METHODS IN MOLECULAR BIOLOGY

Series Editor
John M. Walker
School of Life and Medical Sciences
University of Hertfordshire
Hatfield, Hertfordshire, UK

For further volumes:
http://www.springer.com/series/7651

For over 35 years, biological scientists have come to rely on the research protocols and methodologies in the critically acclaimed *Methods in Molecular Biology* series. The series was the first to introduce the step-by-step protocols approach that has become the standard in all biomedical protocol publishing. Each protocol is provided in readily-reproducible step-by step fashion, opening with an introductory overview, a list of the materials and reagents needed to complete the experiment, and followed by a detailed procedure that is supported with a helpful notes section offering tips and tricks of the trade as well as troubleshooting advice. These hallmark features were introduced by series editor Dr. John Walker and constitute the key ingredient in each and every volume of the *Methods in Molecular Biology* series. Tested and trusted, comprehensive and reliable, all protocols from the series are indexed in PubMed.

Marine Genomics

Methods and Protocols

Edited by

Cinzia Verde

Institute of Biosciences and BioResources (IBBR), National Research Council, Naples, Italy

Daniela Giordano

Institute of Biosciences and BioResources (IBBR), National Research Council, Naples, Italy

Humana Press

Editors
Cinzia Verde
Institute of Biosciences and
BioResources (IBBR)
National Research Council
Naples, Italy

Daniela Giordano
Institute of Biosciences and BioResources (IBBR)
National Research Council
Naples, Italy

ISSN 1064-3745 ISSN 1940-6029 (electronic)
Methods in Molecular Biology
ISBN 978-1-0716-2312-1 ISBN 978-1-0716-2313-8 (eBook)
https://doi.org/10.1007/978-1-0716-2313-8

Cover Illustration Caption: Greenland fjords. Photograph taken by Guido di Prisco.

This Humana imprint is published by the registered company Springer Science+Business Media, LLC, part of Springer Nature.
The registered company address is: 1 New York Plaza, New York, NY 10004, U.S.A.

Dedication

We dedicate this volume to Guido di Prisco, an outstanding and highly ambitious mentor for us both.

Preface

The objective of this book is to provide an overview of recent advances in the application of genomic technologies in several domains of marine biology. Initially, the field of marine genomics was largely driven by scientific curiosity; now genomic approaches are applied to a great variety of questions and organisms: (1) to study evolution and physiological adaptations, (2) to discover novel biomolecules, and (3) to understand ecological interactions among different marine communities/ecosystems within changing environmental conditions.

The aim of the volume is to raise awareness in readers who wish to familiarize themselves with DNA- and RNA-based technologies. These articles also contain essential details for scientists who wish to successfully carry out similar studies to other species or ecosystems.

As a result of recent advances in genomic technologies, a "tool set" is emerging that can contribute to gaining new perspectives on central questions in the field of marine genomics. Targeted studies driven by genomic methods that address physiological and ecological responses of species and their response to changing environmental conditions may be influential for prediction of further changes.

Genomic methods are essential in identifying previously undetected taxonomic (e.g., DNA barcoding), genetic (e.g., sequencing), and functional (e.g., gene expression, analysis of metabolites) diversity as shown in the chapters of this volume. From a resource perspective, tools for marine environmental genomics are developing rapidly. Techniques traditionally reserved for model systems are now applied to the study of the structure and function of genome and transcriptome in non-model marine species.

In this volume, we hope to highlight the utility of some lab protocols and their potential to provide deeper insight into some physiological and ecological mechanisms. High-throughput sequencing (e.g., sequencing DNA at unprecedented speed), more broadly referred to as next-generation sequencing (NGS), has become essential for marine biology. The massive data produced by NGS also presents a significant challenge for data storage, analyses, and management solutions. Advanced bioinformatic tools are essential for the successful application of NGS technology. An efficient combination of many different types of software is necessary. Online forums such as SEQanswers (http://seqanswers.com/) and Biostars (https://www.biostars.org/) provide useful information and guidance.

The first chapters highlight the impact of NGS technologies for novel biological applications and emphasize how choosing the most appropriate method for a specific biological question requires considering the trade-off between benefits and limitations. Chapter 1 (Terraneo et al.) describes a routinely used standardized protocol to obtain mitochondrial genomes of non-model marine organisms by high-throughput next-generation sequencing (NGS). The protocol is composed of five main steps, including DNA extraction, DNA fragmentation, library preparation, high-throughput sequencing, and bioinformatics analyses.

Chapter 2 (Mancia AL) deals with methods to obtain DNA samples ready for NGS sequencing for genome-wide methylation analysis, the major epigenetic modification of DNA in mammalian genomes.

Chapter 3 (De Luca and Lauritano) discusses protocols for transcriptome data mining of genes of interest, from the creation of a protein database to the inference of phylogenetic trees. The chapter looks at transcriptomic analyses in marine protists, which recently received a growing interest in functional ecology and blue biotechnology. The protocol can be used as a general pipeline across different taxa.

Chapter 4 (Sollitto et al.) describes a straightforward and broadly applicable method for the identification of structural variants in fully assembled diploid genomes, leveraging the same reads used for assembly. This chapter also explains the associated gene Presence/Absence Variation (PAV) analysis protocol, broadly applied to any species with a fully sequenced reference genome available. Although the strength of these approaches has been tested and proven in marine invertebrates, which tend to have high levels of heterozygosity, possibly due to their lifestyle traits, they are also applicable to other species across the tree of life, providing a ready means to begin investigations into these potentially widespread phenomena.

Chapter 5 (Cordone et al.) looks at the rapidly evolving field of comparative genomics that allows comparison between genomes of different life forms providing information on the organization of the compared genomes, both in terms of structure and encoded functions. Comparative genomics provides a powerful tool to (1) study and understand the evolutionary changes and adaptation among organisms, (2) compare phylogenetically close marine organisms showing different vital strategies and lifestyles, and (3) obtain information regarding specific adaptations and/or their evolutionary history.

The next three Chapters 6–8 concentrate on the impact of bioinformatics on marine genomics research. Chapter 6 (Achrak et al.) describes a user-friendly automated bioinformatics pipeline via a Galaxy workflow to identify novel venom peptides from raw RNAseq reads of terebrid snails. While designed for venomous terebrid snails, with minor adjustments, this pipeline can be made universal to identify secreted disulfide-rich peptide toxins from any venomous organism.

Chapter 7 (Rivera-Colón and Catchen) provides a detailed protocol to conduct a Restriction site-Associated DNA (RAD) analysis from experimental design to *de novo* analysis—including parameter optimization—as well as reference-based analysis, all in Stacks version 2, which is designed to work with paired-end reads to assemble RAD loci up to 1000 nucleotides in length. The protocol focuses on major points of friction in the molecular approaches and downstream analysis, with special attention given to validating experimental analyses.

Chapter 8 (Cecchetto et al.) proposes a concise bioinformatic pipeline that can be adopted to analyze a metabarcoding dataset. The temporal dynamics of coastal planktic communities can be disclosed through DNA metabarcoding on the filters of reverse osmosis desalination plants. The chapter describes the steps that are necessary to process the filters in order to create the subsamples used for DNA extraction and the bioinformatic pipeline to perform the first exploratory analyses.

DNA barcoding is a powerful and widespread method used to identify large numbers of species collected in the framework of sampling activities in the field. With the exception of big research projects that may count on large teams, in the majority of cases the barcoding effort is handled by a limited number of persons. Chapter 9 (Schiaparelli et al.) focuses on this second case, with special attention paid to field procedures, whose efficiency and smoothness are often overlooked.

Chapter 10 (Núñez-Pons et al.) also deals with DNA barcoding, which is a versatile approach that has revolutionized taxonomy and other akin topics in biology and ecology. The method presented in the chapter consists in the production of one or a few amplicons from informative genetic regions via Sanger sequencing.

Chapters 11 and 12 discuss the relatively recent application of genomic methodologies to the study of marine organisms at the community and population level.

Chapter 11 (Cowart et al.) focuses on the importance of Environmental DNA (eDNA) analysis as a powerful tool for the detection, monitoring, and characterization of aquatic metazoan communities, including vulnerable species. The rapid rate of adopting the eDNA approach across diverse habitats and taxonomic groups attests to its value for a wide array of investigative goals. The aim of this chapter is to familiarize the reader with the eDNA analysis to address questions in marine environments.

Chapter 12 (Ruocco et al.) describes the fundamental steps of metataxonomic analysis of microbial communities associated with marine organisms.

Chapters 13 (Severino et al.), 14 (Giordano and Verde), 15 (Coppola et al.), and 16 (Nuzzo et al.) look at the rapidly growing field of marine biotechnology. Genes from marine environments have a biotechnological potential due to the enormous phylogenetic biodiversity of marine resources and the discovery of novel biological mechanisms. The increasing knowledge of marine genomics has affected the field of marine biotechnology. Marine organisms produce a unique variety of bioactive molecules with a wide structural diversity, potentially valuable for biotechnological applications and for pharmaceutical, nutraceutical, and cosmeceutical sectors. The advent of *omics* techniques is providing new access to the metabolic diversity of the oceans, facilitating the development of new compounds derived from marine biotechnology.

Chapter 13 describes the heterologous expression, which refers to the expression of a gene into a host organism (*Escherichia coli* in this protocol) that does not naturally express the gene. Heterologous expression methods are well established in molecular biology, and it is achieved by cloning the DNA fragment, previously isolated from the organism or metagenome of interest, into a suitable expression vector and introducing the recombinant plasmid into a host that will provide the biochemical apparatus to generate and express the "foreign" protein. The production of recombinant proteins in bacteria made possible to obtain large quantities of proteins essential for basic and applied research. *Escherichia coli* remains one of the organisms of choice for recombinant proteins because of its ability to grow at high density and availability of a vast catalog of cloning vectors and mutant host strains.

In Chapter 14, protocols for the expression of marine cold-adapted (hemo)globins in *Escherichia coli* are described in detail. The next two chapters provide examples of the isolation of bioactive natural products from marine resources. Chapter 15 describes the selection strategy to isolate UV-resistant marine bacteria in marine samples thanks to their ability to produce photoprotective molecules. Chapter 16 focuses on the development of an automated solid phase extraction (SPE)-based method to desalt marine extracts and recover metabolites with biological functions, adapted to be integrated in platform of high-throughput screening.

Diatoms are examples of ecologically important eukaryotic microbes with limited information on their biodiversity and ecology. Chapters 17 (Rogato and Falciatore) and 18 (Russo et al.) describe in detail optimized protocols for the detection and quantification of small regulatory noncoding RNA expression and the CRISPR/Cas9 system coupled with proteolistics as a DNA-free nuclear transformation method in diatoms, respectively.

The last chapters are all dedicated to different methods successfully applied in fish. Chapter 19 (Ametrano and Coscia [a]) looks at CRISPR/Cas9 technology to produce site specifically modified antibody conjugates. The chapter describes the production of a chimeric mouse-fish monoclonal antibody engineered according to CRISPR/Cas9 protocol. Chapter 20 (Ametrano and Coscia [b]) describes different applications of the Polymerase Chain Reaction (PCR) method to amplify up to 2.5-kb DNA regions of the immunoglobulin genes from teleost fishes and protocols to evaluate their expression in different tissues. Chapter 21 (Ghigliotti et al.) provides a step-by-step protocol for mapping genes and noncoding DNA sequences on the chromosomes by fluorescence *in situ* hybridization (FISH). The use of FISH allowed significant advances in our understanding of the fish genome architecture, especially when applied to the study of the repetitive component of the genome that is generally underestimated in the bioinformatic assembly. Chapter 22 (Nikinmaa and Crespel) deals with functional genomics. The aim of any functional genomics study should associate transcriptional change to protein levels and activities. The authors present a methodology for carrying out functional genomics studies with fish erythrocytes.

Chapter 23 (Xue and Corti) discusses the method to analyze mutant zebrafish embryos at different stages by western blot using the Stain-Free technology for correct normalization by taking into account variations not intrinsically dependent on the sample such as different experimental conditions and type of samples. Zebrafish researchers often face the challenge of comparing embryos at different developmental stages or from different strains.

The last two chapters (24 and 25) deal with proteins and proteomics. Chapter 24 (Anjos et al.) presents a detailed workflow for preparation from teleost fish white muscle of extracts for proteomics analysis. The protocol generates samples that can be analyzed by SWATH (Sequential Window data independent Acquisition of the Total High-Resolution-Mass Spectra), a modern mass spectrometry-based quantitative label-free technology.

Chapter 25 (Fusco et al.), the last chapter of the volume, describes *in vitro* assays able to detect the activities of Acylpeptide Hydrolase (APEH) enzyme involved in important metabolic processes in fish.

Naples, Italy *Cinzia Verde*
Naples, Italy *Daniela Giordano*

Contents

Contributors

ELEONORA ACHRAK • *Department of Biology, Hunter College of the City University of New York, New York, NY, USA*
BRUNA AGRILLO • *Institute of Biosciences and BioResources, National Research Council (IBBR-CNR), Naples, Italy; Materias Srl, Naples, Italy*
MARIA CHIARA ALVARO • *Italian National Antarctic Museum (MNA, Section of Genoa), University of Genoa, Genoa, Italy; Department of Earth, Environmental and Life Science (DISTAV), University of Genoa, Genoa, Italy*
ALESSIA AMETRANO • *Institute of Biochemistry and Cell Biology—National Research Council of Italy, Naples, Italy; Department of Environmental, Biological and Pharmaceutical Sciences and Technologies, University of Campania Luigi Vanvitelli, Caserta, Italy*
LILIANA ANJOS • *Centro de Ciências do Mar (CCMAR), Universidade do Algarve, Faro, Portugal*
ROBERTO ARRIGONI • *Department of Biology and Evolution of Marine Organisms (BEOM), Stazione Zoologica Anton Dohrn, Naples, Italy*
JULIETTE AUVINET • *Department of Marine and Environmental Sciences, Northeastern University, Marine Science Center, Nahant, MA, USA*
BARBARA CALCINAI • *DiSVa Department of Life and Environmental Science, Polytechnic University of Marche, Ancona, Italy*
ANDREW D. CALCINO • *Department of Evolutionary Biology, Integrative Zoology, University of Vienna, Vienna, Austria*
ANTONIO CASCONE • *Department of Biology, University of Naples Federico II, Naples, Italy*
JULIAN CATCHEN • *Department of Evolution, Ecology, and Behavior, University of Illinois at Urbana-Champaign, Urbana, IL, USA*
MATTEO CECCHETTO • *Italian National Antarctic Museum (MNA, Section of Genoa), University of Genoa, Genoa, Italy; Department of Earth, Environmental and Life Science (DISTAV), University of Genoa, Genoa, Italy*
C.-H. CHRISTINA CHENG • *Department of Evolution, Ecology, and Behavior, University of Illinois at Urbana – Champaign, Urbana, IL, USA*
ENNIO COCCA • *Institute of Biosciences and BioResources, National Research Council (IBBR-CNR), Naples, Italy*
ALESSANDRO COPPOLA • *Department of Biology, University of Naples Federico II, Naples, Italy*
DANIELA COPPOLA • *Department of Marine Biotechnology, Stazione Zoologica Anton Dohrn (SZN), Villa Comunale, Naples, Italy*
ANGELINA CORDONE • *Department of Biology, University of Naples Federico II, Naples, Italy*
MONICA CORREGGIA • *Department of Biology, University of Naples Federico II, Naples, Italy*
PAOLA CORTI • *Heart, Lung, Blood, and Vascular Medicine Institute, University of Pittsburgh, Pittsburgh, PA, USA; Division of Cardiology, Department of Medicine, University of Pittsburgh, Pittsburgh, PA, USA*
MARIA ROSARIA COSCIA • *Institute of Biochemistry and Cell Biology—National Research Council of Italy, Naples, Italy*
MARIA COSTANTINI • *Stazione Zoologica Anton Dohrn, Department of Ecosustainable Marine Biotechnology, Naples, Italy*

DOMINIQUE A. COWART • *Company for Open Ocean Observations and Logging (COOOL), Saint Leu, La Réunion, France*
AMÉLIE CRESPEL • *Department of Biology, University of Turku, Turku, Finland*
GIULIANA D'IPPOLITO • *National Research Council of Italy, Institute of Biomolecular Chemistry, Pozzuoli, Naples, Italy*
TRAMI DANG • *Bioinformatics and Computational Genomics Laboratory, Hunter College, City University of New York, New York, NY, USA*
DANIELE DE LUCA • *Department of Biology, University of Naples Federico II, Botanic Garden of Naples, Naples, Italy*
ANDREA DI CESARE • *National Research Council of Italy, Water Research Institute (CNR-IRSA), Verbania Pallanza, Italy*
ESTER ECKERT • *National Research Council of Italy, Water Research Institute (CNR-IRSA), Verbania Pallanza, Italy*
JANA EFREMOVA • *Dept. Integrated Marine Ecology (EMI) Stazione Zoologica "Anton Dohrn", Naples, Italy*
ROBERTA ESPOSITO • *Stazione Zoologica Anton Dohrn, Department of Ecosustainable Marine Biotechnology, Naples, Italy; Department of Biology, University of Naples Federico II, Complesso Universitario di Monte Sant'Angelo, Naples, Italy*
ANGELA FALCIATORE • *Laboratoire de Biologie du chloroplaste et perception de la lumière chez les micro-algues, UMR7141, CNRS, Sorbonne Université, Institut de Biologie Physico-Chimique, Paris, France*
JENNIFER FERD • *Department of Chemistry, Hunter College of the City University of New York, New York, NY, USA*
MARIA IMMACOLATA FERRANTE • *Department of Integrative Marine Ecology, Stazione Zoologica Anton Dohrn, Naples, Italy*
ANGELO FONTANA • *National Research Council of Italy, Institute of Biomolecular Chemistry, Pozzuoli, Naples, Italy*
DIEGO FONTANETO • *National Research Council of Italy, Water Research Institute (CNR-IRSA), Verbania Pallanza, Italy*
ZAC FORSMAN • *Hawaii Institute of Marine Biology, Kaneohe, HI, USA*
CARMELA FUSCO • *Institute of Biosciences and BioResources, National Research Council (IBBR-CNR), Naples, Italy; Department of Medicine and Health Sciences, University of Molise, Campobasso, Italy*
CARMELA GALLO • *National Research Council of Italy, Institute of Biomolecular Chemistry, Pozzuoli, Naples, Italy*
MARCO GERDOL • *Department of Life Sciences, Università degli Studi di Trieste, Trieste, Italy*
LAURA GHIGLIOTTI • *National Research Council of Italy (CNR), Institute for the Study of the Anthropic Impacts and the Sustainability of the Marine Environment (IAS), Genoa, Italy*
DANIELA GIORDANO • *Institute of Biosciences and BioResources (IBBR), CNR, Naples, Italy; Department of Marine Biotechnology, Stazione Zoologica Anton Dohrn (SZN), Villa Comunale, Naples, Italy*
DONATO GIOVANNELLI • *Department of Biology, University of Naples Federico II, Naples, Italy; Department of Marine and Coastal Sciences, Rutgers University, New Brunswick, NJ, USA; National Research Council—Institute of Marine Biological Resources and Biotechnologies CNR-IRBIM, Ancona, Italy; Earth-Life Science Institute, Tokyo Institute*

of Technology, Tokyo, Japan; Marine Chemistry & Geochemistry Department, Woods Hole Oceanographic Institution, Falmouth, MA, USA

MARTA GOGLIETTINO • *Institute of Biosciences and BioResources, National Research Council (IBBR-CNR), Naples, Italy*

SAMUELE GRECO • *Department of Life Sciences, Università degli Studi di Trieste, Trieste, Italy*

ALICE GUZZI • *Italian National Antarctic Museum (MNA, Section of Genoa), University of Genoa, Genoa, Italy; Department of Earth, Environmental and Life Science (DISTAV), University of Genoa, Genoa, Italy*

MANDE HOLFORD • *Department of Chemistry & Biochemistry, Hunter College of the City University of New York, New York, NY, USA; The American Museum of Natural History, New York, NY, USA; PhD Programs in Biology, Biochemistry, and Chemistry at the CUNY Graduate Center, New York, NY, USA*

NATHAN J. KENNY • *Faculty of Health and Life Sciences, Oxford Brookes, Oxford, UK; Department of Biochemistry, University of Otago, Dunedin, New Zealand*

KONSTANTINOS KRAMPIS • *Bioinformatics and Computational Genomics Laboratory, Hunter College, City University of New York, New York, NY, USA*

CHIARA LAURITANO • *Department of Ecosustainable Marine Biotechnology, Stazione Zoologica Anton Dohrn, Naples, Italy*

ARSENIOS-ZAFEIRIOS LOUKISSAS • *Centro de Ciências do Mar (CCMAR), Universidade do Algarve, Faro, Portugal*

ANNALAURA MANCIA • *Department of Life Science and Biotechnology, University of Ferrara, Ferrara, Italy*

EMILIANO MANZO • *National Research Council of Italy, Institute of Biomolecular Chemistry, Pozzuoli, Naples, Italy*

KIRUTHIGA G. MARIAPPAN • *Red Sea Research Centre, Division of Biological and Environmental Science and Engineering, King Abdullah University of Science and Technology, Thuwal, Saudi Arabia*

VALERIO MAZZELLA • *Dept. Integrated Marine Ecology (EMI) Stazione Zoologica "Anton Dohrn", Naples, Italy*

ISABELLA MORO • *Department of Biology, University of Padova, Padua, Italy*

KATHERINE R. MURPHY • *Laboratories of Analytical Biology, National Museum of Natural History, Smithsonian Institution, Washington, DC, USA*

MIKKO NIKINMAA • *Department of Biology, University of Turku, Turku, Finland*

LAURA NÚÑEZ-PONS • *Dept. Integrated Marine Ecology (EMI) Stazione Zoologica "Anton Dohrn", Naples, Italy*

GENOVEFFA NUZZO • *National Research Council of Italy, Institute of Biomolecular Chemistry, Pozzuoli, Naples, Italy*

GIANNA PALMIERI • *Institute of Biosciences and BioResources, National Research Council (IBBR-CNR), Naples, Italy*

EVA PISANO • *National Research Council of Italy (CNR), Institute for the Study of the Anthropic Impacts and the Sustainability of the Marine Environment (IAS), Genoa, Italy*

DEBORAH MARY POWER • *Centro de Ciências do Mar (CCMAR), Universidade do Algarve, Faro, Portugal; Shanghai Ocean University, International Center for Marine Studies, Shanghai, China*

FRANCESCA RISPO • *Dept. Integrated Marine Ecology (EMI) Stazione Zoologica "Anton Dohrn", Naples, Italy; DISTAV Dipartimento di Scienze della Terra dell'Ambiente e della Vita. Universitá degli studi di Genova, Genoa, Italy*

Angel G. Rivera-Colón • *Department of Evolution, Ecology, and Behavior, University of Illinois at Urbana-Champaign, Urbana, IL, USA*
Alessandra Rogato • *Institute of Biosciences and BioResources, CNR, Naples, Italy; Department of Integrative Marine Ecology, Stazione Zoologica Anton Dohrn, Naples, Italy*
Nadia Ruocco • *Stazione Zoologica Anton Dohrn, Department of Ecosustainable Marine Biotechnology, Naples, Italy*
Monia Teresa Russo • *Department of Integrative Marine Ecology, Stazione Zoologica Anton Dohrn, Naples, Italy*
Anna Santin • *Department of Integrative Marine Ecology, Stazione Zoologica Anton Dohrn, Naples, Italy*
Stefano Schiaparelli • *Italian National Antarctic Museum (MNA, Section of Genoa), University of Genoa, Genoa, Italy; Department of Earth, Environmental and Life Science (DISTAV), University of Genoa, Genoa, Italy*
Jessica Schulman • *Department of Bioinformatics, New York University Tandon School of Engineering, Brooklyn, NY, USA*
Matteo Selci • *Department of Biology, University of Naples Federico II, Naples, Italy*
Angelica Severino • *Department of Biology, University of Naples Federico II, Naples, Italy*
Marco Sollitto • *Department of Life Sciences, Università degli Studi di Trieste, Trieste, Italy*
Tullia I. Terraneo • *Red Sea Research Centre, Division of Biological and Environmental Science and Engineering, King Abdullah University of Science and Technology, Thuwal, Saudi Arabia*
Carmen Federica Tucci • *Department of Life Sciences, Università degli Studi di Trieste, Trieste, Italy*
Cinzia Verde • *Institute of Biosciences and BioResources (IBBR), CNR, Naples, Italy; Department of Marine Biotechnology, Stazione Zoologica Anton Dohrn (SZN), Villa Comunale, Naples, Italy*
Costantino Vetriani • *Department of Marine and Coastal Sciences, Rutgers University, New Brunswick, NJ, USA; Department of Biochemistry and Microbiology, Rutgers University, New Brunswick, NJ, USA*
Jianmin Xue • *Heart, Lung, Blood, and Vascular Medicine Institute, University of Pittsburgh, Pittsburgh, PA, USA*
Valerio Zupo • *Stazione Zoologica Anton Dohrn, Department of Ecosustainable Marine Biotechnology, Naples, Italy*

Chapter 1

Mitochondrial Genome of Nonmodel Marine Metazoans by Next-Generation Sequencing (NGS)

Tullia I. Terraneo, Kiruthiga G. Mariappan, Zac Forsman, and Roberto Arrigoni

Abstract

Mitochondrial genomes (mtgenome) represent an important source of information for addressing fundamental evolutionary, phylogeographic, systematic, and ecological questions in marine organisms. In the last two decades the advent of high-throughput next-generation sequencing (NGS) has provided an unprecedented possibility to access large amount of genomic data and, as such, there has been a rapid growth in mtgenome resources and studies. In particular, NGS strategies represent a great advantage for investigating nonmodel marine organisms for which no or limited genomic resources are available. Here, we describe a routinely used standardized protocol to obtain mtgenome of nonmodel marine organisms by NGS. The protocol is composed of five main steps, including DNA extraction, DNA fragmentation, library preparation, high-throughput sequencing, and bioinformatic analyses. Each of the first three steps is followed by size/quality and concentration validations. The advantages of the described protocol rely on the assumption that no a priori information on mtgenome of the studied organism is needed and on its versatility as researchers may choose several kits for DNA extraction and library preparation and adopt different methods for DNA fragmentation depending on their needs, experience, and suppliers.

Key words DNA extraction, DNA fragmentation, Gene annotation, High-throughput sequencing, Library preparation, mtgenome assembly

1 Introduction

Mitochondrial DNA (mtDNA) currently represents the most abundant genetic resource deposited in public nucleotide sequence databases. It is widely used for addressing fundamental evolutionary, phylogeographic, systematic, and ecological questions in marine metazoan groups for a number of reasons, including rates of mtDNA nucleotide substitutions up to 10 times faster than nuclear DNA (nDNA), rare recombination, predominant homoplasy, and haploidy via maternal inheritance (e.g., [1–3]). Exceptions to this general scenario exist among marine organisms. For example, mtDNA of Porifera and Anthozoa displays low levels of

Cinzia Verde and Daniela Giordano (eds.), *Marine Genomics: Methods and Protocols*,
Methods in Molecular Biology, vol. 2498, https://doi.org/10.1007/978-1-0716-2313-8_1,

nucleotide variation at both intra- and interspecific levels, resulting in evolutionary rates extremely slower than those of the other metazoans [4]. It has been observed that mussels of the superfamily Unionoidea and clams of the order Veneroida are characterized by a doubly uniparental inheritance of mtDNA, where distinct gender-associated mtDNAs are inherited either maternally or paternally [5]. Apart from these exceptions, the combination of the above mentioned factors results in mtDNA effective population size of approximately four times smaller than nDNA, which results in rapid lineage sorting [6, 7]. As such, mtDNA genes are routinely investigated as source of genetic information for resolving phylogenetic relationships at both deep and shallow nodes, and also for exploring phylogeography, population structure, and demography within species [8–11]. It is noteworthy to indicate that studies based solely on mtDNA may suffer from the effects of mutational saturation, selective sweeps, rate and compositional heterogeneity, and codon-usage bias [12].

Metazoan mitochondrial genomes (mtgenome) are relatively short compared to nuclear and chloroplast genomes, being typically 15–20 kilobase pair in size for the principal groups of marine metazoans (e.g., [13–15]). Metazoan mtgenomes are circular and display a relatively constant gene content, with a total of 13 protein coding genes, two rRNAs, 22 tRNAs, and one noncoding control region [1, 3, 16]. Notable exceptions in terms of tRNA gene number and the presence of introns are found in the basal metazoans Porifera and Cnidaria [11, 13]. Mtgenome sequencing has traditionally relied on polymerase chain reaction (PCR) amplifications and cloning of PCR products from either a set of universal primers or two to three long range PCRs followed by conserved primer walking sequencing strategy [17–19]. Nevertheless, these approaches require a priori information of a similar mtgenome, such as gene order and/or suitable primers, and are laborious and costly. During the last two decades, the advent of high-throughput next-generation sequencing (NGS) has enabled the unprecedented opportunity to acquire huge amount of genomic resources [20]. NGS technologies are rapidly transforming biological studies, with costs continuously decreasing and outstanding progresses in terms of methodological and analytical analyses [21]. A fundamental advantage of NGS approaches is that they can be applied to nonmodel organisms for which no mtgenome information is available. As such, NGS has become the most common strategy also to generate complete mtgenome sequences, resulting in a rapid growth of mtgenome resources and related studies [22–24]. As described in detail by [23], two main laboratory procedures are adopted to obtain mtgenome by NGS strategies, namely, long-range PCR amplicon sequencing and shotgun sequencing of genomic DNA. Moreover, recent studies have also demonstrated the potential of retrieving and assembling high-depth mtgenomes from off-target reads of reduced-genome representation approaches, such as transcriptomics [25], Restriction-site Associated DNA [26], and ultraconserved elements sequencing [27], whose applications in phylogenomics are expected to greatly increase in the coming years.

Here, we describe a routinely used protocol to obtain mtgenome of nonmodel metazoans by NGS. The protocol is composed of five main steps, including DNA extraction, DNA fragmentation, library preparation, high-throughput sequencing, and bioinformatic analyses, with less than 10 days for samples preparation before sequencing. The advantages of the present standardized protocol rely on the assumption that no a priori information on the targeted organism mtgenome sequence is needed and on its versatility as researchers may choose alternative commercial kits for DNA extraction and library preparation and different sequencing platforms depending on their needs, suppliers, and experience. An exhaustive protocol about assembly of mtgenome by NGS strategies has been previously published [23], which we recommend for further reading. During the last 5 years, a rapid increase of mtgenomes obtained through shotgun sequencing of genomic DNA has occurred because of multiple factors, including for example the cost reduction of library preparation and the advantages in the bioinformatic software and pipelines to assemble mitochondrial data from genomic data. As such, here we present a protocol to generate mtgenomes using this laboratory strategy and we include indications and notes about numerous open-source software and pipelines recently developed for mtgenome assembly and annotation and for in silico enzymatic fragmentation.

2 Materials

All steps require 20, 200, and 1000 μL pipettes and tips. Most steps require 0.2 mL PCR tubes, 1.5 mL microcentrifuge tubes, a mini vortex mixer, and mini spin centrifuges for 0.2 and 1.5 mL tubes. Working with numerous samples requires 96-well full skirted and half skirted plates (or 0.2 mL strip tubes) and a 96-well plate shaking incubator. Specific instruments, reagents, and consumables required for each step are given below.

2.1 DNA Extraction

1. Heating block or heated water bath.
2. Laboratory centrifuge for 1.5 mL tubes (minimum 20,000 × *g* required).
3. Forceps, scalpel blades, mortar, and pestle.
4. Timer.

5. DNA extraction kit.
6. Molecular grade ethanol (≥99.8%).
7. Paper towels.

2.2 DNA Fragmentation

2.2.1 Mechanical Fragmentation

1. Sonication system.
2. Nuclease-free water.
3. Tubes for sonication system.

2.2.2 Enzymatic Fragmentation

1. Restriction enzymes. *See* Subheading 3.3.2 for restriction enzyme choice.
2. PCR thermal cycler.
3. Ice bucket.
4. Ice.
5. Nuclease-free water.

2.2.3 Fragmented DNA Purification

1. Magnetic bead separator (either for 1.5 mL tubes or 96-well plate depending on the number of samples).
2. Timer.
3. Magnetic beads for DNA purification. Library preparation kit may be supplied with their own magnetic beads (*see* Subheading 3.3.3).
4. Molecular grade ethanol (≥99.8%).
5. Nuclease-free water.
6. Reagent troughs/reservoirs.
7. 15 mL centrifuge tube.

2.3 Library Preparation

1. Magnetic bead separator (either for 1.5 mL tubes or 96-well plate depending on the number of samples).
2. PCR thermal cycler.
3. Ice bucket.
4. Ice.
5. Library preparation kit.
6. Fleshy prepared 80% molecular grade ethanol.
7. Nuclease-free water.
8. 10 nM Tris–HCl pH 8.5 with 0.1% Tween 20.
9. Timer.
10. Reagent troughs/reservoirs.
11. 15 mL centrifuge tube.

2.4 DNA Extraction, DNA Fragmentation, and Library Validation

2.4.1 Size/Quality Validation Using an Automated Capillary Electrophoresis System

1. High-resolution automated capillary electrophoresis instrument.
2. Timer.
3. Automated capillary electrophoresis kit.
4. Nuclease-free water.
5. Paper towels.

2.4.2 Size/Quality Validation Using Agarose Gel Electrophoresis

1. Horizontal gel electrophoresis system, including casting platform, combs, tank, DC power supply, and gel imaging system (optionally portable UV lamp).
2. Microwave oven.
3. Cylinder.
4. Glass flask.
5. Electrophoresis grade agarose.
6. TAE (50×): 2 M Tris–HCl, 50 mM ethylenediaminetetraacetic acid (EDTA), 1 M glacial acetic acid. Dissolve 242 g Tris–HCl in 700 mL deionized water. Add 57.1 mL 100% glacial acetic acid and 100 mL 0.5 M EDTA. Add water to a final volume of 1 L. Mix and adjust pH to 8.5. Dilute 20 mL TAE (50×) into 980 mL deionized water to make TAE (1×). Store at room temperature.
7. Nontoxic DNA gel stain.
8. 1 kb DNA ladder for DNA extraction validation.
9. 6× DNA gel loading dye.
10. Parafilm.
11. Paper towels.

2.4.3 Double-Stranded DNA (dsDNA) Quantification

1. Fluorometer instrument for dsDNA quantification.
2. Fluorometer kit for dsDNA quantification.
3. Nuclease-free water.
4. 0.5 mL transparent PCR tubes.
5. 15 mL centrifuge tube.

2.5 Pool Validation

2.5.1 Pool Size/Quality Validation

1. High-resolution automated capillary electrophoresis instrument.
2. Timer.
3. Automated capillary electrophoresis kit.
4. Nuclease-free water.
5. Paper towels.

2.5.2 Pool Concentration Quantification

1. Real-time quantitative PCR (qPCR) system.
2. Library quantification qPCR Kit.
3. Nuclease-free water.
4. Ice.
5. Ice bucket.

2.6 Bioinformatic Analyses

You will need a computer with Unix-based environment or access to a remote server with the following software installed.

1. FastQC [28] (http://www.bioinformatics.babraham.ac.uk/projects/fastqc, http://github.com/s-andrews/FastQC) for read quality control.
2. Trimmomatic [29] (http://www.usadellab.org/cms/?page=trimmomatic, https://github.com/timflutre/trimmomatic) for trimming adapter sequences.
3. NOVOPlasty [24] (http://github.com/ndierckx/NOVOPlasty) for de novo assembly of the mtgenome.
4. Internet access to the following three on-line web servers: MITOS [30] (http://mitos.bioinf.uni-leipzig.de/index.py) or its updated version MITOS2 [31] (http://mitos2.bioinf.uni-leipzig.de/index.py), tRNAscan [32] (http://lowelab.ucsc.edu/tRNAscan-SE), Rfam [33] (http://rfam.xfam.org) for gene annotation.

3 Methods

Wear appropriate personal protective equipment and follow correct procedures for waste disposal. The below laboratory protocol (from Subheadings 3.1–3.6) can be roughly completed for ≤96 samples/person in ~10 working days, unless any step repetition for optimization, as follows.

1. Days 1–2: DNA extraction (Subheading 3.1) and validation (Subheading 3.2).
2. Day 3: DNA fragmentation and purification (Subheading 3.3) and validation (Subheading 3.4).
3. Days 4–6: library preparation (Subheading 3.5).
4. Day 7: library validation, normalization, and pooling (Subheading 3.5).
5. Day 8: pool validation (Subheading 3.6).
6. Day 9: high-throughput sequencing (Subheading 3.7).

3.1 DNA Extraction

Numerous commercial kits and customized protocols are available to extract total genomic DNA from a wide variety of marine organisms for NGS projects (e.g., [34, 35]). In particular, for a detailed list of protocols to extract sufficient yields of high-quality DNA from numerous marine organisms, including invertebrates, algae, and yeasts, *see* [35]. Samples can be preserved in either ≥70% ethanol, RNAlater, salt-saturated dimethyl sulfoxide buffer (20% dimethyl sulfoxide, 0.25 M EDTA, pH 8.0 saturated with NaCl), or frozen at −80 °C [36]. DNA degradation and loss of the sample will occur if the tissues are not stored in a sufficient quantity of preservative, or if the samples are not preserved in a timely manner (*see* **Note 1**). We recommend to strictly follow manufacturer's instructions for the DNA extraction protocol.

3.2 DNA Extraction Validation

Validate the DNA samples in terms of both size/quality and concentration. We suggest to control the DNA integrity using a high-resolution automated capillary electrophoresis system (Subheading 3.2.1) because an accurate DNA estimation is needed for the next step. Nevertheless, in case this system is not available, an alternative DNA size/quality check can be carried out using agarose gel electrophoresis and comparison with a DNA ladder of known quantification (Subheading 3.2.2). Concerning DNA quantification, we are interested in measuring the dsDNA concentration and, as such, we strongly suggest to carry out the quantification using a fluorometer instrument (at least 1 μL of high molecular weight DNA is ideal, results may vary with lower quantities or purity) (Subheading 3.2.3).

3.2.1 DNA Size/Quality Control Using an Automated Capillary Electrophoresis System

The samples are verified in terms of size and quality using a high-resolution automated capillary electrophoresis system. The required sample volume is typically 1 μL. We recommend to strictly follow manufacturer's instructions for the samples' preparation and for running the analysis on the preferred instrument (*see* **Note 2**). We aim for DNA samples with high molecular weights.

3.2.2 DNA Size/Quality Control Using Agarose Gel Electrophoresis

For safety reasons we encourage to use a nontoxic DNA gel stain, although it is more expensive than ethidium bromide.

1. Add electrophoresis grade agarose and TAE (1×) sufficient for preparing the 1.5% agarose gel in the glass flask, for example 1.5 g agarose and 100 mL TAE (1×). Plug the glass flask with paper towels and gently mix.
2. Heat in microwave oven and watch carefully and stop the heating process if bubbles begin to form. With a glove to avoid burns, gently mix the solution to avoid agarose deposition and allow bubbles to stop forming. Repeat heating and mixing until the solution is clear and all the agarose is dissolved (~100 °C).

3. Once agarose is melted, gently mix and cool to 55 °C (the process can be speed up by running cool water over the outside of the flask).
4. Add 1.5 μL nontoxic DNA gel stain. Gently mix by pipetting up and down five times and gently stirring.
5. Most gel casting platforms require adhesive tape or gaskets to be sealed, insure a proper seal.
6. Pour in the gel casting platform, add properly cleaned combs. Use pipette tips to remove any bubbles from the gel surface. Ensure to have the agarose gel with 0.5–1 cm thickness.
7. Incubate at room temperature until the gel is solidified (about 20–30 min).
8. Remove any adhesive tape or sealing gaskets from the gel casting platform. Carefully remove the combs. Place in the electrophoresis tank. Add sufficient TAE (1×) to ensure that gel is completely immersed.
9. On a small piece of parafilm, for each DNA sample prepare one drop composed of 3 μL DNA sample and 1 μL 6× DNA gel loading dye. Prepare then one drop composed of 3 μL 1 kb quantitative DNA ladder and 1 μL 6× DNA gel loading dye. Gently mix by pipetting up and down ten times.
10. Carefully load the DNA gel loading dye into the first gel well. Carefully load the samples into the remaining gel wells.
11. Connect the electrophoresis tank to the DC power supply cables. DNA is negatively charged and will run from the cathode (negative) to the anode (positive).
12. Set DC power supply to the desired voltage and amperage, routinely 100 V and 360 A respectively, for 40 min. Monitor the gel run either by checking the migration of the DNA loading dye or with a portable UV lamp. Run for additional time if the migration of the DNA loading dye is still insufficient.
13. Once the run is ended, visualize the DNA run with a gel imaging system or a portable UV lamp. A single and intense high molecular weight banding with absent to limited smearing can be considered as indicator of high-quality DNA, while a large smearing indicates degraded/fragmented DNA. Although high molecular weight banding is ideal, some degradation and smearing may be acceptable.

3.2.3 dsDNA Extract Quantification

The samples are quantified in terms of dsDNA using a fluorometer instrument. It is essential to use a fluorescence-based technology because we aim to selectively quantify only the dsDNA. The required sample volume is typically 1 μL. We recommend to follow manufacturer's instructions for the samples' preparation and for running the analysis on the preferred fluorometer instrument.

3.3 DNA Fragmentation

Prior to library preparation, DNA samples are mechanically (Subheading 3.3.1) or enzymatically (Subheading 3.3.2) fragmented to a desired insert size averaging approximately 350 bp or 550 bp. The fragmented DNA samples are then purified through a standard magnetic bead cleanup with a DNA–beads ratio of 1:1.6 (Subheading 3.3.3).

3.3.1 DNA Mechanical Fragmentation

The DNA mechanical fragmentation is obtained using a sonication system, aiming at a size of either 350 bp or 550 bp depending on the desired sequencing platform. We recommend to strictly follow manufacturer's instructions for the samples' preparation (including indications about the required quantity and volume of dsDNA) and for running the DNA shearing analysis. Moreover, to add a sufficient amount of DNA consider also the DNA quantity required for the library preparation step (Subheading 3.5) and the potential loss of DNA occurring in the DNA purification (Subheading 3.3.3).

3.3.2 DNA Enzymatic Fragmentation

DNA enzymatic fragmentation is obtained through the use of one or two restriction enzymes and ~1–1.2 μg dsDNA initial quantity. In the presence of mtgenome resources of the targeted organisms or their relatives, it is possible to run an in silico approach to predict the number and size of mtgenome loci recovered by restriction enzymes that researcher desires to test [37]. This in silico digestion prediction allows the researcher to choose the restriction enzymes' combination that maximizes the recovery of mtgenome loci in terms of both number and lengths. For a detailed description of commands and software *see* https://github.com/tkchafin/fragmatic. In the absence of mtgenome resources of the targeted organisms or their relatives, the use of frequent cutter enzymes results in considerable horizontal coverage, allowing mtgenome assembly [38]. Frequent cutter enzymes, such as DpnII, MboI, or Sau3AI, recognise GATC sites, but have different methylation sensitivities (see manufacturer's instructions). Based on our experience, the double digestion with MboI and Sau3AI allows the assembly of complete mtgenomes in corals, molluscs, and sponges [38–43]. Once the restriction enzymes are chosen, follow the below steps for digestion.

1. Save the following program on the PCR thermal cycler (without the preheat lid option).
   (a) 37 °C for 3 h; 65 °C for 20 min; hold at 15 °C.
2. The typical restriction digestion mix (50 μL) is as follows.
   (a) 43 μL DNA + nuclease-free water (about 1.2 μg dsDNA) (*see* **Note 3**);

(b) 5 μL supplied buffer (10x) (*see* **Note 4**);

(c) 1 μL each enzyme.

3. Add the requested 43 μL DNA + nuclease-free water in a 0.2 μL PCR tube.
4. Thaw restriction enzyme buffer at room temperature. Place buffer and restriction enzymes on ice. Return to −20 °C storage after use.
5. Add 5 μL buffer (10×), then add 1 μL each enzyme, for a final volume of 50 μL. Gently mix by pipetting up and down ten times and then briefly spin down.
6. Place the samples on the PCR thermal cycler and run the program.
7. Proceed immediately to Subheading 3.3.3 or store at 4 °C for up 24 h.

3.3.3 Fragmented DNA Purification

The fragmented DNA samples obtained from either Subheadings 3.3.1 or 3.3.2 are purified and cleaned through a standard magnetic bead clean-up procedure, with a DNA–beads ratio of 1:1.6 (*see* **Note 5**). Library preparation kits may be supplied with their own magnetic beads, so there is no need to buy separate magnetic beads.

1. In a 15 mL centrifuge tube prepare fresh 80% ethanol using molecular grade ethanol (≥99.8%) and nuclease-free water. Prepare 420 μL 80% ethanol for each sample.
2. Briefly spin the samples down by mini spin centrifuge.
3. Mix the magnetic beads' solution well for 1 min by mini vortex mixer until it is homogeneous and consistent in colour.
4. Add 80 μL magnetic beads' solution to each sample, for a total volume of 130 μL. Gently mix by pipetting up and down ten times.
5. Incubate at room temperature for 5 min.
6. Place the samples onto the magnetic bead separator and wait until the solution is clear (~5 min).
7. Aspirate out all solution, without disturbing the magnetic beads.
8. With the samples onto the magnetic bead separator, wash with fresh 80% ethanol two times:

   (a) Dispense 200 μL fresh 80% molecular grade ethanol.

   (b) Wait for 30 s.

   (c) Carefully aspirate out ethanol without disturbing the magnetic beads.
9. Air-dry onto the magnetic bead separator until ethanol is no longer visible (~5 min).

10. Remove the samples from the magnetic bead separator and add 38 μL elution buffer for resuspending. Gently mix by pipetting up and down ten times until beads are completely resuspended.
11. Incubate at room temperature for 2 min.
12. Place the samples onto the magnetic bead separator and wait until the solution is clear (~5 min).
13. Transfer 33 μL eluate to a new 0.2 mL microcentrifuge tube. Carefully avoid transferring beads.
14. Store at −20 °C.

3.4 Fragmented DNA Validation

Validate the fragmented DNA samples in terms of both size/quality and concentration as described in Subheading 3.2.

3.5 Library Preparation

Numerous kits of different commercial companies are available for NGS library preparation and, as such, researchers may choose the preferred one depending on their needs, experience, sequencing platform, and suppliers. For example, there are PCR-free kits that are less-error prone but require a relatively high amount of DNA and PCR-based kits that require small quantities of initial DNA. Alternatively, a researcher may also choose not to use a commercial kit for the library preparation and buy separate reagents. The use of a library preparation kit allows for a standardized protocol, including also multiple advantages such as (1) no additional reagents are typically required to complete the library preparation protocol; (2) the protocol is typically easy to follow using manufacturer's instructions even for a researcher with moderate laboratory experience; (3) uniquely dual-indexed libraries are typically generated and, as such, they can be eventually pooled prior to the sequencing (*see* **Note 6**). To carry out the library preparation and to properly store the supplied reagents we recommend to strictly follow the manufacturer's instructions. Alternatively, follow [44] to half volumes of all reagents per sample.

1. At the end of the library preparation, validate the libraries in terms of both size/quality and concentration as described in Subheading 3.2.
2. For the library normalization, to convert ng/μL (as reported by the fluorometer) to nM use the following formula (the molecular weight of 1 bp is 660 Da).

$$\begin{aligned}\text{Concentration(nM)} &= [\text{concentration(ng/}\mu\text{L)}\\ &\quad \times 1{,}000{,}000]/[660\text{(g/mol)}\\ &\quad \times \text{final average fragment size(bp)}]\end{aligned}$$

3.6 Pool Validation

It is essential to accurately define the size and concentration of the libraries' pool by using a high-resolution automated capillary electrophoresis system (Subheading 3.2.1) and a qPCR (Subheading 3.6.1), respectively, to obtain the highest yield and quality of sequencing data.

3.6.1 qPCR

The libraries' pools are quantified using a qPCR system. Several qPCR commercial kits are available depending on the needs of the researcher. We recommend strictly following manufacturer's instructions for the samples' preparation and for running the analysis on the qPCR instrument (*see* **Note 2**).

3.7 High-Throughput Sequencing

Libraries are sequenced using NGS technologies. There are several bench-top high-throughput sequencing instruments currently available and researchers may choose the preferred one based on multiple factors such as availability, cost, and output (*see* **Note 6**).

3.8 Bioinformatic Analyses

The sequencing facility usually carries out the demultiplexing of the raw sequence reads using the unique sample combination of sequence adapters and, in case of paired-end sequencing, provides two separated files per sample in compressed FASTQ format, including forward and reverse sequence reads. Usually ~1–2% of the generated sequence reads will belong to mtgenome [45]. Below we provide indications for read quality control, trimming, mtgenome assembly, and gene annotation using the open-source software listed in Subheading 2.6, installed in a computer with a Unix-based environment or access to a remote server. In particular, we select NOVOPlasty [24] for the de novo assembly of mtgenome because is among the most used open-source software for conducting this analysis, outperforms most assemblers in terms of accuracy and coverage, and is user-friendly even for a researcher with limited bioinformatics experience. NOVOPlasty includes a fast and straightforward algorithm to generate a single circular high quality mtgenome sequence by extending a provided seed that can be either a single sequence read of the dataset belonging to the mtgenome, a published partial/complete conserved gene, or a partial/complete mtgenome from a related or distant taxon. As such, the seed can be chosen from a wide range of sequences and is used not to start the assembly but only to retrieve a raw sequence read belonging to the mtgenome. For a list of alternative software and pipelines for the mtgenome assembly (*see* **Note 7**).

1. Download the raw sequence data ("sample_name_forward.fastq.gz" and "sample_name_reverse.fastq.gz") into a folder called "data." All software should be installed in our path and the profile should be edited so that all commands can be accessible when run from the folder "data." $ indicates that we are working in the Unix/Linux-based environment.

2. Check the quality of the raw sequence data using FastQC [28] (http://www.bioinformatics.babraham.ac.uk/projects/fastqc, http://github.com/s-andrews/FastQC).

```
$ fastqc sample_name_forward.fastq.gz sample_name_re-
verse.fastq.gz
```

Open the two HTML files using a web server and visualize the plots. See the FastQC documentation to interpret the plots and assess the quality of sequencing reads.

3. Trim raw sequence reads for Illumina adapter sequences using Trimmomatic [29] (http://www.usadellab.org/cms/?page=trimmomatic, https://github.com/timflutre/trimmomatic).

```
$ java -jar trimmomatic-0.35.jar PE -phred33 sample_na-
me_forward.fastq.gz sample_name_reverse.fastq.gz output_sam-
ple_name_forward_paired.fq.gz output_sample_name_forward_un-
paired.fq.gz output_sample_name_reverse_paired.fq.gz output_
sample_name_reverse_unpaired.fq.gz ILLUMINACLIP:TruSeq3-PE.
fa:2:30:10 MINLEN:36
```

It will generate four output files (forward and reverse paired files where both reads survive, forward and reverse unpaired files where only one read survives). The command ILLUMINACLIP (followed by the library preparation kit information) indicates Illumina adapters while, for trimming non Illumina adapters, the command HEADCROP (followed by the number indicating the adapter's length) can be used. See the Trimmomatic documentation for additional explanations. Numerous other programs are available for trimming adapter sequences and we encourage to explore their specifics and performances (e.g., [46, 47]).

4. The mtgenome assembly is carried out using NOVOPlasty [24] (http://github.com/ndierckx/NOVOPlasty). Prior to running the analysis, two files need to be provided: a) a seed sequence in FASTA format ("seed_sequence.fas") that can be a single sequence read of the dataset that belongs to the mtgenome, a published partial/complete conserved gene or even a partial/complete mtgenome from a related or distant taxon; b) a configuration file ("config.txt") that is provided by NOVOPlasty (see the documentation for the explanation of all parameters). For example, this file contains the details of sequencing platform (read length, insert size, single/paired, Illumina/Ion Torrent) and the path and name of seed sequence, reference sequence (if any), forward and reverse reads (generated with Trimmomatic). Select "Mito" as

"Type" and then use the default settings. Move both files ("seed_sequence.fas" and "config.txt") into the folder called "data."

```
$ perl NOVOPlasty4.3.pl -c config.txt
```

It will generate a single circular high quality mtgenome sequence. See NOVOPlasty documentation for troubleshooting.

5. The gene annotation of the assembled mtgenome can be carried out using the on-line web server MITOS [30] (http://mitos.bioinf.uni-leipzig.de/index.py) or its updated version MITOS2 [31] (http://mitos2.bioinf.uni-leipzig.de/index.py). MITOS has a very user-friendly interface, including a step-by-step tutorial. For running the mtgenome annotation, upload the assembled mtgenome in FASTA format and select the appropriate genetic code. The analysis takes about 1.5 h for a single mtgenome of average length. Alternatively, the user-friendly software MOSAS can be used for mtgenome annotation [48] (http://mosas.byu.edu).
6. In case there are complete or partial mtgenomes or mitochondrial genes of taxa related to the targeted organism, belonging for example to the same family, class, order, or phylum, we suggest to further manually inspect the gene annotation generated in MITOS. Perform separated alignments for each gene and manually verified whether the genes annotated in MITOS correspond exactly to the genes deposited in public databases.
7. Additionally scan tRNAs for both primary and secondary structures using two on-line web servers: tRNAscan [32] (http://lowelab.ucsc.edu/tRNAscan-SE) and Rfam [33] (http://rfam.xfam.org). Both tRNAscan and Rfam are user-friendly web servers and detailed instructions can be found in the "help" sections of both web servers. For tRNAscan, upload the assembled mtgenome or potential tRNA sequence previously annotated by MITOS in FASTA format, and select the default "search mode" and the appropriate "sequence source" and "genetic code." For Rfam, select "search by sequence" and enter the tRNA sequence. Both searches take a few minutes to scan for tRNAs in the uploaded sequence, providing also their secondary structures.

4 Notes

1. For increasing the DNA yield and quality we suggest to: (1) for marine invertebrates with mucopolysaccharides (such as corals), carefully remove mucous by blotting with a chem-wipe if

possible because mucous will coprecipitate with DNA and may inhibit some reactions. It is also recommended to carefully remove the preservative from the tissue sample, for example drying ethanol-preserved samples; (2) for DNA extraction protocols including an enzymatic lysis, incubate the samples at 56 °C overnight using a heating block or a heated water bath.

2. Since we work with small μL amounts of reagents and samples, carefully and slowly pipette and mix the reagents. Ensure to avoid bubbles that would create bias in the instrument analyses.
3. Add the corresponding DNA sample μL to include ~1–1.2 μg dsDNA total quantity. Add nuclease-free water to a final volume of 43 μL. For example, if dsDNA concentration is 0.12 μg/μL, add 10 μL DNA sample and 33 μL nuclease-free water. In case dsDNA concentration is >1.2 μg/μL we recommend to dilute 1:10 the DNA sample to avoid pipetting errors when using volumes <1 μL. In case dsDNA concentration is <24 ng/μL, we recommend to concentrate the DNA samples using either vacuum concentrator (for samples eluted in water) or a typical ethanol/isopropanol precipitation method (for samples eluted in buffer solution). Based on our experience, the digestion can work well even if the dsDNA input is 0.7–0.8 μg. If the total yield of the DNA extraction is <0.7 μg dsDNA, we strongly recommend to either repeat the DNA extraction or use alternative commercial kits or customized protocols.
4. Restriction enzymes may be supplied with multiple buffers. Select the buffer that maximizes the enzyme activity. In case the enzymatic fragmentation is carried out with two enzymes, select the buffer that allows the most activities for both restriction enzymes.
5. The magnetic bead purification is performed numerous times during the entire protocol and it is essential to try to minimize the potential loss of beads during the cleanup steps. Follow these simple tips: homogenize magnetic beads by mini vortex mixer for 1 min before use; slowly pipette the supernatant; carefully avoid to disturb beads when removing/adding the supernatant and ethanol; elute beads only when ethanol is completely evaporated. The magnetic beads are typically stored at 4 °C.
6. The samples can be pooled only by using a unique combination of adapters for each library. Carefully annotate on your laptop/laboratory book the unique combination of adapters that is used for each sample. This information is essential for demultiplexing sequencing reads. The number of samples to be pooled and sequence in a single lane depends on a number of factors

including for example the desired sequencing output, the sequencing platform availability and cost, and the adapters used for library preparation.

7. Several other open-source software and pipelines can be used to assemble and annotate mtgenomes from NGS reads obtained following the presented protocol. For example, [49] developed MitoZ, an open-source software for de novo assembly, annotation, and visualization of animal mtgenomes with one-command solution. MitoZ works in a Unix-based environment and is designed to assemble mtgenome without the need of closely related references. For documentation and commands *see* http://github.com/linzhi2013/MitoZ. Alternatively, another frequently used software for mtgenome assembly from NGS reads of genomic DNA libraries is MITObim [22] (http://github.com/chrishah/MITObim). This program is open-source and requires a closely or distantly related mtgenome reference or mitochondrial barcode sequences and is easy to use. Additionally, GRAbB [50] (http://github.com/b-brankovics/grabb) can be used to assemble mtgenomes and other multi copy regions, by specifying a bait reference sequence. Finally, [23] described the commands and steps to assemble mtgenomes using the software Geneious (http://www.geneious.com). Apart from these software, several other programs are available and we suggest to explore their specifics and performances.

References

1. Brown WM, George M, Wilson AC (1979) Rapid evolution of animal mitochondrial DNA. Proc Natl Acad Sci U S A 76: 1967–1971
2. Avise JC, Arnold J, Ball RM et al (1987) Intraspecific phylogeography: the mitochondrial DNA bridge between population genetics and systematics. Annu Rev Ecol Syst 18:489–522
3. Chen XJ, Butow RA (2005) The organization and inheritance of the mitochondrial genome. Nat Rev Genet 6:815–825
4. Hellberg ME (2006) No variation and low synonymous substitution rates in coral mtDNA despite high nuclear variation. BMC Evol Biol 6:24
5. Breton S, Beaupre HD, Stewart DT et al (2007) The unusual system of doubly uniparental inheritance of mtDNA: isn't one enough? Trends Genet 23:465–474
6. Birky CW Jr, Fuerst P, Maruyama T (1989) Organelle gene diversity under migration, mutation, and drift: equilibrium expectations, approach to equilibrium, effects of heteroplasmic cells, and comparison to nuclear genes. Genetics 121:613–627
7. Bazin E, Glémin S, Galtier N (2006) Population size does not influence mitochondrial genetic diversity in animals. Science 312: 570–572
8. Moritz C, Dowling TE, Brown WM (1987) Evolution of animal mitochondrial DNA: relevance for population biology and systematics. Annu Rev Ecol Syst 18:269–292
9. Harrison RG (1989) Animal mitochondrial DNA as a genetic marker in population and evolutionary biology. Trends Ecol Evol 4:6–11
10. Hebert PD, Cywinska A, Ball SL et al (2003) Biological identifications through DNA barcodes. Proc R Soc Lond B Biol Sci 270: 313–321
11. Osigus HJ, Eitel M, Bernt M et al (2013) Mitogenomics at the base of Metazoa. Mol Phylogenet Evol 69:339–351
12. Liu Y, Cox CJ, Wang W et al (2014) Mitochondrial phylogenomics of early land plants: mitigating the effects of saturation, compositional

heterogeneity, and codon-usage bias. Syst Biol 63:862–878

13. Kayal E, Roure B, Philippe H et al (2013) Cnidarian phylogenetic relationships as revealed by mitogenomics. BMC Evol Biol 13:5
14. Arnason U, Gullberg A, Janke A (2004) Mitogenomic analyses provide new insights into cetacean origin and evolution. Gene 333: 27–34
15. Tan MH, Gan HM, Lee YP et al (2019) Comparative mitogenomics of the Decapoda reveals evolutionary heterogeneity in architecture and composition. Sci Rep 9:10756
16. Boore JL (1999) Animal mitochondrial genomes. Nucleic Acids Res 27:1767–1780
17. Boore JL, Macey JR, Medina M (2005) Sequencing and comparing whole mitochondrial genomes of animals. Methods Enzymol 395:311–348
18. Hu M, Jex AR, Campbell BE et al (2007) Long PCR amplification of the entire mitochondrial genome from individual helminths for direct sequencing. Nat Protoc 2:2339–2344
19. Lin MF, Luzon KS, Licuanan WY et al (2011) Seventy-four universal primers for characterizing the complete mitochondrial genomes of scleractinian corals (Cnidaria; Anthozoa). Zool Stud 50:513–524
20. Van Dijk EL, Auger H, Jaszczyszyn Y et al (2014) Ten years of next-generation sequencing technology. Trends Genet 30:418–426
21. Schuster SC (2008) Next-generation sequencing transforms today's biology. Nat Methods 5: 16–18
22. Hahn C, Bachmann L, Chevreux B (2013) Reconstructing mitochondrial genomes directly from genomic next-generation sequencing reads—a baiting and iterative mapping approach. Nucleic Acids Res 41:e129
23. Briscoe AG, Hopkins KP, Waeschenbach A (2016) High-throughput sequencing of complete mitochondrial genomes. In: Bourlat SJ (ed) Marine genomics. Springer, Berlin, pp 45–64
24. Dierckxsens N, Mardulyn P, Smits G (2017) NOVOPlasty: de novo assembly of organelle genomes from whole genome data. Nucleic Acids Res 45:e18
25. Nabholz B, Jarvis ED et al (2010) Obtaining mtDNA genomes from next-generation transcriptome sequencing: a case study on the basal Passerida (Aves: Passeriformes) phylogeny. Mol Phylogenet Evol 57:466–470
26. Stobie CS, Cunningham MJ, Oosthuizen CJ et al (2019) Finding stories in noise: mitochondrial portraits from RAD data. Mol Ecol Resour 19:191–205
27. Amaral FR, Neves LG, Resende MF et al (2015) Ultraconserved elements sequencing as a low-cost source of complete mitochondrial genomes and microsatellite markers in non-model amniotes. PLoS One 10:e0138446
28. Andrews S (2010) FastQC: a quality control tool for high throughput sequence data. Available from: http://www.bioinformatics.babraham.ac.uk/projects/fastqc
29. Bolger AM, Lohse M, Usadel B (2014) Trimmomatic: a flexible trimmer for Illumina sequence data. Bioinformatics 30:2114–2120
30. Bernt M, Donath A, Jühling F et al (2013) MITOS: improved de novo metazoan mitochondrial genome annotation. Mol Phylogenet Evol 69:313–319
31. Donath A, Jühling F, Al-Arab M et al (2019) Improved annotation of protein-coding genes boundaries in metazoan mitochondrial genomes. Nucleic Acids Res 47:10543–10552
32. Chan PP, Lowe TM (2019) tRNAscan-SE: searching for tRNA genes in genomic sequences. Methods Mol Biol 1962:1–14
33. Kalvari I, Nawrocki EP, Ontiveros-Palacios N et al (2021) Rfam 14: expanded coverage of metagenomic, viral and microRNA families. Nucleic Acids Res 49:D192–D200
34. Nishiguchi MK, Doukakis P, Egan M et al (2002) DNA isolation procedures. In: DeSalle R, Giribet G, Wheeler W (eds) Techniques in molecular systematics and evolution. Methods and tools in biosciences and medicine. Springer, Berlin, pp 249–287
35. Panova M, Aronsson H, Cameron RA et al (2016) DNA extraction protocols for whole-genome sequencing in marine organisms. In: Bourlat SJ (ed) Marine genomics. Springer, Berlin, pp 13–44
36. Gaither MR, Szabó Z, Crepeau MW et al (2011) Preservation of corals in salt-saturated DMSO buffer is superior to ethanol for PCR experiments. Coral Reefs 30:329–333
37. Chafin TK, Martin BT, Mussmann SM et al (2018) FRAGMATIC: in silico locus prediction and its utility in optimizing ddRADseq projects. Conserv Genet Resour 10:325–328
38. Toonen RJ, Puritz JB, Forsman ZH et al (2013) ezRAD: a simplified method for genomic genotyping in non-model organisms. PeerJ 1:e203
39. Tisthammer KH, Forsman ZH, Sindorf VL et al (2016) The complete mitochondrial genome of the lobe coral *Porites lobata* (Anthozoa: Scleractinia) sequenced using ezRAD. Mitochondrial DNA B Resour 1:247–249

40. Terraneo TI, Arrigoni R, Benzoni F et al (2018a) The complete mitochondrial genome of *Porites harrisoni* (Cnidaria: Scleractinia) obtained using next-generation sequencing. Mitochondrial DNA B Resour 3:286–287
41. Terraneo TI, Arrigoni R, Benzoni F et al (2018b) Using ezRAD to reconstruct the complete mitochondrial genome of *Porites fontanesii* (Cnidaria: Scleractinia). Mitochondrial DNA B Resour 3:173–174
42. Price M, Knapp ISS, Toonen RJ et al (2021) Phylogenomics of an adaptive radiation in endangered Hawaiian tree snails (Achatinellidae). PeerJ 9:e10993
43. Price MR, Forsman ZH, Knapp ISS et al (2018) A comparison of mitochondrial genomes from five species in three genera suggests polyphyly in the subfamily Achatinellinae (Gastropoda: Pulmonata: Stylommatophora: Achatinellidae). Mitochondrial DNA B Resour 3: 611–612
44. Knapp IS, Puritz JB, Bird CE et al (2016) ezRAD-an accessible next-generation RAD sequencing protocol suitable for non-model organisms v3. Protocols io Life Sciences Protocol Repository. https://doi.org/10.17504/protocols.io.e9pbh5n
45. Ekblom R, Smeds L, Ellegren H (2014) Patterns of sequencing coverage bias revealed by ultra-deep sequencing of vertebrate mitochondria. BMC Genomics 15:467
46. Martin M (2011) Cutadapt removes adapter sequences from high-throughput sequencing reads. EMBnet J 17:10–12
47. Schubert M, Lindgreen S, Orlando L (2016) AdapterRemoval v2: rapid adapter trimming, identification, and read merging. BMC Res Notes 9:1–7
48. Sheffield NC, Hiatt KD, Valentine MC et al (2010) Mitochondrial genomics in Orthoptera using MOSAS. Mitochondrial DNA 21: 87–104
49. Meng G, Li Y, Yang C et al (2019) MitoZ: a toolkit for animal mitochondrial genome assembly, annotation and visualization. Nucleic Acids Res 47:e63
50. Brankovics B, Zhang H, van Diepeningen AD et al (2016) GRAbB: selective assembly of genomic regions, a new niche for genomic research. PLoS Comput Biol 12:e1004753

Chapter 2

Genome-Wide DNA Methylation Protocol for Epigenetics Studies

Annalaura Mancia

Abstract

Epigenetic modification are heritable changes in gene expression not encoded by the DNA sequence therefore playing a significant role in a broad range of biological processes and diseases.

A key player of the epigenetic modifications is the DNA methylation, a process involving the covalent transfer of a methyl group to the C-5 position of the cytosine ring of DNA forming 5-methylcytosine (5mC), catalyzed by DNA methyltransferases. Altering the structure of the chromatin, DNA methylation has the potential to down-regulate gene expression.

The here presented protocol shows a method to obtain DNA samples ready for NGS sequencing for genome-wide methylation analysis.

Key words Methylated cytosine, Bisulfite conversion, Reduced Representation Bisulfite Sequencing, CpG island

1 Introduction

The major epigenetic modification of DNA in mammalian genomes is the methylation of cytosine (5mC), namely, DNA methylation. The process of DNA methylation involves the transfer of a methyl group (CH_3) to the C-5 position of a cytosine in the DNA sequence [1]. The primary target sequence of methylation in the DNA of mammals is at 5′-CpG-3′ dinucleotides. Regions with a high frequency of CpG sites are known as CpG islands and are often found in gene promoters [2]. The main known mechanism by which DNA methylation can regulate gene expression is by preventing the binding of factors to the DNA that promotes transcriptional activity thereby repressing gene expression [3]. However, DNA methylation can also induce gene expression by preventing the binding of proteins that repress gene expression, thus inducing gene expression under the regulatory activity of nearby enhancers. Alterations to the DNA methylation profile (or methylome) can

Cinzia Verde and Daniela Giordano (eds.), *Marine Genomics: Methods and Protocols*,
Methods in Molecular Biology, vol. 2498, https://doi.org/10.1007/978-1-0716-2313-8_2,

occur at specific loci or more broadly across the genome and can also promote chromosome stability, silence the transcription of transposons [4, 5] playing a critical role in genome function both in health and disease.

In the past 30 years the technology to investigate epigenetic marks has substantial improved. One of the most powerful approaches described to interrogate for regional DNA methylation changes at a genome-wide scale is based on the chromatin immune-precipitation of methylated fragments (MeDIP), where the selected methylated DNA is coupled to microarray chips hybridization [6]. Other methods are also array-based and some take advantage of the restriction activity by the methylation-dependent enzymes.

The most used method to detect 5mC with a resolution at the level of CpG sites, is the bisulfite sequencing, a method that employs sodium bisulfite to chemically convert the DNA prior to DNA sequencing. First described by Frommer et al. (1992) [7] then optimized by Clark (1994) [8] to detect local DNA methylation changes, bisulfite sequencing involves the differential conversion of cytosine or methylated cytosine to uracil when DNA is treated with sodium bisulfite. Further PCR amplification and sequencing allow for this differential conversion to be detected. While methylated cytosines are sequenced as cytosines, unmethylated cytosines are sequenced as thymines.

To enrich for CG-rich parts of the genome, thereby reducing the amount of sequencing required while capturing most promoters and other relevant genomic regions, the treatment of DNA with the restriction endonuclease *MspI* has been applied prior to converting with bisulfite treatment and sequencing. This method, first proposed in 2005 [9], is called reduced representation bisulfite sequencing (RRBS) [10–12].

Described here is an adjusted method for RRBS. To increase coverage, while limiting the increase in sequencing cost, the combination of *MspI* and *ApeKI* is employed [13]. The method yields data with an appropriate coverage per CpG site for differential methylation site and region determinations and for epigenetic polymorphism detection.

Briefly, the main steps are the following: digestion of genomic DNA by restriction enzymes *MspI* and *ApeKI*; end-repair, A-tailing, and ligation to methylated Illumina adapters; size selection and bisulfite conversion of CpG-rich DNA fragments; PCR amplification and sequencing on an Illumina Genome Analyzer. A schematic timetable of the procedure described below is shown in Fig. 1. A schematic diagram of the critical steps of the short fragment generation, bisulfite genomic conversion and PCR amplification protocol is outlined in Fig. 2.

Detecting genome-wide 5-hmC at single-base resolution: 5mC vs *5hmC.* Sodium bisulfite sequencing cannot discriminate between 5-methylcytosine, 5mC, and 5-hydroxymethylcytosine, 5hmC, an

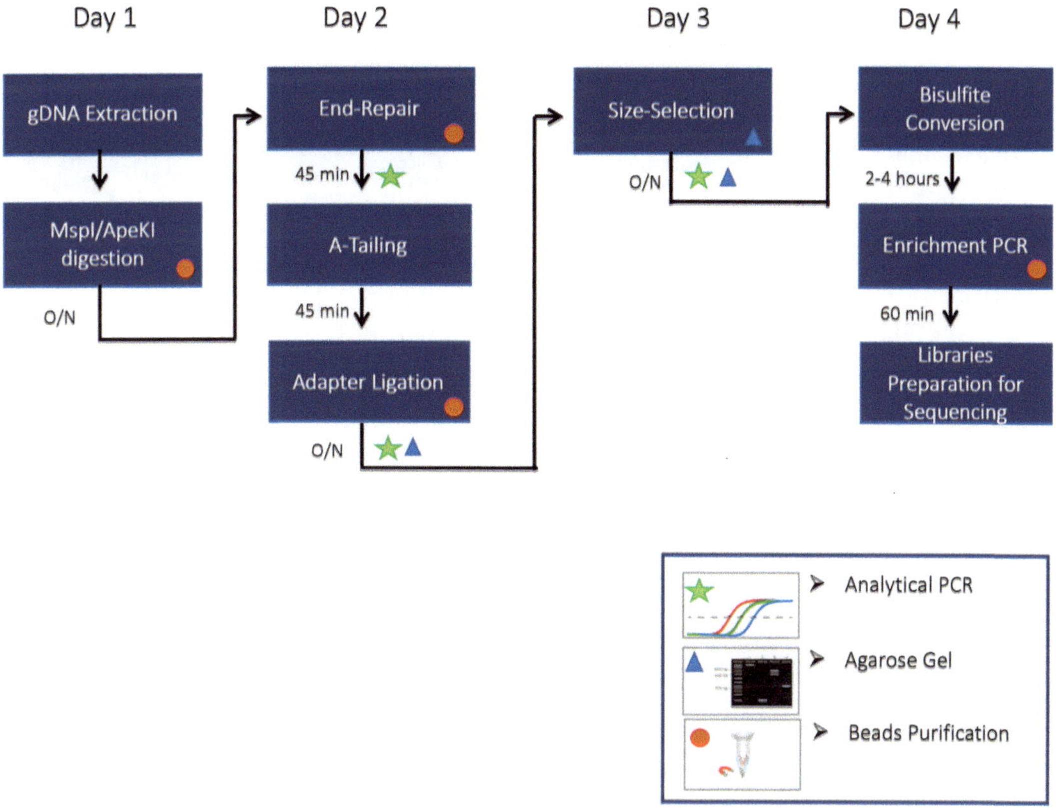

Fig. 1 Timetable of the complete DNA methylation procedure with bisulfite conversion. In Day 1 the gDNA is extracted and digested with *MspI* and *ApeKI*. In Day 2, the digested DNA goes through end-repair, A-tailing, and adapter ligation. In Day 3 fragments are size selected. In Day 4 size selected libraries are subject to enrichment PCR. Resulting libraries are quantified and prepared for sequencing. Purification steps with bead are indicated with red bullet points; checking points by analytical PCR are indicated by green stars; checking points with electrophoresis agarose gels are indicated by blue triangles (*see* legend in the bottom left corner)

intermediate metabolite in active DNA demethylation. However, in the past few years, there has been increasing evidence that 5hmC is a stable derivate of 5mC, with emerging roles in defining cellular fate [14–18]. Researchers have developed methods such as oxidative bisulfite sequencing (oxBS-seq) to allow discrimination between 5mC and 5hmC [19] (*see* **Note 1**).

2 Materials

Prepare all solutions using ultrapure water (prepared by purifying deionized water, to attain a sensitivity of 18 MΩ-cm at 25 °C) and analytical grade reagents. Prepare and store all reagents at the temperature indicated. Meticulously follow all waste disposal regulations when disposing of waste materials.

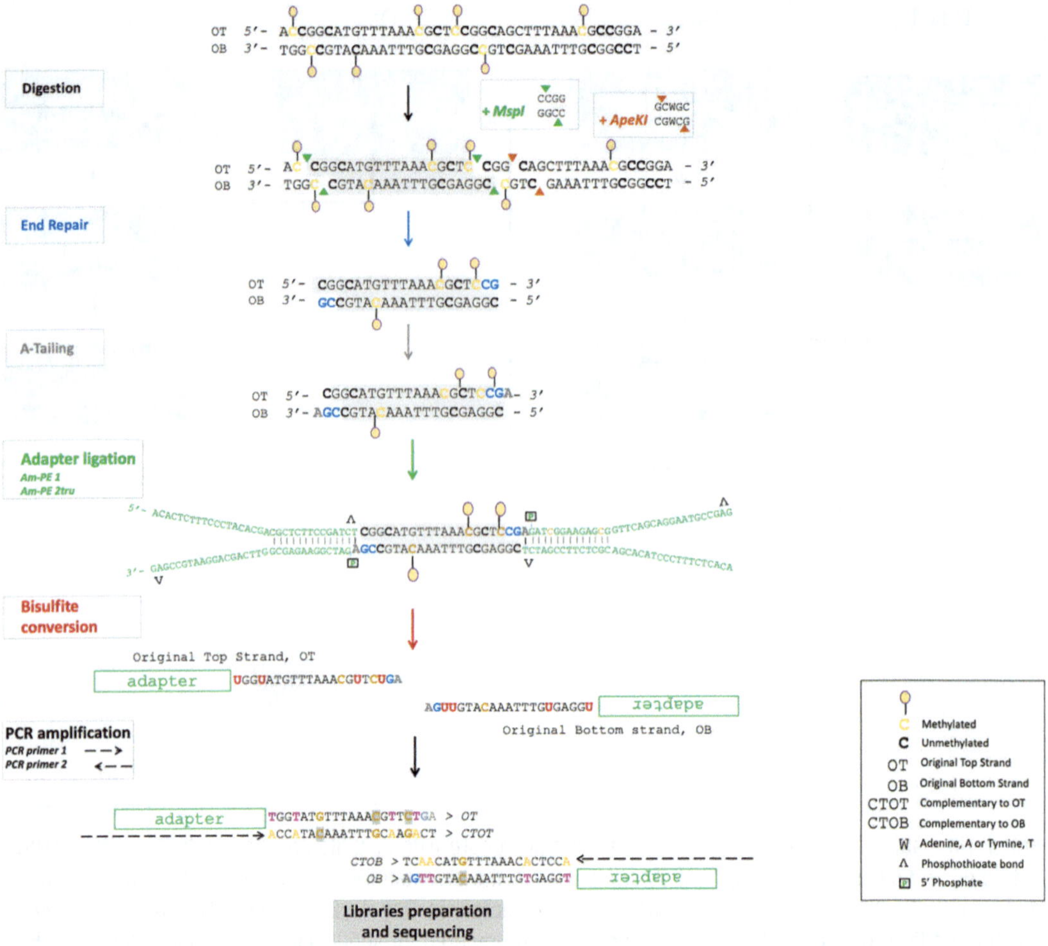

Fig. 2 Schematic diagram of the critical steps in the bisulfite conversion and PCR amplification reaction. In this example, a DNA sequence, 5′ to 3′ orientation, with the complementary top (OT) and bottom (OB) DNA strands is shown. Methylated C are indicated by balloons (*see* legend in the bottom left corner)

2.1 Equipment

1. Pipettes (P2, P10, P20, P100, P200, P1000) and corresponding sterile filter tips, multichannel pipette eight-channel (0.5–10 μL, 5–100 μL).
2. 1.5 mL microcentrifuge tubes, PCR tubes, PCR eight-strip tubes and PCR 96-well plates and sealing films; 15 mL and 50 mL tubes.
3. Benchtop centrifuges for 1.5/2.0 mL tubes.
4. Microplate centrifuge.
5. Microvolume spectrophotometer.
6. Vortex.
7. Incubator.
8. Heating block.

9. Electrophoresis equipment.
10. Microwave oven.
11. Thermal cycler PCR system 96-well block format (with heated lid).
12. Real time PCR detection system.
13. Magnetic stand.
14. Microplate Spectrophotometer.
15. Microchip automated electrophoresis instruments.

2.2 Reagents

1. *MspI, ApeKI,* NEBuffer r3.1—used to digest genomic DNA.
2. Low melting Agarose—used for fine resolution of small DNA and PCR products $\leq$1 kb on agarose gel electrophoresis.
3. Tris–borate–EDTA, TBE, buffer.
4. Tris–acetate–EDTA, TAE, buffer.
5. Magnetic beads.
6. DNase/RNase-free distilled water.
7. Absolute ethanol.
8. 1 M Tris–HCl.
9. PEG 8000.
10. Acrylamide–bis 19:1, 40% (w/v) solution, ammonium persulfate (APS), TEMED, 6× Agarose Gel-Loading Dye—used for 5% Native Polyacrylamide DNA gel.
11. Nanosep Centrifugal Devices with wwPTFE membrane—0.45 μm.
12. DNA minicolumns.
13. Glycogen.
14. 3 M sodium acetate solution, pH 5.2.
15. SybrGreen PCR Master Mix—used for qPCR of glucosyl-5hmC optional protocol.
16. 2′-deoxyadenosine 5′-triphosphate (dATP) and deoxynucleotide triphosphate (dNTP) solution set.
17. 10× T4 ligase reaction buffer (-ATP), T4 DNA Polymerase, DNA Polymerase I Large (Klenow) Fragment, T4 Polynucleotide Kinase, Quick Ligase—used to end repair digested DNA and to anneal adapters.
18. Klenow Fragment (3′–5′ exo-), 10× T4 ligase reaction buffer (-ATP), 10 mM 2′-deoxyadenosine 5′-triphosphate (dATP)—used to A-tail digested end repaired DNA.
19. 1 kb DNA ladder and 50 bp DNA ladder.
20. SybrGold Nucleic Acid Gel Stain.

21. Phusion Hot Start II DNA polymerase (2 U/μL), 5× Phusion HF Buffer—used for Analytical PCRI.
22. Platinum Taq DNA polymerase, $MgCl_2$, 10× PCR buffer -$MgCl_2$—used for Analytical PCRII and Enrichment PCR.
23. Bisulfite Reaction kit; in alternative, if bisulfite reaction is performed manually:
    (a) Degassed distilled water (degassed dH_2O) used to make the following solutions: 3 N NaOH, 100 mM hydroquinone.
    (b) Sodium bisulfite/sodium metabisulfite:
    (c) 0.5 M Na_2EDTA, pH 8.0.
    (d) Minicolumn-based DNA purification kit.
    (e) Desulfonation buffer.
    (f) 1× Tris-EDTA buffer, pH 8.0.

3 Methods

Carry out all procedures at the temperature indicated. Ensure that pipettes are correctly calibrated and that they are not used at the volume extremes of their performance specifications.

Day 1.

3.1 Digestion of Genomic DNA

1. Prepare 500–1000 ng of high-quality genomic DNA (gDNA, *see* **Note 2**) as starting material in 44 microliter (μL) of DNase-free water (*see* **Note 1**).
2. Mix gDNA with 1 μL of *MspI* (100 U/μL), 5 μL of 10× NEB r3.1 reaction buffer, to a total volume of 50 μL, mix, spin down.
3. Incubate the reaction at 37 °C in a thermal cycler for 2 h (lid temperature set: 60 °C). After 2 h add 1 μL of *MspI* (100 U/μL) directly to all samples. Mix, spin down.
4. Incubate the reaction at 37 °C in a thermal cycler for 2 more hours (lid temperature set: 60 °C) (*see* **Note 3**).
5. Add 4 μL of *ApeKI* 5 U/μL, 5 μL of 10× NEB r3.1 reaction buffer, 1 μL of 5 M NaCl, 4.5 μL 1 M Tris–HCl and 33 μL DNase-free water to bring the reaction to a total volume of 100 μL.
6. Mix and spin briefly. Incubate reaction mixtures at 75 °C in a thermal cycler for 8 h (lid temperature set: 100 °C).
7. Proceed to digested gDNA (dgDNA) purification using magnetic beads (1.8×) (*see* **Note 4**):

(a) Transfer samples to newly labeled 1.5 mL tubes.
(b) Vortex the beads thoroughly (20–30 seconds, sec).
(c) Add 180 μL of beads to each tube containing 100 μL of dgDNA.
(d) Incubate the tubes at RT for 15 minutes (min).
(e) Place the tubes on magnetic stand at RT for 2 min or until the liquid appears clear.
(f) Remove and discard 270 μL of supernatant from each tube using P200. Keep the plate on the magnetic stand and to not disturb the beads.
(g) Wash the beads by adding 190 μL of freshly prepared 80% EtOH to each tube.
(h) Incubate at RT for 30 s.
(i) Remove and discard all the supernatant from each tube, 200 μL in 2 steps.
(j) With tubes still on magnet, add 190 μL of freshly prepared 80% EtOH to each tube.
(k) Let stand for 1 min.
(l) Remove all of supernatant (in 2 steps) and discard.
(m) Let the tubes stand at RT for 5 min to dry and then remove the tubes from the magnetic stand.
(n) Resuspend the dried pellet in 22.5 μL resuspension buffer (according to the manufacturer's instructions) and pipet mix up and down 10 times.
(o) Incubate the tubes at RT for 2 min and then place the tubes on magnetic stand at RT for 2 min.
(p) Transfer 20 μL of clear supernatant from each tube to new labeled tubes. Keep 5 μL for OD (and qPCR, if GSD reaction is performed—*see* **Note 1**). Use 14 μL in End Repair.

Day 2.

3.2 End-Repair (ER) Reaction

1. On ice, prepare ER mix: 2.5 μL 10× T4 Ligase Rxn Buffer (-ATP), 2.5 μL 10 mM ATP, 1 μL 10 mM dNTP solution mix, 1.25 μL 3000 U/mL T4 DNA Polymerase, 0.25 μL 3000 U/ mL DNA Polymerase I Large (Klenow) Fragment, 1.25 μL 10,000 U/mL T4 Polynucleotide Kinase, and 2.25 μL DNase-free water.
2. Transfer the 14 μL of dgDNA (from Subheading 3.1, **step** 7.p) into a PCR tube on ice (4 °C) and add ER mix for a final volume of 25 μL.

3. Incubate on the thermal cycler with the heat lid on for 30 min at 20 °C.
4. Perform purification using magnetic beads (1.8×).
   (a) Transfer samples to newly labeled 1.5 mL tubes.
   (b) Vortex the magnetic beads thoroughly (20–30 s).
   (c) Add 180 μL of beads to each tube containing 100 μL of dgDNA.
   (d) Incubate the tubes at RT for 15 min.
   (e) Place the tubes on magnetic stand at RT for 2 min until the liquid appears clear.
   (f) Remove and discard 270 μL of supernatant from each tube using P200. Do not disturb the beads.
   (g) Wash the beads by adding 190 μL of freshly prepared 80% EtOH to each tube.
   (h) Incubate at RT for 30 s.
   (i) Remove and discard all the supernatant from each tube, 200 μL in 2 steps.
   (j) With tubes still on magnet, add 190 μL of freshly prepared 80% EtOH to each tube.
   (k) Let stand for 1 min.
   (l) Remove all of supernatant (in 2 steps) and discard.
   (m) Let the tubes stand at RT for 5 min to dry and then remove the tubes from the magnetic stand.
   (n) Resuspend the dried pellet in 22.5 μL resuspension buffer (according to the manufacturer's instructions) and pipet-mix up and down 10 times.
   (o) Incubate the tubes at RT for 2 min and then place the tubes on magnetic stand at RT for 2 min.
   (p) Transfer 20 μL of clear supernatant from each tube to newly labeled tubes. Store in −20 °C, overnight (o/n) (*see* **Note 5**).

3.3 A-tailing (A-t) Reaction

1. On ice (4 °C), add 20 μL of DNA solution from Subheading 3.2, **step 4.p** to 5 μL of A-t mix: 2.5 μL 10× T4 Ligase Rxn Buffer (-ATP), 0.5 μL 10 mM dATP, 1.5 μL 5000 units/mL Klenow Fragment (3′ → 5′ exo), 0.5 μL DNase-free water.
2. Incubate the 25 μL reaction at 37 °C for 30 min followed by 20 min at 75 °C with the heat lid temperature set at 100 °C.
3. Go immediately into Adapter Ligation Reaction (Subheading 3.5).

Table 1
Adapter sequences used in RRBS: Am-PE1 and Am-PE2tru

Am-PE 1	5'-ACACTCTTTCCCTACACGAC GCTCTTCCGATC*T A/iMe-dC/A/iMe-dC/T/iMe-dC/TTT/iMe-dC//iMe-dC//iMe-dC/TA/iMe-dC/A/iMe-dC/GA/iMe-dC/G/iMe-dC/T/iMe-dC/TT/iMe-dC//iMe-dC/GAT/iMe-dC/*T All C's are methylated,* = phosphorothioate bond
Am-PE 2tru	P-GATCGGAAGAGC ACACGTCTGAACTCCAGTCAC-3'-/3SpC3/ /5Phos/GAT/iMe-dC/GGAAGAG/iMe-dC/A/iMe-dC/A/iMe-dC/GT/iMe- All C's are methylated, 5' phosphate, C3 spacer

3.4 Adapter Annealing

1. Anneal methylated adapters at 10 μM in 100 μL final volume starting from stock solutions (*see* **Note 6**).
2. Thaw adapters (adapt methyl PE1, Am-PE1 and adapt methyl PE2tru, Am-PE2tru) on ice. Set heat block at 80 °C. Am-PE1 and Am-PE2tru info are in Table 1.
3. Mix reagents, vortex, spin down and move in heat block for 10 min at 80 °C.
4. Cool down in the same block at RT.
5. Store annealed adapters at −20 °C.

3.5 Adapter Ligation (AL) Reaction

(*See* **Note 7**, before starting).

1. Add to 1.5 mL tube: 25 μL DNA (from Subheading 3.3), 6 μL 50% PEG 8000, 4 μL ATP 10 mM, 1.5 μL 10× T4 Ligase Rxn Buffer (-ATP). Mix and spin down.
2. Add 3 μL Annealed adapters 10 μM (from Subheading 3.4, **step 5**) vortex gently and spin down.
3. Add 1.5 μL 2,000,000 units/mL Quick ligase (*see* **Note 8**).
4. Incubate on the thermal cycler for 30 min at 25 °C.
5. Perform purification using magnetic beads (1.0×)—2 rounds.
6. Place beads at RT 30 min before starting.
   (a) Transfer samples to newly labeled 1.5 mL tubes.
   (b) Vortex the magnetic beads thoroughly (20–30 s).
   (c) Add 41 μL of beads to each tube containing DNA.
   (d) Incubate the tubes at RT for 15 min.
   (e) Place the tubes on magnetic stand at RT for 2 min until the liquid appears clear.
   (f) Remove and discard 65 μL of supernatant from each tube using P200. Do not disturb the beads.

(g) Wash the beads by adding 190 μL of freshly prepared 80% EtOH to each tube.
(h) Incubate at RT for 30 s.
(i) Remove and discard all supernatant from each tube, 180 μL in 2 steps.
(j) With tubes still on magnet, add 190 μL of freshly prepared 80% EtOH to each tube.
(k) Let stand for 1 min.
(l) Remove all of supernatant (in 2 steps) and discard.
(m) Let the tubes stand at RT for 5 min to dry and then remove the tubes from the magnetic stand.
(n) Resuspend the dried pellet in 42 μL resuspension buffer (according to the manufacturer's instructions) and pipet-mix up and down 10×.
(o) Incubate the tubes at RT for 2 min and then place the tubes on magnetic stand at RT for 2 min.
(p) Transfer 40 μL of clear supernatant from each tube to a new 1.5 mL tube containing 40 μL beads.
(q) Repeat **steps a–n** and then resuspend in 14 μL to take 12 μL (*see* **Note 9**).

3.6 Analytical PCRI

After AR reaction, check 1 μL of each library (ALigDNA) with 17 cycles of analytical PCR along with positive and negative controls (*see* **Note 10**). The PCR product will be run on low melting agarose gel (*see* below **step 5**).

1. Prepare the PCR master-mix (MM1) on ice. For one reaction add reagents in the following order: 6.44 μL H_2O, 2 μL 5× Phusion HF Buffer, 0.16 μL 25 μM PCR PE primer 1, 0.16 μL 25 μM PCR PE primer 2 indexl (Table 2), 0.08 μL 2 U/μL Phusion Hot Start II DNA Polymerase, 0.16 μL 10 mM dNTP. Mix by pipetting.
2. Add 9 μL MM1 to 1 μL of ALigDNA for a final volume of 10 μL.
3. Mix gently, but thoroughly, and centrifuge briefly.
4. Amplify using the following PCR process with a heated lid (set the heat lid to 100 °C): 5 min at 98 °C, 17 × (20 s at 98 °C, 30 s at 65 °C, 1 min at 72 °C), hold at 4 °C.
5. Prepare 3% low melting agarose DNA gel (Table 3) as follows:
   (a) Choose a beaker that is 2–4 times the volume of the solution. Add chilled 1× or 0.5× electrophoresis buffer and a stir bar to a beaker.
   (b) Slowly sprinkle in the agarose powder while the solution is rapidly stirred.

Table 2
Primer sequences used in Analytical PCRI: PCR PE primer 1 and PCR PE primer 2 index1

PCR PE 1	5'-AATGATACGGCGACCACCGAGATCT ACACTCTTTCCCTACACGA*C * = phosphorothioate bond
PCR PE 2 index1	5'-CAAGCAGAAGACGGCATACGAGAT CGTGATGTGACTGGAGTTCAGACGTG*T * = phosphorothioate bond

Table 3
Suggested agarose concentration for Low melting Agarose Gels

Suggested Agarose Concentrations		
Size Range (bp)	Final Agarose Concentration (%)	
	1X TAE Buffer	1X TBE Buffer
500-1,000	3	2
100-500	4	3
10-100	6	5

3% Agarose Gel		
	Low melting Agarose	1X TAE Buffer
Mini	1.5	50
Midi	3	100

(c) Remove the stir bar if not Teflon coated.

(d) Soak the agarose in the buffer for 15 min before heating. This reduces the tendency of the agarose solution to foam during heating.

(e) Weigh the beaker and solution before heating.

(f) Cover the beaker with plastic wrap. Pierce a small hole in the plastic wrap for ventilation. For agarose concentrations >4%, the following additional steps will further help prevent the agarose solution from foaming during melting/dissolution.

- Heat the beaker in the microwave oven on medium power for 1 min.
- Remove the solution from the microwave oven.
- Allow the solution to sit on the bench for 15 min.

(g) Heat the beaker in the microwave oven on medium power for 2 min.

(h) Remove the beaker from the microwave oven (*see* **Note 11**).

(i) Gently swirl the beaker to resuspend any settled powder and gel pieces.

(j) Reheat the beaker on high power until the solution comes to a boil.

(k) Hold at boiling point for 1 min or until all the particles are dissolved.

(l) Remove the beaker from the microwave oven.

(m) Gently swirl the beaker to thoroughly mix the agarose solution.

(n) After dissolution, add sufficient hot distilled water to obtain the initial weight. Mix thoroughly.

(o) Cool the solution to 50–60 °C prior to casting.

Prepare Samples

(p) Add 2 μL 6× DNA loading dye to DNA amplified and purified in Subheading 3.5.

(q) Add required volume of 6× DNA loading dye to 300 ng of 50 bp DNA ladder in 12 μL final volume. Vortex gently, spin down.

(r) Load samples and DNA ladder into wells.

Run and Stain

(s) Run gel at 85 V (constant voltage) for 1 h/1 h 30 min in 1× TAE.

(t) Add 12 μL of 10,000× SYBRGold stain to 100 mL 1× TAE buffer to submerge the gel in a staining tray. Mix stain well.

(u) Place gel in tray and stain for 1 h in the dark.

(v) Photograph gel and save file.

Day 3.

3.7 Size Selection

1. Prepare and run a 5% native polyacrylamide cDNA gel:

(a) Wash plates, spacer and comb with soap, water then rinse with ethanol. Assemble casting apparatus.

(b) Prepare fresh 10% APS mix.

(c) Prepare 5% acrylamide mix in 50 mL tubes. For one gel: 4 μL 5× TBE, 2.5 μL 40% 19:1 Acryl amide, 13.5 μL DNase-free water to a final volume of 20 μL.

(d) Degas the mix in side-arm flask under vacuum for 10 min with intermittent gentle swirling.

(e) Add 30 μL of TEMED, mix well.

(f) Add 200 μL of 10% APS, mix well.

(g) Cast the gel using a 10 mL pipette and insert the comb.

(h) Allow the gel to polymerize for 30 min.

(i) Store the excess acryl amide mix in 50 mL tube to check for polymerization.

(j) Setup to run the gels in 1× TBE buffer.

(k) Prerun the gel for 5 min.

Prepare Samples

(l) Add 7 μL Acrylamide DNA loading dye 2.5× (6× Ficoll-BB DNA loading dye, Fresh 5× TBE, DNase-free water) to 10 μL ALigDNA (from Subheading 3.5, **step q**).

(m) Add required volume of 2.5× Acrylamide DNA loading dye to 50 bp DNA ladder in 17 μL final volume. Vortex gently, spin down.

(n) Load the samples on 5% TBE-PAGE Gel.

Run and Stain

(o) Run gel at 200 V, 20 m AMPS/gel. Run for 55 min.

(p) Add a 1/10,000 dilution of SybrGold stain to enough 1× TBE buffer to submerge the gel in a staining tray (40 mL/ one gel).

(q) Mix stain well. Place gel in tray and stain for 10 min in the dark.

(r) Photograph gel with ruler, save file and put gel back in staining solution.

(s) Calculate the region to excise for 150–450 bp.

(t) Excise using ruler following protocol below (**step 2**).

2. Perform gel-based DNA purification (2 days protocol).

(a) Use small spatula to slightly lift gel in staining tray, grab by edges and transfer to light box.

(b) Slice off the ladders using a clean razor blade and visualize them on a UV transilluminator. Do not expose the samples to UV light. Mark the 150 bp, 250 bp and 450 bp bands of the ladders using a razor blade or pipette tips (this will allow the excision of samples without exposing them to UV light).

(c) Align the ladder slices with the rest of the gel. Using the marked bands as reference, excise one slice containing the 150–250 bp library fraction (lower) and another slice containing the 250–450 bp library fraction (higher). Place each of the two slices into different tubes (*see* **Note 12**).

(d) Repeat the excision for all samples prepared using a clean razor blade for each gel lane used (*see* **Note 13**).

NOTE.

(e) Add 600 μL of Acrylamide Elution buffer, cap the tubes and seal well with parafilm.

(f) Elute the cDNA by shaking tube on side on a rotary orbital shaker at 200 rpm at 37 °C o/n.

Day 4

1. Spin samples at full speed for 5 min.
2. Filter the sample through a nanosep 0.45 μm filter into a 2 mL tube by centrifuging at 8000 × *g* for 2 min.
3. Add 3 μL of 5 mg/mL Glycogen and 1/10 the volume (60 μL) of 3 M Na Acetate pH 5.2 and vortex well.
4. Add 2 volumes (1300 μL) 100% EtOH, vortex well.
5. Immediately, spin tubes at 15,000 × *g* for 30 min at RT.
6. Carefully remove the supernatant without disturbing the pellet and save it in separate collection tubes.
7. Add 500 μL of 75% EtOH and vortex the pellet until it breaks up and come off from the bottom.
8. Spin again at full speed for 5 min.
9. Pour off the supernatant to the collection tube and spin briefly.
10. Remove remaining supernatant with P200 and dry the pellets in tubes on bench for 5 min at RT.
11. Resuspend in 20 μL of RNase-free water.
12. Use 2 μL for Analytical PCRII.

3.8 Analytical PCRII

Check 2 μL of the size selected libraries with 20 cycles of analytical PCR along with positive and negative controls. The PCR product will be run on low melting agarose gel (*see* below **step 5**).

1. Prepare the PCR master-mix (MM2) on ice. For one reaction add reagents in the following order: 6.3 μL H_2O, 1 μL 10× PCR Buffer -$MgCl_2$, 0.3 μL 50 mM $MgCl_2$, 0.08 μL 25 μM PCR PE primer 1, 0.08 μL 25 μM PCR PE primer 2 index1 (Table 2), 0.04 μL Platinum TaqDNA Pol, 0.2 μL 10 mM dNTP. Mix by pipetting.
2. Add 8 μL MM2 to 2 μL of size selected libraries for a final volume of 10 μL.
3. Mix gently, but thoroughly, and centrifuge briefly.
4. Amplify using the following PCR process with a heated lid (set the heat lid to 100 °C): 2 min at 94 °C, 20× (30 s at 94 °C, 30 s at 65 °C, 1 min at 72 °C), hold at 4 °C.
5. Prepare 3% low melting agarose DNA gel (Table 3):
   (a) Choose a beaker that is 2–4 times the volume of the solution. Add chilled 1× or 0.5× electrophoresis buffer and a stir bar to a beaker.

(b) Slowly sprinkle in the agarose powder while the solution is rapidly stirred.

(c) Remove the stir bar if not Teflon coated.

(d) Soak the agarose in the buffer for 15 min before heating. This reduces the tendency of the agarose solution to foam during heating.

(e) Weigh the beaker and solution before heating.

(f) Cover the beaker with plastic wrap. Pierce a small hole in the plastic wrap for ventilation. For agarose concentrations >4%, the following additional steps will further help prevent the agarose solution from foaming during melting/dissolution:

- Heat the beaker in the microwave oven on medium power for 1 min.
- Remove the solution from the microwave oven.
- Allow the solution to sit on the bench for 15 min.

(g) Heat the beaker in the microwave oven on medium power for 2 min.

(h) Remove the beaker from the microwave oven (*see* **Note 11**).

(i) Gently swirl the beaker to resuspend any settled powder and gel pieces.

(j) Reheat the beaker on High power until the solution comes to a boil.

(k) Hold at boiling point for 1 min or until all the particles are dissolved.

(l) Remove the beaker from the microwave oven.

(m) Gently swirl the beaker to thoroughly mix the agarose solution.

(n) After dissolution, add sufficient hot distilled water to obtain the initial weight. Mix thoroughly.

(o) Cool the solution to 50–60 °C prior to casting.

Prepare Samples

(p) Add 2 μL 6× DNA loading dye to DNA amplified in Subheading 3.7.

(q) Add required volume of 6× DNA loading dye to 300 ng of 50 bp DNA ladder (100 ng/μL) in 12 μL final volume. Vortex gently, spin down.

(r) Load samples and DNA ladder into wells.

Run and Stain

(s) Run gel at 85 V (constant voltage) for 1 h/1 h 30 min in 1× TAE.

(t) Add a 1/10,000 dilution of SybrGold stain to enough 1× TBE buffer to submerge the gel in a staining tray (40 mL/one gel). Mix stain well.

(u) Place gel in tray and stain for 1 h in the dark.

(v) Photograph gel and save file.

3.9 Bisulfite Conversion

Unmethylated cytosines are deaminated into uracil; methylated cytosines remain unchanged, due to the methyl group protecting them from the reaction. The main stages of bisulfite conversion are (1) DNA denaturation; (2) incubation with bisulfite at high temperature; (3) desalting for bisulfite removal; (4) desulfonation of sulfonyl uracil adducts with basic pH; (5) removal of the desulfonation solution.

1. (A) Perform bisulfite conversion using a commercial kit protocol per manufacturer's recommendations. Elute the samples in 10 μL of DNase-free water (*see* **Note 14**).

   (B) Perform bisulfite conversion manually using the following protocol.

   (a) Set a water bath to 50 °C.

   (b) Prepare fresh sample denaturation buffer by gently mixing 0.5 μL 0.5 M EDTA, pH 8.0; 3 μL 3 N NaOH; and 0.7 μL 3 mg/mL glycogen with degassed dH_2O to bring volume to 10 μL (*see* **Note 15**).

   (c) Add 10 μL sample denaturation buffer to each DNA sample in 20 μL total volume.

   (d) Prepare fresh saturated sodium bisulfite solution (*see* **Note 16**). Pipet 7 mL degassed dH_2O, 100 μL 100 mM hydroquinone into a 20 mL glass vial and stir gently on the stir plate. Quickly add 1 mL 3 N NaOH and adjust the pH to 5.0 while stirring by adding more 3 N NaOH (may require 200–300 μL). Cap the vial and preheat the saturated sodium metabisulfite solution to 50 °C in a beaker containing water in a water bath.

   (e) Place DNA samples in a thermocycler with a heated lid and denature 5 min at 98 °C.

   (f) Add 200 μL of preheated saturated sodium bisulfite solution from step d to the sample, vortex, and place in a water bath preheated to 50 °C for 4–6 h in the dark.

   (g) Immobilize DNA by binding to a minicolumn and remove bisulfite solution by centrifugation. Wash with 80% EtOH (v/v) to remove residual bisulfite (a commercially available kit is recommended).

   (h) To desulfonate DNA, incubate with 0.3 N NaOH (preheated to 30 °C) for 15–25 min at 30 °C. This can be done on-column by placing the columns in a 30 °C incubator.

(i) Wash DNA twice with 80% EtOH to desalt before eluting in 20 μL TE preheated to 90 °C.

2. Use and treat positive and negative spike-in controls in the same manner as the samples throughout the bisulfite conversion protocol. Controls are essential for monitoring the bisulfite conversion efficiency.
3. Proceed with the enrichment PCR (Subheading 3.10) on the same day that the bisulfite conversion is completed.
4. Use 1–4 μL of eluted DNA for enrichment PCR.

3.10 Enrichment PCR

1. Prepare the PCR master-mix (MM3) on ice in 1.5 mL tube. For one reaction add reagents in the following order: 34.3 μL H_2O, 5 μL 10× PCR Buffer -$MgCl_2$, 1.5 μL 50 mM $MgCl_2$, 0.2 μL 5 U/μL Platinum TaqDNA Pol, 1 μL 10 mM dNTP.
2. Add 22 μL MM3 to 6 μL of bisulfite converted (BSC) libraries (from Subheading 3.9) and 2 μL of primer mix 1–12 (PE1tru/PE2tru index1-12, 10 μM each) (Table 4) (*see* **Note 17**).
3. Add 3 μL of BSC-gDNA (controls from kits per manufacturer's recommendations).
4. Mix gently, but thoroughly, and centrifuge briefly.
5. Amplify using the following PCR process with a heated lid (set the heat lid to 100 °C): 2 min at 94 °C, 18× (20 s at 94 °C, 30 s at 65 °C, 1 min at 72 °C), hold at 4 °C (*see* **Note 18**).
6. Perform purification using magnetic beads (1.0×).
7. Place beads at RT 30 min before starting.
   (a) Transfer samples to newly labeled 1.5 mL tubes.
   (b) Vortex the magnetic beads thoroughly (20–30 s).
   (c) Add 50 μL of beads to each tube containing DNA.
   (d) Incubate the tubes at RT for 15 min.
   (e) Place the tubes on magnetic stand at RT for 2 min until the liquid appears clear.
   (f) Remove and discard 70 μL of supernatant from each tube using P200. Do not disturb the beads.
   (g) Wash the beads by adding 190 μL of freshly prepared 80% EtOH to each tube.
   (h) Incubate at RT for 30 s.
   (i) Remove and discard all the supernatant from each tube, 180 μL in 2 steps.
   (j) With tubes still on magnet, add 190 μL of freshly prepared 80% EtOH to each tube.

Table 4
Primer sequences used in Enrichment PCR for Illumina sequencing

PCR PE1tru primer	5'-AATGATACGGCGACCACCGAGATCT ACACTCTTTCCCTACACGA*C
PCR PE2tru index1	5'-CAAGCAGAAGACGGCATACGAGAT CGTGATGTGACTGGAGTTCAGACGTG*T
PCR PE2tru index2	5'-CAAGCAGAAGACGGCATACGAGAT ACATCGGTGACTGGAGTTCAGACGTG*T
PCR PE2tru index3	5'-CAAGCAGAAGACGGCATACGAGAT GCCTAAGTGACTGGAGTTCAGACGTG*T
PCR PE2tru index4	5'-CAAGCAGAAGACGGCATACGAGAT TGGTCAGTGACTGGAGTTCAGACGTG*T
PCR PE2tru index5	5'-CAAGCAGAAGACGGCATACGAGAT CACTGTGTGACTGGAGTTCAGACGTG*T
PCR PE2tru index6	5'-CAAGCAGAAGACGGCATACGAGAT ATTGGCGTGACTGGAGTTCAGACGTG*T
PCR PE2tru index7	5'-CAAGCAGAAGACGGCATACGAGAT GATCTGGTGACTGGAGTTCAGACGTG*T
PCR PE2tru index8	5'-CAAGCAGAAGACGGCATACGAGAT TCAAGTGTGACTGGAGTTCAGACGTG*T
PCR PE2tru index9	5'-CAAGCAGAAGACGGCATACGAGAT CTGATCGTGACTGGAGTTCAGACGTG*T
PCR PE2tru index10	5'-CAAGCAGAAGACGGCATACGAGAT AAGCTAGTGACTGGAGTTCAGACGTG*T
PCR PE2tru index11	5'-CAAGCAGAAGACGGCATACGAGAT GTAGCCGTGACTGGAGTTCAGACGTG*T
PCR PE2tru index12	5'-CAAGCAGAAGACGGCATACGAGAT TACAAGGTGACTGGAGTTCAGACGTG*T

(k) Let stand for 1 min.

(l) Remove all of supernatant (in 2 steps) and discard.

(m) Let the tubes stand at RT for 5 min to dry and then remove the tubes from the magnetic stand.

(n) Resuspend the dried pellet in 22.5 μL of resuspension buffer (according to the manufacturer's instructions) and pipet-mix up and down 10×.

(o) Incubate the tubes at RT for 2 min and then place the tubes on magnetic stand at RT for 2 min.

(p) Transfer 20 μL of clear supernatant from each tube to a new 1.5 tube.

8. Use 2 μL for microplate spectrophotometer quantification and 1 μL for high sensitivity DNA microchip automated electrophoresis run to calculate concentration and molarity of BSC libraries (*see* below Subheading 3.11).

3.11 Prepare Libraries for Sequencing and for Data Analysis

1. Calculate the molarity (nanomolar concentration, nM) of each library using the following formula:

$$\text{nM DNA} = [(\text{ng}/1000)/(\text{bp} \times 660)] \times 10^9$$

where ng is the concentration expressed in ng/μL (as measured Subheading 3.10, **step 8** using microplate spectrophotometer), bp is the average size of the library fraction, and 660 is the weight of a standard double stranded DNA base pair.

2. Prepare 10 μL of a 2 nM solution of each library fraction. Dilute the library with DNase-free water.
3. Pool the library fractions by combining 10 μL of the 2 nM lower fraction with 10 μL of the 2 nM higher fraction for each sample library prepared. This pool is the final library for sequencing.
4. Sequence the library:
   (a) load the libraries with a goal of optimal densities of 600,000–650,000 clusters per millimeter squared (loading concentration of 7–8 picomolar).
   (b) sequence the libraries using a minimum of single-read 51 cycles sequencing on a HiSeq 2500 sequencer in high output mode.
5. Once the sequencing is complete, the final step in the protocol is the bioinformatic analysis, which includes sequence alignment and methylation identification (using Bismark), requiring the application of computational tools, such as Bismark [20] (not addressed specifically here) (*see* **Note 19**).
6. The post-processed output can be converted to formats easily accessible in a genome browser such as the UCSC Genome Browser [21] or IGV [22] for data analysis and representation of the results.

4 Notes

1. If the protocol to discriminate between 5mC vs. 5hmC is included, remember, before starting to double the amounts of each DNA to be tested into each of the separate reaction setups: to be treated without (control, *) or with β-glucosyltransferase, to allow final sequenced product to be compared for 5hmC identifications. The enzyme β-glucosyltransferase adds glucose to 5hmC but not 5mC (thus yielding a modified base, glucosyl-5hmC). In further steps, glucosyl-5hmC sensitive restriction endonucleases (GSRE, such as *MspI*, *GlaI*, *Csp6I*, *HaeIII*, *MbolI*, *TaqI*) cleaves DNA when cytosine, 5mC, or 5hmC is located at the inner C position (CCGG) within its recognition sequence (i.e., CCGG for *MspI*). Glucosylation at this position (glucosyl-5hmC) will inhibit cleavage by the enzyme. Glucosylation should be tested using hypermethylated genes targeted qPCR designing primer sets that flank the GSRE recognition site of interest. To detect 5hmC by qPCR, test DNA must be added to the qPCR that has not been processed by glucosylation and digestion (no treatment, NT). This NT DNA will establish the level that is representative of complete hydroxymethylation at

the site, while the unglucosylated sample (*) will be a control indicating the level of no hydroxymethylation at the site. An example for a gene to be used in locus-specific 5-hmC detection via qPCR is Galanin [23]. Otherwise, Tet enzymes oxidize 5mC to 5-carboxylcytosine, 5caC, and converts to U on bisulfite treatment but glucosyl-5hmC cannot be oxidized [24]. hmC will be read as C and mC will be read as T in the subsequent step of Tet assisted bisulfate sequencing.

2. Use only good quality genomic DNA (gDNA); gDNA should be free of contaminants such as phenol, chloroform, alcohol, EDTA, detergents or excessive salts. Check gDNA on agarose gel before starting. Start the protocol including controls **to monitor the downstream steps of bisulfite conversion and amplification.**

3. By definition, one unit of restriction enzyme will completely digest 1 μg of DNA in a 50 μL reaction in 60 min. However, it is best to use 5–10 units of enzyme per μg DNA, and 10–20 units for genomic DNA; to overcome variability in DNA source, quantity and purity, a ten-fold overdigestion is often recommended.

4. Always place beads at RT 30 min before starting magnetic beads purification (beads purification is performed also in Subheading 3.2, **step 4**; Subheading 3.5, **step 5**; Subheading 3.10, **step 6**. *See* also timetable in Fig. 1). Immediately prior to use, vortex the beads until they are well dispersed, and the color of the liquid is homogeneous. Prepare fresh 80% ethanol for the wash steps for optimal results. Remove all ethanol from the bottom of the tubes because it may contain residual contaminants. Allow for the complete evaporation of residual ethanol, or it will impact the performance of the subsequent reactions.

5. The samples from end-repair reaction can be stored at −20 °C for up to 7 days if do not plan to proceed to 3′-adenylation (A-tailing) immediately.

6. Allow 2 h to anneal adapters. Adapters used in this protocol are compatible with Illumina sequencing. It is critical that the barcoded adapters are methylated so they will not be converted during bisulfite treatment. It is also important to ensure that the barcoded adapters chosen for the ligation reaction in any given library are compatible with other barcoded adapters chosen for libraries planned for sequencing in the same lane. The information below is optimized for TruSeq adapters. The reader is advised to carefully review the manufacturer's guidelines for index combinations for use in the downstream sequencing. Adapters allowing for multiplexed sequencing have a structure that may lead to migrations differently than expected. Prior to use, allow the beads to come to room temperature.

7. Remember to warm up PEG 8000.
8. It is critical to use a high concentration ligase (2,000,000 units/mL) and methylated adapters to ensure that the ligation reaction is efficient, and that the bisulfite conversion does not influence the adapter sequences essential for accurate data alignment in downstream data analysis. Inefficient ligation reaction could result in failed libraries. Adaptors used in excess (concentration higher than suggested) will interfere with the sequencing. To aid the elimination of excess adapters and primers, bead purification (rather than column-based assays) is recommended for ligation and PCR product isolations.
9. Final volume of adapter-ligated DNA (ALigDNA) is 12 μL (2 μL for PCR and 10 μL for size selection of libraries). ALigDNA can be stored at −20 °C before proceeding with size selection.
10. Analytical PCRI on libraries (digested, end repaired, A-tailed, adapter ligated DNA) will be run to confirm successful ligation of the methylated adapters. 17 cycles of PCR are recommended for first time users of the protocol, but the protocol can be optimized to 14 amplification cycles.
11. *Caution:* Any microwaved solution may become superheated and foam over when agitated.
12. Size selection has been optimized to keep small fragments of interest and to remove adaptor dimers, resulting in a better coverage. The lower and higher library fractions must be maintained as separate samples throughout the bisulfite conversion and PCR enrichment steps to ensure efficient PCR amplification of the libraries. If combined there could be the risk of unequal representation of the genomic loci covered in each library fraction (lower and higher) during the sequencing step.
13. Take a UV snapshot of the gel post excision. Limit exposure to 3 s.
14. If using a commercial kit, use one with high bisulfite conversion rates of unmethylated cytosines while keeping low the DNA degradation. Commercial kits usually have very simple protocols giving consistent results in less time.
15. Degassed dH_2O day must be prepared the day before starting. Boil 200 mL dH_2O for 20 min in a glass beaker containing a stir bar. Carefully pour the boiling water into a 125 mL glass bottle until it is completely full above the lip using surface tension and screw the cap on tightly to ensure an air-tight seal. Cool o/n at RT. Recap the bottle tightly after each use and discard after bisulfite conversion is completed.

16. The described recipe provides enough bisulfite solution for at least 30 deamination reactions. It is suggested to incubate some unused solution at 50 °C in in parallel to the samples undergoing bisulfite conversion to make sure the pH does not increase significantly. An increase in pH of more than 0.2 units indicates potential decreased conversion efficiency.
17. Multiplexed sequencing offer a lower cost protocol per sample, but also result in reduced coverage per CpG site represented in the data. Do not choose multiplexing if performing downstream analyses requiring high coverage per CpG site.
18. Bisulfite PCR primers need to be between 26–30 bases; primers should not contain CpG sites, however, if they are necessary locate them at the 5′-end of the primer with a mixed base at the cytosine position. The amplicon size should be between 150 and 300 bp; annealing temperature gradient should be run with every new primer set to ensure optimal amplification of the specific target. The number of PCR cycles required to amplify bisulfite converted DNA varies with the enzyme used. 18 cycles of PCR are recommended for first time users of the protocol, but the protocol can be optimized for fewer amplification cycles: cycle number is lower with high efficiency enzymes. Use multiple controls to monitor conversion and amplification. In order to generate high quality data, it is important to ensure efficient bisulfite conversion.
19. When working with a marine organism make sure to have access to a reference genome or a reference genome of a closely related species. Filtered adapter trimmed sequence reads (bisulfite treated reads) will be aligned to the bisulfite converted reference genome of the species analyzed to determine the methylation context for each cytosine. For example, Bismark [20]—cited in the main text—is a customized short read mapping tool that aligns bisulfite treated reads to a bisulfite converted genome (where all C are converted to T) and returns methylation calls for cytosines in CpG, CHG, and CHH context. Mapping and bioinformatics analysis will be performed with tools that can align reads first by chromosome, then start position, and finally strand.

References

1. Allis D, Caparros ML, Jenuwein T, Reinberg D (eds) (2015) Epigenetics, 2nd edn. Cold Spring Harbor, New York
2. Gardiner-Garden M, Frommer M (1987) CpG islands in vertebrate genomes. J Mol Biol 196: 261–282
3. Barter MJ, Bui C, Young DA (2012) Epigenetic mechanisms in cartilage and osteoarthritis: DNA methylation, histone modifications and microRNAs. Osteoarthr Cartil 20:339–349
4. Rodriguez J, Frigola J, Vendrell E et al (2006) Chromosomal instability correlates with genome-wide DNA demethylation in human primary colorectal cancers. Cancer Res 66: 8462–9468

5. Di Giacomo M, Comazzetto S, Saini H et al (2013) Multiple epigenetic mechanisms and the piRNA pathway enforce LINE1 silencing during adult spermatogenesis. Mol Cell 50: 601–608
6. Weber M, Davies JJ, Wittig D et al (2005) Chromosome-wide and promoter-specific analyses identify sites of differential DNA methylation in normal and transformed human cells. Nat Genet 37:853–862
7. Frommer M, McDonald LE, Millar DS et al (1992) A genomic sequencing protocol that yields a positive display of 5-methylcytosine residues in individual DNA strands. Proc Natl Acad Sci U S A 89:1827–1831
8. Clark SJ, Statham A, Stirzaker C et al (2006) DNA methylation: bisulphite modification and analysis. Nat Protoc 1:2353–2364
9. Meissner A, Gnirke A, Bell GW et al (2005) Reduced representation bisulfite sequencingfor comparative high-resolution DNA methylation analysis. Nucleic Acids Res 33:5868–5877
10. Gu H, Bock C, Mikkelsen TS, Jäger N et al (2010) Genome-scale DNA methylation mapping of clinical samples at single-nucleotide resolution. Nat Methods 7: 133–136
11. Gu H, Smith ZD, Bock C et al (2011) Preparation of reduced representation bisulfite sequencing libraries for genome-scale DNA methylation profiling. Nat Protoc 6:468–481
12. van Gurp TP, Wagemaker NC, Wouters B et al (2016) epiGBS: reference-free reduced representation bisulfite sequencing. Nat Methods 13:322–324
13. Martinez-Arguelles DB, Lee S, Papadopoulos V (2014) In silico analysis identifies novel restriction enzyme combinations that expand reduced representation bisulfite sequencing CpG coverage. BMC Res Notes 7:534
14. Huang Y, Pastor WA, Shen Y et al (2010) The behaviour of 5-hydroxymethylcytosine in bisulfite sequencing. PLoS One 5:e8888
15. Wang J, Tang J, Lai M et al (2014) 5-Hydroxymethylcytosine and disease. Mutat Res Rev Mutat Res 762:167–175
16. Song CX, Yin S, Ma L et al (2010) 5-Hydroxymethylcytosine signatures in cell-free DNA provide information about tumor types and stages. Cell Res 27:1231–1242
17. Li FJ, Li LM, Zhang RH et al (2017) The role of 5-hydroxymethylcytosine in melanoma. Melanoma Res 27:175–179
18. Ecsedi S, Rodríguez-Aguilera JR, Hernandez-Vargas H (2018) 5-Hydroxymethylcytosine (5hmC), or how to identify your favorite cell. Epigenomes 2:3
19. Booth MJ, Ost TW, Beraldi D et al (2013) Oxidative bisulfite sequencing of 5-methylcytosine and 5-hydroxymethylcytosine. Nat Protoc 8: 1841–1851
20. Krueger F, Andrews SR (2011) Bismark: a flexible aligner and methylation caller for bisulfite-Seq applications. Bioinformatics 27: 1571–1572
21. Kent WJ, Sugnet CW, Furey TS et al (2002) The human genome browser at UCSC. Genome Res 12:996–1006
22. Thorvaldsdóttir H, Robinson JT, Mesirov JP (2013) Integrative genomics viewer (IGV): high-performance genomics data visualization and exploration. Brief Bioinform 14:178–192
23. Yoon D, Bae K, Lee MK et al (2018) Galanin is an epigenetically silenced tumor suppressor gene in gastric cancer cells. PLoS One 13: e0193275
24. Tahiliani M, Koh KP, Shen Y et al (2009) Conversion of 5-methylcytosine to 5-hydroxymethylcytosine in mammalian DNA by MLL partner TET1. Science 324:930–935

Chapter 3

Transcriptome Mining to Identify Genes of Interest: From Local Databases to Phylogenetic Inference

Daniele De Luca and Chiara Lauritano

Abstract

The advancement in next-generation sequencing technologies and the dropping of sequencing costs have seen an increase in the amount of transcriptome data generated each year. These data are of big potential for identifying genes and molecular pathways of interest across a plethora of organisms. However, navigating these resources requires some bioinformatics and evolutionary skills. Here, we describe a protocol of transcriptome data mining for genes of interest, from the creation of a protein database to the inference of phylogenetic trees, which was used for marine protists, but can be used as general pipeline across different taxa.

Key words Local databases, Marine protists, Phylogenies, Sequence alignment, Transcriptome mining, Genes of interest, Biosynthetic pathways, Bioinformatics

1 Introduction

In the last decade, the amount of sequencing data produced has increased exponentially, prompted by the development of new next-generation sequencing (NGS) technologies and falling of price [1]. For marine protists, the generation of NGS data, especially in the form of transcriptomes (Fig. 1), has also benefited from the growing interest in functional ecology (e.g., the MMETSP project [2] or the study of toxic dangerous species [3]) or blue biotechnology studies, such as for the discovery of bioactive compounds or enzymes with potential industrial applications [4–9]. These transcriptomes are generally deposited in public sequence databases (e.g., GenBank) or in dedicated webpages, as AlgaePath (http://algaepath.itps.ncku.edu.tw) [10] and MMETSP (https://github.com/dib-lab/dib-MMETSP), where often both raw reads and assemblies are available. AlgaePath integrates gene information, biological pathways and NGS data for green algae, while MMETSP provides more than 650 functionally annotated and

Cinzia Verde and Daniela Giordano (eds.), *Marine Genomics: Methods and Protocols*,
Methods in Molecular Biology, vol. 2498, https://doi.org/10.1007/978-1-0716-2313-8_3,

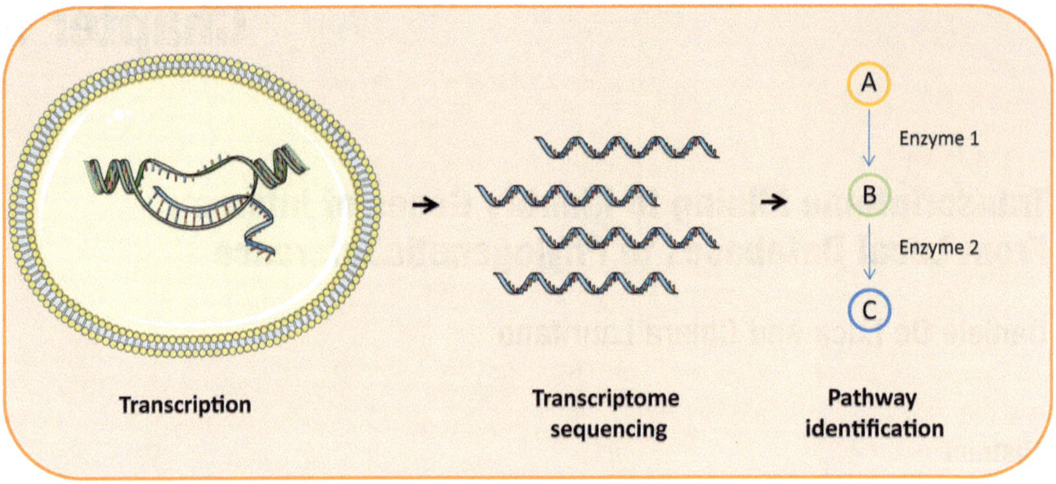

Fig. 1 Flowchart showing RNA sequencing and pathway analysis

assembled transcriptomes. These data provide a valuable tool for identifying enzymes or biosynthetic pathways responsible for the synthesis of bioactive compounds. However, such "in silico" identification of genes of interest requires some bioinformatic and evolutionary biology knowledge, which are not always at hand for some biologists. Here we describe a protocol of transcriptome data mining for genes of interest, from the creation of local databases for homologs retrieval to sequence alignment and phylogenetic inference (Fig. 2).

2 Bioinformatic Pipeline

2.1 Creating a Protein Database

The first step for transcriptome mining of protein of interest is the creation of a local database. This can contain one or a collection of transcriptomes to be searched; the only important thing is that each transcriptome has a unique sequence identifier (seqID) for all the contigs, allowing tracing the source of each blast hit. For this exercise, we will use BLAST+ [11], the command line version of BLAST [12].

1. On your Linux/Ubuntu OS, open a new shell and run the command below to install a ready-to-use version of BLAST+:

```
sudo apt-get install ncbi-blast+
```

2. Create a protein database in a directory of your choice by typing (*see* **Note 1**):

```
makeblastdb -in data_for_database.fasta -dbtype prot -out mydatabase -parse_seqids
```

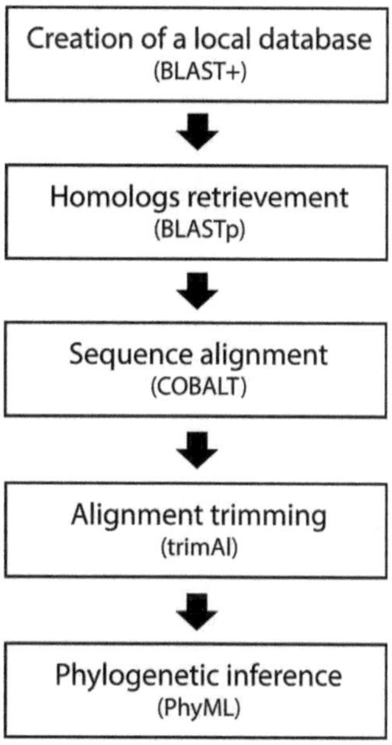

Fig. 2 Example of pipeline for transcriptome mining

3. To carry out a blastp search using all the sequences (or just one) in "myseqs.fasta" as queries against your database type:

```
blastp -query myseqs.fasta -db mydatabase -out blastp_results.
txt -outfmt 6 -evalue 1e-10
```

This command will output the results to the file "blastp_results.txt" in a tabular format and with a similarity threshold of 1e-10 (*see* **Notes 2** and **3**).

4. To get the blast hits in fasta format, type in the shell (*see* **Note 4**):

```
blastdbcmd -db mydatabase -dbtype prot -entry_batch hits.txt
-outfmt %f -out hits.fasta
```

Optional: If you want to check what are the sequences in GenBank with the highest similarity with your query sequence or with the blast hits just obtained, you can run a remote blast against the nonredundant (nr) protein sequence database held on the NCBI website by typing (*see* **Note 5**):

```
blastp -query hits.fasta -remote -db nr -out blastp_re-
sults_nr.txt -outfmt 6
```

2.2 Sequence Alignment

The next step is to build up an alignment with the blast hits just obtained. A multiple sequence alignment (MSA) is the result of a process of arranging three or more sequences so that identical or similar nucleotides/amino acids are located in the same column. MSAs are useful at identifying evolutionarily conserved (and therefore putatively functionally important) sites and homologous sequences, and constitute the first step toward the inference of molecular phylogenies. Most of the popular multisequence alignment algorithms, such as ClustalW [13], MUSCLE [14, 15] and ProbCons [16] use a progressive alignment technique [17] which starts aligning the most similar pairs of sequences and then continues with multiple pairs until all sequences have contributed to the alignment. These algorithms do not take into account other information that could improve the quality of the multiple alignment, as secondary structure or conserved domains and are time demanding. To overcome these factors, we will use COBALT [18], a multiple sequence alignment tool that uses pairwise constraints derived from the conserved domain database (CCD) [19], the protein motif database (PROSITE) [20], and sequence similarity, using RPS-BLAST, BLASTP, and PHI-BLAST. The resulting pairwise constraints are then incorporated into a progressive multiple alignment. COBALT is included in the NCBI C++ toolkit, but in this protocol we will use the online version.

1. Access COBALT via the webserver available at https://www.ncbi.nlm.nih.gov/tools/cobalt/re_cobalt.cgi. All you need to do is to upload your sequences (single file) in fasta format and run the analysis. The default parameters are generally good for most analyses.
2. When the analysis is completed (one hundred sequences generally require less than 5 min to be aligned), you can visually inspect the alignment through NCBI Multiple Sequence Alignment Viewer. In the default visualisation mode, amino acid residues are coloured by conservation, with red colour indicating highly conserved positions and blue colour less conserved ones; red lines indicate gaps. You can download the alignment in fasta format by clicking on the button "Download" (*see* **Note 6**).

2.3 Trimming of Spurious Regions

The accuracy of an alignment of protein sequences depends on their sequence similarity, which generally correlates to evolutionary distance. If the target and the query proteins are closely related, the alignment is often of good quality (e.g., few gaps). On the contrary, protein alignments of distantly related taxa are made of many gappy or poorly aligned regions and few, accurately aligned regions, often corresponding to highly conserved domains [21, 22]. Sequence alignments become poor with sequence similarity lower than 40%

and even worse when sequence similarity enters the "twilight zone" of <25% of identical residues [23]. Various trimming methods can be used to identify and remove these unreliable regions in the alignment to reduce noise and improve downstream phylogenetic inferences. Here we will use trimAl [24], a tool for rapid automated alignment trimming. Given an input alignment file in different file formats, trimAl can trim poorly aligned regions using either a user-defined threshold or applying an automated method. Unless there is a particular reason to use a cut-off threshold for quality trimming, the best option is to take advantage of the automated options, which will automatically select the most appropriate thresholds for your alignment. These optimal cut-off thresholds are based on gap distribution (*-gappyout* option) or similarity scores (*-strict* and *-strictplus* options). Here we will launch trimAl with the option *-automated1*, which uses a heuristic method to choose the best automated method among the ones abovementioned to trim the alignment.

1. Download trimAl by typing the following:

```
wget http://trimal.cgenomics.org/_media/trimal.v1.2rev59.tar.
gz
```

2. Unzip the archive to the "trimAl" folder:

```
tar zxvpf trimal.v1.2rev59.tar.gz
```

3. Enter in the directory containing the package's source code:

```
cd source
```

4. Type "make" to compile the package:

```
make
```

5. In this folder ("source") launch trimal as follows (*see* **Note 7**):

```
./trimal -in <inputfile> -out <outputfile> -phylip -auto-
mated1
```

2.4 Evolution Models Selection and Phylogenetic Inference

Now that poorly aligned regions have been removed from the alignment, we are ready to select the proper evolution model of sequences and infer a phylogenetic tree. Users can choose between different software to compute sequence model selection (e.g., jModelTest 2 [25], ModelFinder, [26]), different methods of phylogenetic tree inference (maximum likelihood or Bayesian inference) and different methods for calculating support values of nodes on phylogenies (e.g., bootstrap procedure and approximate likelihood-ratio tests, aLRTs). Different tutorials are available for running phylogenetic analysis from command line [27]. Here, we will use the PhyML webserver available at http://www.atgc-montpellier.fr/phyml/ for both model selection and phylogenetic inference.

1. Access the website at the above mentioned link.
2. Load the input file in phylip format in the "Input Data" menu and select the "Amino-Acids" data type option.
3. Calculate the model selection using SMS [28] with the Akaike Information Criterion (AIC).
4. Leave the default options in the menu "Tree Searching."
5. In the "Branch Support" menu, choose the aLRT-SH test (approximate likelihood-ratio tests with the nonparametric Shimodaira-Hasegawa).
6. Name your analysis and insert an email account to receive a message once the analysis is completed, with a link for downloading the results (*see* **Note 8**).

The resulting tree can be visualised and graphically edited using several tools as FigTree (downloadable at https://github.com/rambaut/figtree/releases) and iTOL (https://itol.embl.de/).

3 Notes

1. If you get the error message "BLAST options error: File does not exist," this means that you are trying to run the program in a different directory from the one that contains your files. To solve this, create a new folder in Ubuntu and then copy in it the fasta file containing the sequences you want to include in your database. Then, launch the *makeblastdb* command from this folder (it will create the database in it).
2. The output format "6" (-outfmt 6) is a tabular format with the following default headers: *qseqid* = query (e.g., your sequence) sequence id; *sseqid* = subject (e.g., target sequence) sequence id; *pident* = percentage of identical matches; *length* = sequence overlap; *mismatch* = number of mismatches; *gapopen* = number of gaps in the alignment; *qstart* = start of alignment in query; *qend* = end of alignment in query; *sstart* = start of alignment in subject; *send* = end of alignment in subject; *evalue* = expect value; *bitscore* = bit score. If you want to retain only some information in your output file (e.g., qseqid, sseqid, and bitscore), you can modify the option as follows: *-outfmt "6 qseqid sseqid bitscore"*.
3. A common rule of thumb for similarity thresholds is that two sequences are homologous if they share more than 30%

identical sites/residues over their entire alignment length [29]. However, unless there is a specific reason to conduct a blast search based on a similarity threshold, evalue and bitscore values are better predictors of homology. Indeed, many homologs can be missed with this threshold because similarity percentage is a reasonable indicator of evolutionary distance only once homology has been established [29]. The e-value represents the number of expected hits of similar quality (score) that could be found just by chance; the smaller the e-value, the better the match. The e-value depends on the size of the sequence database used for the search; this means that, when searching for homologs in large databases, smaller e-values should be preferred (e.g., 1e-10) over bigger ones (e.g., 1e-5 for small-medium size databases). On the contrary, the bitscore is a database size-independent measure of hits that could be found by chance. Consequently, bit-score values can be compared across differently sized databases and, for this reason, bit-score is a better predictor of homology. For sequences of comparable length, bit-scores ≥200 indicate high sequence similarity, but also values up to 50 are generally significant. However, below 50, it is of good practice to compare this value to the significance of e-value, which should be <0.001 [29].

4. The -entry_batch option requires a text file (here called "hits.txt") containing the list of the sequence IDs (as they appear in the BLAST results file), in consecutive rows, to be retrieved. The option *-outfmt* specifies the format of output sequences ("%f" = "fasta" format).

5. If you do not specify the -evalue option, the default value of "1e-30" will be applied. You can also modify the option -outfmt 6 by adding "salltitles" if you want, alongside the default entries, the description of the sequence. The full option will be *-outfmt "6 qseqid sseqid pident length mismatch gapopen qstart qend sstart send evalue bitscore salltitles"*.

6. COBALT renames your sequences progressively as "Query_1000x"; therefore, when you download the alignment, you have to restore the original names. Linux users can rename their sequences with the original IDs using the command: *awk 'NR%2==0' myalignment.fasta | paste -d'\n' mynames.txt - > myoutput.fasta*. The text file "mynames" includes the original sequence names, preceded by ">", on consecutive rows. Please note that fasta sequences should not be interleaved (on multiple rows), but on single lines.

7. The input file (extension included in the name) must be in the "source" folder where you are running trimAl. You can either copy the input file to this folder or add the path to the input file

in the –in option (e.g., -in /home/myfolder/input.fasta). In alternative to the command line version of trimAl, there is also a web server available at http://phylemon2.bioinfo.cipf.es/.

8. The online server can only perform 100 bootstrap replicates, which are often not sufficient to provide reliable statistic support; therefore, we use the aLRT-SH test, which has been proven as accurate as other support methods [30]. In general, aLRT-SH values >0.80 are considered robust.

Acknowledgments

The authors thank Servier Medical Art (SMART) website (https://smart.servier.com/) by Servier for the elements of Fig. 1.

References

1. Giani AM, Gallo GR, Gianfranceschi L et al (2020) Long walk to genomics: history and current approaches to genome sequencing and assembly. Comput Struct Biotechnol J 18:9–19
2. Keeling PJ, Burki F, Wilcox HM et al (2014) The marine microbial eukaryote transcriptome sequencing project (MMETSP): illuminating the functional diversity of eukaryotic life in the oceans through transcriptome sequencing. PLoS Biol 12(6):e1001889
3. Vingiani GM, Stalberga D, De Luca P et al (2020) *De novo* transcriptome of the non-saxitoxin producing *Alexandrium tamutum* reveals new insights on harmful dinoflagellates. Mar Drugs 18:386
4. Lauritano C, De Luca D, Amoroso M et al (2019) New molecular insights on the response of the green alga *Tetraselmis suecica* to nitrogen starvation. Sci Rep 9:3336
5. De Luca D, Lauritano C (2020) In silico identification of type III PKS chalcone and stilbene synthase homologs in marine photosynthetic organisms. Biology 9:110
6. Vingiani GM, De Luca P, Ianora A et al (2019) Microalgal enzymes with biotechnological applications. Mar Drugs 17:459
7. Elagoz AM, Ambrosino L, Lauritano C (2020) De novo transcriptome of the diatom *Cylindrotheca closterium* identifies genes involved in the metabolism of anti-inflammatory compounds. Sci Rep 10:4138
8. Riccio G, De Luca D, Lauritano C (2020) Monogalactosyldiacylglycerol and sulfolipid synthesis in microalgae. Mar Drugs 18:237
9. Lauritano C, Ferrante MI, Rogato A (2019) Marine natural products from microalgae: an-omics overview. Mar Drugs 17(5):269
10. Zheng HQ, Chiang-Hsieh YF, Chien CH et al (2014) AlgaePath: comprehensive analysis of metabolic pathways using transcript abundance data from next-generation sequencing in green algae. BMC Genomics 15(1):1–12
11. Camacho C, Coulouris G, Avagyan V et al (2009) BLAST+: architecture and applications. BMC Bioinformatics 10(1):1–9
12. Altschul SF, Gish W, Miller W et al (1990) Basic local alignment search tool. J Mol Biol 215(3):403–410
13. Thompson JD, Higgins DG, Gibson TJ (1994) CLUSTAL W: improving the sensitivity of progressive multiple sequence alignment through sequence weighting, position-specific gap penalties and weight matrix choice. Nucleic Acids Res 22(22):4673–4680
14. Edgar RC (2004) MUSCLE: a multiple sequence alignment method with reduced time and space complexity. BMC Bioinformatics 5(1):1–19
15. Edgar RC (2004) MUSCLE: multiple sequence alignment with high accuracy and high throughput. Nucleic Acids Res 32(5):1792–1797
16. Do CB, Mahabhashyam MS, Brudno M et al (2005) ProbCons: probabilistic consistency-based multiple sequence alignment. Genome Res 15(2):330–340
17. Feng DF, Doolittle RF (1987) Progressive sequence alignment as a prerequisite to correct phylogenetic trees. J Mol Evol 25(4):351–360

18. Papadopoulos JS, Agarwala R (2007) COBALT: constraint-based alignment tool for multiple protein sequences. Bioinformatics 23(9):1073–1079
19. Marchler-Bauer A, Anderson JB, Cherukuri PF et al (2005) CDD: a conserved domain database for protein classification. Nucleic Acids Res 33(suppl_1):D192–D196
20. Hulo N, Bairoch A, Bulliard V et al (2006) The PROSITE database. Nucleic Acids Res 34 (suppl_1):D227–D230
21. Edgar RC, Batzoglou S (2006) Multiple sequence alignment. Curr Opin Struct Biol 16(3):368–373
22. Thompson JD, Linard B, Lecompte O et al (2011) A comprehensive benchmark study of multiple sequence alignment methods: current challenges and future perspectives. PLoS One 6(3):e18093
23. Vogt G, Etzold T, Argos P (1995) An assessment of amino acid exchange matrices in aligning protein sequences: the twilight zone revisited. J Mol Biol 249(4):816–831
24. Capella-Gutiérrez S, Silla-Martínez JM, Gabaldón T (2009) trimAl: a tool for automated alignment trimming in large-scale phylogenetic analyses. Bioinformatics 25(15):1972–1973
25. Darriba D, Taboada GL, Doallo R et al (2012) jModelTest 2: more models, new heuristics and parallel computing. Nat Methods 9(8): 772–772
26. Kalyaanamoorthy S, Minh BQ, Wong TK et al (2017) ModelFinder: fast model selection for accurate phylogenetic estimates. Nat Methods 14(6):587–589
27. Stamatakis A (2015) Using RAxML to infer phylogenies. Curr Protoc Bioinformatics 51(1):6–14
28. Lefort V, Longueville JE, Gascuel O (2017) SMS: smart model selection in PhyML. Mol Biol Evol 34(9):2422–2424
29. Pearson WR (2013) An introduction to sequence similarity ("homology") searching. Curr Protoc Bioinformatics 42(1):3–1
30. Anisimova M, Gil M, Dufayard JF et al (2011) Survey of branch support methods demonstrates accuracy, power, and robustness of fast likelihood-based approximation schemes. Syst Biol 60(5):685–699

Chapter 4

Detecting Structural Variants and Associated Gene Presence–Absence Variation Phenomena in the Genomes of Marine Organisms

Marco Sollitto, Nathan J. Kenny, Samuele Greco, Carmen Federica Tucci, Andrew D. Calcino, and Marco Gerdol

Abstract

As complete genomes become easier to attain, even from previously difficult-to-sequence species, and as genomic resequencing becomes more routine, it is becoming obvious that genomic structural variation is more widespread than originally thought and plays an important role in maintaining genetic variation in populations. Structural variants (SVs) and associated gene presence–absence variation (PAV) can be important players in local adaptation, allowing the maintenance of genetic variation and taking part in other evolutionarily relevant phenomena. While recent studies have highlighted the importance of structural variation in Mollusca, the prevalence of this phenomenon in the broader context of marine organisms remains to be fully investigated.

Here, we describe a straightforward and broadly applicable method for the identification of SVs in fully assembled diploid genomes, leveraging the same reads used for assembly. We also explain a gene PAV analysis protocol, which could be broadly applied to any species with a fully sequenced reference genome available. Although the strength of these approaches have been tested and proven in marine invertebrates, which tend to have high levels of heterozygosity, possibly due to their lifestyle traits, they are also applicable to other species across the tree of life, providing a ready means to begin investigations into this potentially widespread phenomena.

Key words Hemizygosity, Genome, Presence–Absence Variation, Structural Variation, Pangenome

1 Introduction

It is becoming apparent from whole genome resequencing that intraspecific genome variation, once mostly thought to be a prerogative of prokaryotes, is more widespread than previously expected in eukaryotes. These phenomena, collectively referred to as structural variations (SV), encompass any large scale difference in genome architecture and, in addition to translocations, duplications, inversions, and copy number variations, they also include

Cinzia Verde and Daniela Giordano (eds.), *Marine Genomics: Methods and Protocols*,
Methods in Molecular Biology, vol. 2498, https://doi.org/10.1007/978-1-0716-2313-8_4,

insertions and deletions [1]. Insertions and deletions are particularly consequential because they include regions of a species' genome which are subject to presence–absence variation (PAV) either between homologous chromosomes within an individual or between individuals. The outcome of such variation is that the genetic complement of two individuals belonging to a single species or population may differ from one another. This deviates from the commonly held idea that variation between individuals is primarily the result of allelic variation of a common genomic repertoire. In genome assemblies of diploid organisms, PAV can be determined by the observation of large genomic regions present in a hemizygous state, that is, only present in one of the two homologous chromosomes. These loci can be readily identified by their reduced read coverage which exists at approximately half that of flanking homozygous regions. On a broader population level, any genomic region affected by PAV could be present in zero, one, or two copies (found in a nullizygous, hemizygous, or homozygous state) across different individuals.

Unexpectedly, not only do genomic regions subject to PAV include intergenic or intronic sequence, but they may also include protein-coding genes which retain full functionality, with a potential impact on phenotypic diversity [2] and adaptation [3, 4]. In this respect, the gene repertoire of any given species can be divided into two different categories: (1) *core* genes, which are thought to carry out functions indispensable for survival and are shared by all individuals, and (2) *dispensable* genes, which may provide accessory functions and are only found in a subgroup of individuals. Altogether, the full complement of *core* and *dispensable* genes found across all the individuals belonging to the same species define the pangenome [5]. While this concept has been long applied to microbial and viral genomes [6, 7], broad scale genomic PAV has also been observed in a number of plants, microalgae and fungi [3, 8–11], where the accessory functions provided by *dispensable* genes have been linked with an improved ability to colonize new ecological niches, to withstand biotic and abiotic stress, or to escape host immune response in the case of pathogenic organisms. Even though pangenomic studies have been only very recently extended to metazoans, gene PAV has already been noted in a number of lineages, including in humans [12–14]. Following early observations collected for single genes [15], a subsequent genome-wide approach allowed for the first description among marine animals of the presence of a pangenome in the Mediterranean mussel *Mytilus galloprovincialis* [16].

The scale of PAV varies widely between species. In plants [10], and some animal species, variable regions, which can be readily identified thanks to their hemizygous state (i.e., for diploid species, regions with half the expected sequencing coverage), can encompass large proportions of the genome. For example, hemizygous regions account for more than 35% of the reference genome assembly of *M. galloprovincialis*, which results in the presence of more than 20,000 *dispensable* genes [16]. However, in other animals, such as humans, PAV accounts for a much smaller fraction of the genome. Typically, less than 0.1% of the human genome varies between individuals and a small fraction of genes (i.e., 240 out of the ~20,000 genes annotated in the human genome) are potentially subject to homozygous deletions [12]. The rate of *dispensable* to *core* genes is generally used to describe the "openness" of a pangenome. Even though this definition is somewhat arbitrary, genomes such as those of the mussel, plants and bacteria, where a high proportion of *dispensable* genes have been detected, are considered "open" pangenomes. In contrast, genomes with a relatively lower *dispensable* to *core* gene proportion are defined as "closed" pangenomes. While the presence of "open" pangenomes has been previously associated with an improved potential for adaptation in bacteria [4], the possibility that accessory genes may also be associated with neutral or slightly deleterious effects on fitness [17] reveals that a consensus about the functional role of microbial pangenomes is still far from being reached by the scientific community. Although *dispensable* genes are characterized by an enrichment of functions linked with immune defense and survival in *M. galloprovincialis* [16], it is unclear whether this seemingly adaptive role for PAV is common to other marine species that share similar life traits. This remains an open question of great ecological and evolutionary relevance.

Due to the relatively recent inclusion of metazoans among the targets of pangenomic studies, no consolidated method for the analysis of SVs and PAV have been developed and broadly validated to date in these organisms, which (unlike bacteria) have complex diploid genomes. The standard method for detecting genomic regions subject to PAV between individuals is the use of whole genome resequencing (WGR). In this approach, a sample of individuals from a population or species are resequenced and compared to a reference genome assembly of the target species. This allows for the identification of genomic loci that are subject to presence–absence variation between each sequenced individual and the reference assembly. Through this iterative comparative approach, the true genomic complement of the population or species can be approached as more individuals are sampled.

While WGR remains the only method to determine the complete genomic complement of a population or species, it is expensive and time consuming, and is not an efficient option to gain a first insight into the level of PAV that may exist in a population. Moreover, it does not necessarily provide information on PAV between homologous chromosomes within an individual. We have established a pipeline which can be applied to any well-assembled genome, and which leverages the long reads used in the initial assembly to identify genomic loci that are subject to PAV between the homologous chromosomes of diploid individuals rather than between individuals. As previously mentioned, PAV between homologous chromosomes results in stretches of hemizygous DNA that can be identified using preexisting reads. This means that no additional sequencing is required to gain a preliminary understanding of the level of hemizygosity in the sequenced individual [18]. This method accounts for regions of tandem duplication and false positives due to genome assembly artifacts, producing robust yet conservative estimates of hemizygosity through a mapping and coverage estimation-based approach.

The stark differences in PAV raise many interesting questions for evolutionary biology. In an era where genomes are generally generated from single individuals, it is useful to consider whether they are representative of populations as a whole, and the methods described here provide a ready initial assessment scheme. In concert with the method for the detection of hemizygous regions in diploid genome assemblies detailed in this chapter, we also describe here a protocol for the use of Illumina WGR data for the detection of gene PAV among individuals. While the costs of third generation sequencing technologies, such as those offered by Oxford Nanopore and Pacific Biosciences are rapidly dropping, price considerations still prevent the broad applicability of such methods to large scale resequencing projects. For many research groups the use of short reads remains the preferred choice for high-throughput nonmodel species genomics. The reliability of the short-read based method presented here has been proven on mollusks, and in particular in *M. galloprovincialis* [16]. However, this approach will be just as applicable to other species, as long as the quality of both the raw resequencing data and of the annotated reference genome are sufficient.

The ready nature of the approach we describe could be helpful in ascertaining the extent to which phenomena such as genomic and genic PAV are coupled to particular environments or life history traits. This approach could be particularly useful in understanding heterozygosity in marine broadcast spawners. Marine species have been shown to possess higher levels of heterozygosity than other species [19]. This could be due to large effective population sizes, rapid dispersal, high levels of genetic outcrossing and panmixture, coupled with local variations in environment, although factors such as local sperm viability do impact "real world" genetic diversity levels [20]. However, genetic variation is ultimately the result of mutational events, and when many mutations are retained rather than purged from genomes within a population over evolutionary timescales, this will result in high levels of heterozygosity. Mutation rate will be further impacted by the rate of transpositional activity or low fidelity error correction mechanisms, which we have previously shown to be likely involved in PAV in mollusks [18].

This approach provides new means for raising and addressing questions regarding the origin, maintenance and prevalence of PAV in nonmodel species, which may be more suitable for addressing these issues than more typical laboratory organisms. It also provides a means of linking these investigations with gene PAV, the adaptive nature of this phenomenon, and processes of gene family expansion and contraction. As genomic sequences become readily available across the tree of life, and as long read sequencing reduces in price, it will only become easier to gain a more representative understanding of the level of PAV in all species, and to reflect on the contribution of these phenomena to the evolution of marine life as we know it.

2 Materials

2.1 Target Selection

The procedures described in Subheading 3.1 involve the analysis of a near complete or complete (chromosomal) haploid genome assemblies produced from diploid species using both PacBio long reads (*see* **Note 1**) and Illumina short reads. Care should be taken to ensure that the sequenced individual was not the product of selective breeding as this would likely reduce the level of structural variation that would be expected from comparable individuals from outbreeding or wild populations. Target species/individuals should be selected based on the availability of the following.

1. A high-quality genome assembly (*see* **Note 2** for additional recommendations).
2. High quality PacBio long read libraries (good results were attained with >40× coverage, and we have not tested this to determine a minimum required level) derived from the same individual from which the genome assembly was constructed.
3. High quality Illumina short read libraries (good results were attained with >20× coverage) derived from the same individual from which the genome assembly was constructed. Please *see* **Note 3** for additional recommendations.

On the other hand, the analyses described in Subheading 3.2 can be applied to any fully sequenced and annotated genome, regardless of the use of long reads during the de novo assembly process. The choice of the target species should be based on the following criteria.

1. Availability of a full genome assembly with associated gene annotations in a .gff file format. Since the accuracy of gene model annotations will be crucial to allow a proper interpretation of PAV data, we suggest to evaluate the completeness of the set of the annotated genes in the target species with BUSCO [21], selecting the appropriate taxonomic dataset available from the most recent release of OrthoDB [22].

Availability of high quality Illumina WGR short read libraries (good results were attained with >20× coverage) derived from one or more individuals different from the one used for the generation of the genome assembly. Please *see* **Note 2** for additional recommendations.

2.2 Software

All computational procedures should be performed on a Linux-based system and the following software should be installed prior to commencement.

bbduk [23]: https://sourceforge.net/projects/bbmap

bedmap [24]: https://github.com/bedops

bedtools [25]: https://github.com/arq5x/bedtools2

BUSCO [21]: https://busco.ezlab.org

Bwa [26]: http://bio-bwa.sourceforge.net

bwa mem [27]: https://github.com/lh3/bwa

fastp [28]: https://github.com/OpenGene/fastp

FastQC [29]: https://github.com/s-andrews/FastQC

jellyfish [30]: https://github.com/gmarcais/Jellyfish

mosdepth [31]: https://github.com/torfinnnome/mosdepth

Numpy [32]: https://numpy.org

Pandas [33]: https://pandas.pydata.org

pbmm2 [34]: https://github.com/PacificBiosciences/pbmm2

pbsv [35]: https://github.com/PacificBiosciences/pbsv

Python: https://www.python.org

samtools [36]: https://github.com/samtools/samtools

Scipy [37]: https://www.scipy.org

Tandem Repeats Finder [38]: https://tandem.bu.edu/trf/trf.html

3 Methods

3.1 Allelic Structural Variation Detection Within Assembled Genomes

Identify tandem repeats in genome assembly and convert output to zero-based six field bed file.

1. `trf genome.fa 2 7 7 80 10 50,500 -d -h`
2. `TRFdat_to_bed.py --dat genome.fa.2.7.7.80.10.50.500_mod.dat --bed genome.fa.2.7.7.80.10.50.500.bed`
3. `awk '{print $1"\t"$2-1"\t"$3"\t"$4}' genome.fa.2.7.7.80.10.50.500.bed | sed 's/Sequence://' | awk '{print $0"\t"$3-$2"\t1"}' >genome.fa.2.7.7.80.10.50.500_0based_6field.bed`

Align PacBio reads to genome assembly with the minimap2 [39] wrapper pbmm2.

1. `pbmm2 align -j 16 genome.fa pacbio_reads.fofn genome.aligned.bam --sort --median-filter --sample sample1`

Identify structural variants in genome assembly (*see* **Note 4**).

1. `pbsv discover --tandem-repeats genome.fa.2.7.7.80.10.50.500_0based_6field.bed genome.aligned.bam genome.svsig.gz`

2. `pbsv call -j 16 genome.fa genome.svsig.gz genome.var.vcf`

Extract deletions (DELs) and insertions (INSs) for which two alleles can be detected to avoid likely false positives that occur due to genome assembly errors.

1. `grep DEL genome.var.vcf | grep PASS | grep -v '1/1' | awk '{print $1"\t"$2-1"\t"$2+length($4)-1"\t"$3}' | awk '{print $0"\t"$3-$2"\t1"}' >genome.var.DEL.6field.bed`

2. `grep INS genome.var.vcf | grep PASS | grep -v -P '1/1' | awk '{print $1"\t"$2-1"\t"$2+length($5)-1"\t"$3}' | awk '{print $0"\t"$3-$2"\t1"}' >genome.var.INS.6field.bed`

Remove adapters and low quality regions from Illumina libraries. This step could be carried out using several different tools as an alternative to bbduk, reported in the example below (*see* **Note 5** for additional recommendations).

1. `bbduk.sh in1=reads_1.fastq in2=reads_2.fastq out1=reads_clean_trimmed_1.fq out2=reads_clean_trimmed_2.fq ref=adapters.fa ktrim=r k=25 mink=11 hdist=1 qtrim=r trimq=30 tpe tbo threads=8`

Map processed reads to genome assembly, merge resulting bam files if more than one library is independently mapped (**step 2**) and convert to fasta format.

1. `bwa mem -t 16 genome.fasta reads_clean_trimmed_1.fastq reads_clean_trimmed_2.fastq | samtools sort -@16 -o reads_bwa_aligned.bam -`

2. `samtools merge mergedBamFile.bam *.bam`

3. `samtools view -@ 8 -F 4 -h mergedBamFile.bam >all_mapping.sam`

4. `reformat.sh in=all_mapping.sam out=all_mapping.fa`

Produce kmer histogram of mapped reads.

1. `jellyfish count -t 8 -C -m 21 -s 16G all_mapping.fa -o all_reads.jf`

2. `jellyfish histo -o all_reads.histo all_reads.jf`

Extract reads that map to deletions and that are completely contained within a deletion.

1. `samtools view -@ 8 -F 4 -h -b -L genome.var.DEL.6field.bed mergedBamFile.bam >del_mapping.bam`

2. `bedmap --echo --fraction-map 1 <(bam2bed <del_mapping.bam) genome.var.DEL.6field.bed >del_reads.bed`

3. `cut -f1-6 del_reads.bed >del_reads.6field.bed`

4. `fastaFromBed -fi genome.fa -bed del_reads.6field.bed -fo del_reads.fa`

Produce kmer histogram of deletion mapping reads to compare against kmer histogram of all mapped reads.

1. `jellyfish count -t 8 -C -m 21 -s 16G del_reads.fa -o del_reads.jf`

2. `jellyfish histo -o del_reads.histo del_reads.jf`

Produce bam file of reads mapped to deletions.

1. `cut -f4 del_reads.bed | sort -u >del_reads.names`

2. `samtools view mergedBamFile.bam | fgrep -w -f del_reads.names >del_reads.sam`

3. `samtools view -H mergedBamFile.bam >del_reads_header.sam`

4. `cat del_reads.sam >>del_reads_header.sam`

5. `samtools view -@8 -S -b del_reads_header.sam >del_reads_header.bam`

6. `samtools index -@8 del_reads_header.bam`

Determine median coverage of deletion mapped reads.

1. `mosdepth -t 8 -m -b del_reads.6field.bed genome del_reads_filtered.bam`

Produce histogram of deletion coverages (rounded to whole number) so that each deletion has its coverage counted only once.

1. `zcat genome.regions.bed.gz | awk 'BEGIN{OFS=FS="\t"}{$6=sprintf("%.0f",$5) }1' | cut -f6 | sort -n | uniq -c | sed -e 's/^ *//' -e 's/\ /\t/' | awk '{print $2"\t"$1}' | head -200 >deletion_coverage.txt`

Determine read coverage at every nucleotide of the genome and split the output file by chromosome.

1. `genomeCoverageBed -d -ibam mergedBamFile.bam >genome.cov`

2. `awk '{print>$1".cov"}' genome.cov`

Calculate median coverage using a 1000 bp sliding window and a step of 1 nt then count how many windows have each coverage value (rounded to whole number). Script (median_sliding_window.gawk) can be found in **Note 6**.

1. `for i in *cov ; do cat $i | sh median_sliding_window.gawk | sed 's/\ /\t/g' >`echo $i | sed 's/cov/median/'` ; done`

2. `cat *median | awk 'BEGIN{OFS=FS="\t"}{$4=sprintf("%.0f", $3) }1' | cut -f4 | sort -n | uniq -c | sed -e 's/^ *//' -e 's/\ /\t/' | awk '{print $2"\t"$1}' >coverage_count.txt`

3.2 Gene Presence–Absence Variation Detection and Analysis

Some useful scripts to run the pipeline of analysis described below can be found online at the following link: https://github.com/Carmen-Tuc/PAV_pipeline.

Recover all the short read Illumina libraries derived from WGR experiments carried out on the target species of interest (*see* **Note 1** for further recommendations about the selection of the data to be analyzed). Although the procedure described below could be applied, with some modifications, to other types of sequencing reads, the reliability of this analysis pipeline has only been tested so far with Illumina short reads (*see* **Note 7** for further details).

Perform quality control on reads and trim them accordingly. This task can be performed with a number of different tools, including FastQC [40], that we are using in the example below. *See* **Note 5** for further recommendations.

```
# perform quality control
fastqc read1.fq read2.fq

# perform trimming
fastp -i read1 file.fq -I read2 file.fq -o read1.trimmed.fq -O
read2.trimmed.fq --detect_adapter_for_pe -V -f 5 -t 3 -F 5 -T
3 -h report.html -w 16 -x -n f -5 -3 -p -l 75 -M 24
```

Map the reads on the reference genome, making sure to allow the nonunique mapping of reads (*see* **Note 8**). Each sequencing library should be mapped separately from the others. We recommend the use of bwa [27], paired with samtools [41], for this task.

```
# index the genome
bwa index genome.fna
```

```
# map trimmed reads on the reference, piping the result into
# samtools view for bam conversion
bwa mem -M -t 64 genome.fna read1.trimmed.fq read2.trimmed.fq
| samtools view -bS - > mapping.bam

# sort the bam file
samtools sort -@ 64 -O bam -o mapping.sorted.bam mapping.bam
```

Produce a coverage file for each set of Illumina reads mapped. This file should link each position of the genome with read mapping coverage. Make sure to include sites that display coverage = 0.

```
# produce the depth file
samtools depth -aa mapping.sorted.bam > genome.depth
```

Extract the coordinates for all genes and corresponding exons from the genome annotation.

```
# extract meaningful rows and columns from the gff annotation
awk '{ if ($3 == "exon") print ($0) }' genome.gff | cut -f
1,4,5,9 > filtered_annotation.tsv

# clean the last column, reducing it to the exon id only.
# this command may vary depending on the format of the
annotation

cut -d "|" -f 1,7 filtered_annotation.tsv | sed 's/ID=exon-
gnl|//g' | cut -d ";" -f 1 > exons_coordinates

# input row example:
MTYJ01000001.1  Genbank exon    13344   13541   .   +   .
ID=exon-gnl|WGS:MTYJ|mrna.BV898_00003.1-1;Parent=rna-gnl|WGS:
MTYJ|mrna.BV898_00003.1;gbkey=mRNA;locus_tag=BV898_00003;or-
ig_protein_id=gnl|WGS:MTYJ|BV898_00003.1;orig_transcrip-
t_id=gnl|WGS:MTYJ|mrna.BV898_00003.1;partial=true;product=hy-
potheticalprotein;start_range=.,13344

# output row example:
MTYJ01000001.1   13344   13541 BV898_00003.1-1
```

Run a BUSCO analysis [21] on the extracted transcript sequences obtained from the genome, making sure to select the most appropriate dataset available in OrthoDB, and create a list of genes flagged as "complete".

```
busco -m transcriptome -l $lineage -c 64 -i transcript.fa -o
busco_exons -f
```

```
awk '$2 == "Complete"' full_table.tsv | cut -f 3 | cut -d ":"
-f 1 > complete_busco.list
```

Load the coverage file in a data analysis environment such as R or python. In the following example, python will be used with the numpy [32] and pandas [33] modules. Estimate the average coverage for the exon regions of each gene (if alternatively spliced exons are present, make sure to include the longest exons available. *See* **Note 9** for a simple python code to achieve this computation and **Note 10** for an explanation about the need to exclude intronic regions from these calculations.

Plot the distribution of the exon coverages of the complete set of BUSCO genes. As this set only includes single-copy genes, the graph produced should display a gaussian shaped curve, with its peak approximately corresponding to the expected genome sequencing depth (*see* **Note 11** for additional details). Estimate the median coverage of the BUSCO genes and divide it by eight to obtain a coverage threshold for calling absent genes (*see* **Note 12** for additional details).

```
def readlist(filename):
    with open(filename, "r") as input_list:
        return([x.rstrip() for x in input_list.readlines()]

complete_busco_genes = read_list("complete_busco.list")

exons_busco_coverage = total_exons[total_exons.ID_gene.isin
(complete_busco_genes)].groupby("ID_gene").apply(flatten_ex-
ons)
sns.distplot(exons_busco_coverage)
threshold = exons_busco_coverage.median()/8
```

Plot the exon coverage for all genes annotated in the reference genome and observe the peaks in the generated graph. The presence of genes with zero coverage and of a hemizygous peak in addition to the homozygous peak indicates the detection of gene PAV (*see* **Note 13** for additional details).

```
# Update xlim accordingly to the sequencing depth
xlim = 100
sns.distplot(coverage_total_exons[coverage_total_exons<xlim])
```

For troubleshooting purposes, we recommend to run the same analysis using the Illumina short reads library originally used for the de novo assembly of the reference genome itself. In this case, no absent gene should be identified (*see* **Note 14** for additional details).

Please note that this protocol has some limitations in PAV detection for genes subjected to copy number variations and, in particular, for genes associated with transposable element activity (*see* **Note 15**).

The users might want to optionally collect unmapped reads resulting from the bwa mem mapping step. As these may belong to regions which are not found in the reference genome, the de novo assembly of unmapped paired-end reads can be used to generate a collection of alternate contigs, building the pangenome of the species of interest. *See* **Note 16** for some suggestions.

3.3 PAV Gene Functional Enrichment Analysis

Perform enrichment analysis by hypergeometric test [42] on the union set of the *dispensable* genes identified in all the analyzed resequenced genomes, using the complete set of genes annotated in the reference genome as a background (this will constitute the "universe" dataset, see below). To achieve this, at least one type of annotation will be needed. While multiple alternative functional annotation resources could be used, we recommend Gene Ontology (GO) terms, which can be further subdivided in three main categories: Biological Process, Molecular Functions, and Cell Component [43, 44]. For functional enrichment analysis in nonmodel species, we suggest using PFAM [45] in addition to GO terms (*see* **Note 17**).

While these annotations are usually already available from the genome annotation file, they can be also obtained with InterProScan [46, 47], which should be run on the proteome resulting from the translation of the coding sequences associated with gene model annotations.

A simple python script to perform functional enrichment tests is available online (https://gitlab.com/54mu/enrichment_test). This analysis requires a "universe" dataset (the complete set of genes annotated in the reference genome) and a subset of genes to test for enrichment (which in this case comprises all the *dispensable* genes identified with the previous steps). The universe dataset needs to be a table matching gene ids and feature ids, one per line, as reported in the example below. It is also important for each match to be unique.

extract from a universe file for GO enrichment.

```
...
LOC105326593    GO:0065003
LOC105326593    GO:0070062
LOC105326593    GO:1903561
LOC105344258    GO:0000212
LOC105344258    GO:0001516
LOC105344258    GO:0001525
...
```

```
Tyr     GO:0042470
Tyr     GO:0046872
Tyr     GO:0050149
Tyr     GO:0055114
Pepck   GO:0000287
Pepck   GO:0003729
Pepck   GO:0004550
Pepck   GO:0004611
Pepck   GO:0004613
...
```

The subset file must contain the list of unique ids of the *dispensable* genes subject to PAV. The last parameter required to run the analysis is the number of unique features in the universe (in this case the total number of genes annotated in the reference genome).

Enriched annotations can be filtered by the user based on arbitrary p-value thresholds, which may be also combined with a threshold of observations for any given annotation in the *dispensable* gene subset, to filter out annotations linked with a very low number of genes.

4 Notes

1. This procedure may potentially also work with long reads generated with Oxford Nanopore technologies (ONT), that is, obtained with MinION, GridION, or PromethION platforms. Since we have not tested the performance of the protocol with this type of data, in case of availability of ONT long reads, we suggest testing alternative SV detection tools in addition to psvb. Among these, NanoSV [48], SVIM [49] and cuteSV [50] have been previously indicated as suitable for SV detection using ONT data.
2. We here define a "high quality" genome as a genome whose assembly approaches a chromosome-scale quality. Ideally, such a genome would display a low amount of genomic sequence located in unplaced scaffolds and a very high completeness, which can be generally estimated either with gene model-centric tools, such as BUSCO [21], or with k-mer based methods, such as Merqury [51].

   Most importantly, a high quality reference genome analyzed with the hemizygous regions and PAV detection protocols must be entirely devoid of exogenous contaminations, that is, contigs and scaffolds deriving from other associated organisms, such as symbionts, parasites, and pathogens, as these may lead to artifacts (i.e., contaminant genes might be wrongly identified as *dispensable* genes part of the pangenome of the species of interest). The overwhelming majority of the reference genomes deposited in public repositories, such as Ensembl, NCBI Genomes and others, are expected to meet these quality requirements. Nevertheless, we recommend extra caution, especially when no other genomes from closely related species are available as a reference, since the discrimination between novel orphan genes, horizontally transferred genes and exogenous contaminants is often not a trivial task [52].

3. Illumina genomic DNA libraries can be obtained with a number of different commercial kits and sequenced using different strategies. These can affect in a significant way both the accuracy of read mapping and the evenness of read coverage across the reference genome assembly. The protocols described in this chapter were extensively tested on regular paired-end libraries sequenced on HiSeq and NovaSeq series platforms with a 2×100, 2×125 or 2×150 strategies, which usually allow the attainment of a relatively uniform distribution of reads on the full length of chromosomes, with the exception of local peaks with very high coverage in highly repetitive regions. The use of particularly short reads (e.g., <50 base pairs) generated with single-end sequencing might lead to a decrease in mapping accuracy, determining an increase in the fraction of reads mapping to multiple genomic sites. This might have a negative effect on the quantification of the sequencing coverage of some loci, most notably those that include paralogous genes.

   At the same time, we discourage the use of paired-end libraries which derive from chromosome conformation capture approaches (such as Hi-C or Dovetail Genomics Omni-C™) and include long-range connectivity information, as these may lead to an uneven distribution of mapped reads, not suitable for downstream PAV analyses. On the other hand, we have successfully tested this protocol on 10× Genomics Illumina libraries [18].

4. Due to the limitations of pbsv, deletions larger than 100 kb will be missed. We recommend checking the length distribution of the insertions identified in the target genome to evaluate whether deletions larger than this threshold are likely to be present. In the case of genomes including a high amount of large hemizygous regions, the output of this protocol might result in an underestimate of total hemizygous genomic DNA content.

5. The trimming parameters may be modified based on the quality of the raw sequencing data available. In particular, note that Illumina short reads generated with different library preparation kits might include different adapter sequences and

barcodes. Make sure to check the technical documentation provided by the manufacturers to identify the most appropriate list of adapter sequences to be used. We recommend using stringent trimming parameters to discard all possible sources of bias. Discarding all the reads whose length, following the trimming procedure, is lower than 50 nucleotides, might be also beneficial in some cases, as these reads may result in ambiguous mappings on multiple sites.

6.

```
median_sliding_window.gawk script
#!/usr/bin/sh
gawk -v wsize=1000 '
   BEGIN {
   if (wsize % 2 == 0) { m1=wsize/2; m2=m1+1; } else {
   m1 = m2 = (wsize+1)/2; }
   }
   function roundedmedian() {
     asort(window, a);
   return (m1==m2) ? a[m1] : int(0.5 + ((a[m1] + a[m2])
   / 2));
   }
   function push(value) {
     window[NR % wsize] = value;
   }
   NR < wsize { window[NR]=$3; next; }
   { push($3);
     $3 = roundedmedian();
     print $0;
   }'
```

7. Short reads generated with other sequencing platforms (e.g., with BGISEQ platforms) may work as well, as long as the sequencing error rate is in line with those expected from Illumina approaches. We do not recommend using reads generated with sequencing methods that are known to suffer from relatively high error rates, in particular in correspondence with homopolymeric sequence stretches, such as 454 Life Sciences pyrosequencing and Thermo Fisher Scientific Ion Torrent, as they may introduce significant biases in read mapping profiles. This protocol might be implemented in the future to also allow the mapping of resequencing data obtained with third generation methodologies (i.e., nanopore sequencing by Oxford Nanopore and SMRT sequencing by Pacific Biosciences), even though significant modifications might be required to take into account their different length and error rates of the reads generated with these approaches.

8. This can be achieved by selecting the -M option in bwa mem. We have previously shown that this is necessary in order to avoid erroneously obtaining long stretches of genomic sequences with coverage = 0 in the presence of repeats. Not selecting this option would therefore result in an inflation of PAV calls.

9.
```
import pandas as pd
import numpy as np
import matplotlib.pyplot as plt
import seaborn as sns

def flatten_exons(gene):

    # by using a coverage map this function returns the average
    # coverage of the exonic regions of a gene (passed as a
    # pandas groupby object)

    global cov_map
    contig = gene.contig.values[0]
    start = min(gene.start.values)-1
    end = max(gene.end.values)
    size = end - start
    mask = np.full([size], False)
    pairs= np.array([gene.start.values, gene.end.values])
    startendarray = pairs.flatten('F')
    for i in range(0, startendarray.size, 2):
        this_exon = np.arange(startendarray[i]-start,starten-
darray[i+1]-start)
        mask[this_exon] = True
    sub_cov_map = cov_map[contig][start:start+size]
    try:
        return sub_cov_map[mask].mean()
    except Exception as e:
        return np.nan

coverage_file = pd.read_csv("/path/genome.depth", sep = "\t",
names = ["contig", "position", "coverage"])
cov_map = coverage_file.groupby("contig")["coverage"].apply
(np.array).to_dict()
total_exons = pd.read_csv("/path/exons_coordinates", sep =
"\t", names = ["contig", "start", "end", "ID_gene"])

coverage_total_exons = total_exons.groupby("ID_gene").apply
(flatten_exons)
```

10. While the fragments included in Illumina paired-end sequencing data derive from the random fragmentation of genomic

DNA, reads are not necessarily expected to be evenly distributed across the genome assembly. In particular, due to the mapping strategy used in this protocol (*see* **Note 8**), a significant number of reads may be aligned to multiple genomic locations, resulting in local mapping spikes. We expect such spikes to be found in genomic regions encompassing repeats, and therefore to be mostly associated with intergenic regions. However, some species are known to include a considerable amount of heterochromatic introns, which may also include short repeats that could lead to the nonspecific mapping of short reads [53]. Since anomalous mapping may lead to an artefactual inflation of read coverage estimates at the gene level, we recommend to perform such calculations based on exonic regions only.

11. While the theoretical sequencing depth of any resequenced genome can be easily calculated by dividing the total amount of sequence data generated (i.e., the total number of reads multiplied by their average read) by the size of the genome, we recommend using the strategy explained in this protocol to obtain a more reliable estimate of the mapping coverage that would be expected for any single-copy *core* gene in the genome of a diploid organism. As a matter of fact, several factors may cause some discrepancies between the theoretical and actual genome sequencing coverage: namely, the presence of a high amount of reads mapping on mitochondrial (and plastidial) genomes, the presence of exogenous contamination and the occurrence of low quality, unmappable reads. By only taking into account the coverage observed in exonic regions of validated single-copy genes, the method we propose disregards the aforementioned confounding factors and provides a much more reliable estimate.

    The distribution of the coverage of BUSCO genes is expected to follow a Gaussian curve, centered on the actual genome-wide sequencing depth, which identifies the "homozygous peak," as shown in the example provided in Fig. 1.

    Note that this could be also achieved by the use of an approach based on genome-wide k-mer distribution, as explained in Subheading 3.1. However, we noted that the identification of a clear homozygous peak may become difficult in genomes characterized by high heterozygosity rates and resequenced with low coverage (i.e., <25×).

12. This is an arbitrary threshold, which has been previously shown to work well for PAV detection in *M. galloprovincialis*, as it displayed high correlation between in silico gene presence–absence calls and PCR confirmation [16]. In any given resequenced genome of a diploid organism, gene coverage

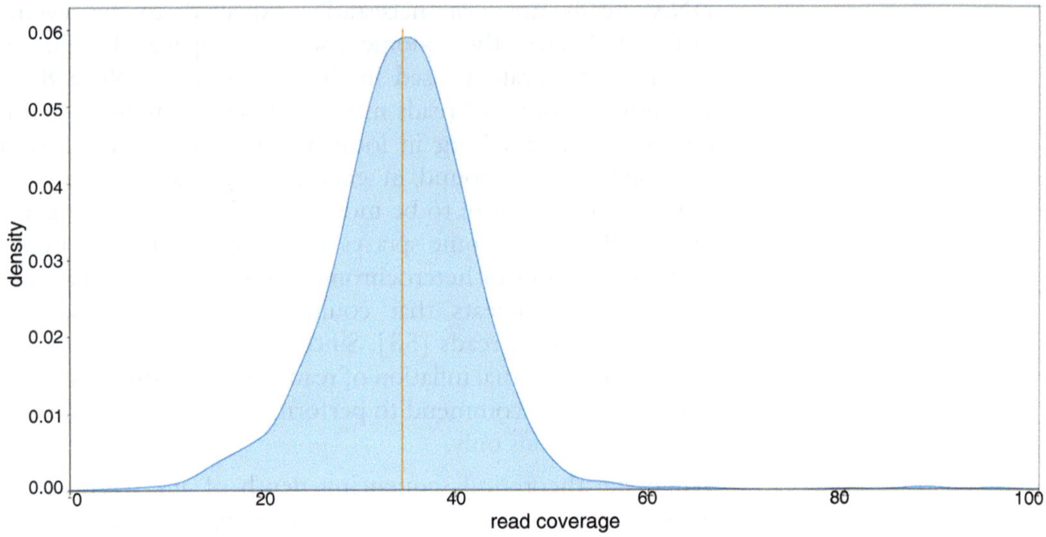

Fig. 1 Expected distribution of the coverage of BUSCO genes in a diploid organism resequenced with a paired-end Illumina short reads library and 36× coverage. The vertical bar indicates the median coverage observed for all BUSCOs. Note the Gaussian distribution around the "homozygous peak"

(calculated on exons only) would be expected to follow a distribution similar to the one shown in the example below (Fig. 2).

In brief, a main "homozygous peak" of coverage, corresponding to the one previously identified from the analysis of BUSCOs, should be observed. This peak indicates genes present with two alleles in the diploid genome. A second "hemizygous peak" should be observed at a coverage equal to exactly half of the "2n" peak. This peak indicates genes present with a single allele in the diploid genome. The relative height of the two peaks can be used to estimate the rate of genes encoded by hemizygous genomic regions in any resequenced individual. Note that species where PAV is very rare are not expected to display a visible hemizygous peak. Finally, a third peak should be observed at zero coverage, marking *dispensable* genes that are present in the reference genome, but absent in the resequenced individual.

We have empirically noted that whenever a relatively low sequencing coverage is used for resequencing (e.g., <50×), the two Gaussian curves cannot be well separated, but rather result in a valley where the upper- and lower-end tails of the two curves are partially overlapping, which does not permit to discriminate with certainty the homozygous or hemizygous state of a given gene (*see* Fig. 2). With sequencing coverage <30×, the hemizygous peak will simply appear as a "shoulder" on the side of the homozygous peak. The same consideration

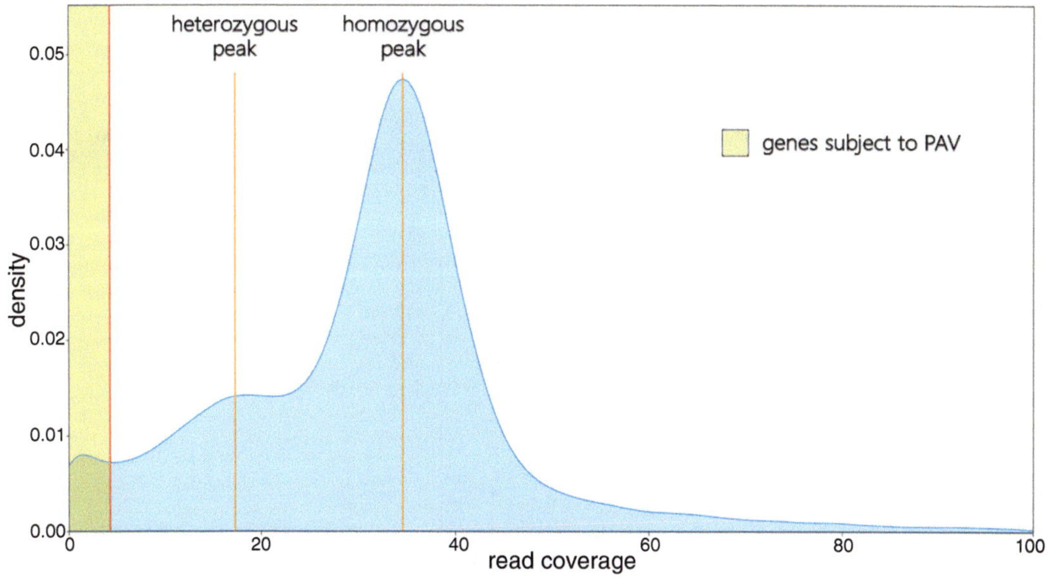

Fig. 2 Expected distribution of the sequencing coverage of all genes in a resequenced genome of a diploid species with widespread PAV. This example shows a genome resequenced with 36× coverage using a paired-end Illumina short read library. The homozygous and hemizygous peaks are placed at 36× and 18×, respectively. The threshold for gene PAV calling is set at 4.5× (i.e., all genes showing a coverage lower than 4.5× are called as absent)

applies to the ability to discriminate between the lower-end tail of the hemizygous peak and the "gene absence peak". This issue is most likely linked with the cross-mapping of a few reads derived from paralogous *core* genes on local regions with high pairwise sequence homology. While higher sequencing depths increase the confidence level of presence–absence calls, we here provide a conservative way of estimating absent genes by using as a threshold 1/8th of the coverage of the homozygous peak. This means that only genes showing an exon sequencing coverage lower than 25% of the expected coverage of a *dispensable* gene found in a hemizygous genomic region will be called as absent.

13. This protocol has been extensively tested on the genome of diploid organisms only. In the case of target genomes with different ploidy levels, multiple peaks might be observed (e.g., four peaks, denoting genes present with one, two, three or four alleles, should be present in a tetraploid species). In such cases, we recommend setting the threshold for PAV detection at 1/4th of the "single allele" peak coverage. Please note that small peaks can be occasionally observed in diploid genomes at coverages which are multiples of the homozygous peak. These may indicate the presence of nearly identical, recently duplicated, paralogous genes. We do not expect this

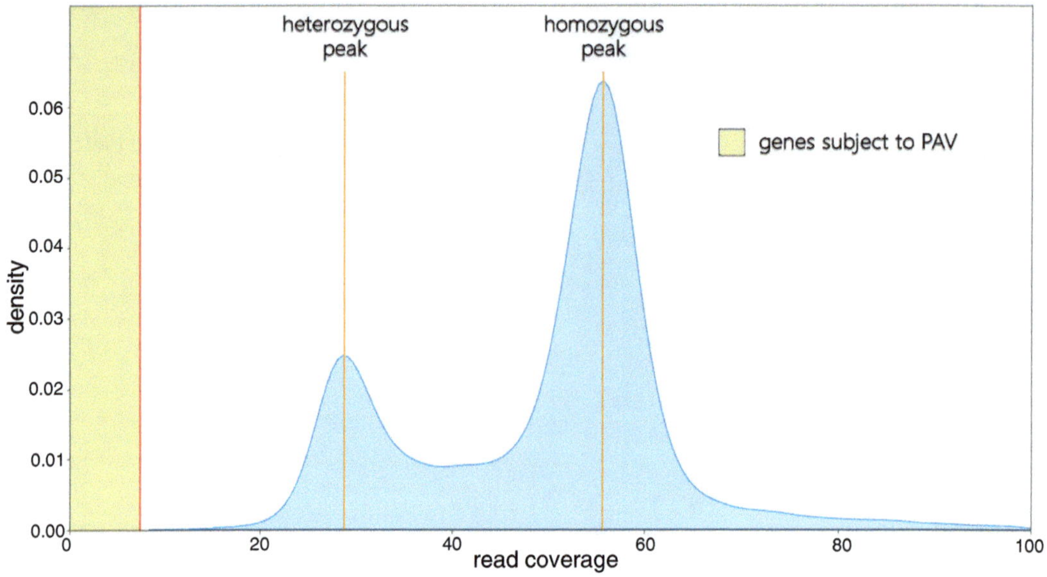

Fig. 3 Expected distribution of the sequencing coverage of all genes in the reference genome of a diploid species with widespread PAV, based on the mapping of a paired-end Illumina short read library generated from the same individual used for the de novo genome assembly. This example shows a genome resequenced with 56× coverage. The homozygous and hemizygous peaks are placed at 56× and 28×, respectively. The threshold for gene PAV calling is set at 7× (i.e., all genes showing a coverage lower than 4.5× are called as absent). Note that no absent genes can be detected in the reference genome

factor to represent a relevant issue in most cases, unless the target species has been subjected to recent whole genome duplication events.

14. We recommend running this test to check the correct setting of all parameters. The mapping of Illumina short read libraries used to perform the de novo assembly of the reference genome against the reference genome itself should, by definition, not produce any gene with zero coverage. On the other hand, depending on the level of hemizygosity of the sequenced individual, the hemizygous peak might be visible. *See* an example of the expected sequencing coverage in Fig. 3.

15. The allowance of read multimapping (*see* **Note 8**), applied in this protocol to avoid the artefactual identification of PAV within highly repetitive genomic regions, may lead to the impossibility of detecting PAV for genes present with multiple identical or nearly identical paralogous gene copies, which may be subject to copy number variation. Although we have previously reported that the activity of transposable elements is likely associated with hemizygous genomic regions in several molluscan reference genomes [18], no significant enrichment of gene families linked with reverse transcriptase, integrase, transposase and other TE-related enzymatic activities has

been identified associated with *dispensable* genes in *M. galloprovincialis* with the protocol explained in Subheading 3.2 [16]. In our interpretation, this discrepancy is linked with the cross-mapping of reads that originated from identical gene copies placed in distinct genomic locations. The protocol described in Subheading 3.1 might be more indicative of PAV detection in these particular cases, thanks to the possibility of exploiting flanking unambiguous sequence to accurately map the reads to the correct genomic locations.

16. Unmapped reads, collected from all the individual resequenced genomes analyzed, can be de novo assembled using several different algorithms, using a recursive reassembly approach [16]. For this task, we recommend paying attention to the assembly of contaminant contigs, which may derive from a series of different sources, in particular in marine filter-feeding organisms. Considering that most pangenomic contigs assembled using unmapped reads are expected to be present in hemizygous genomic regions in the resequenced individuals, the estimated coverage of the hemizygous peak (*see* **Note 12**) can be used as a guidance to set appropriate coverage thresholds to identify contigs which belong to the target species. Both contigs showing excessively high or particularly low coverage compared with expectations should be flagged as suspect and discarded. In addition, nucleotide composition, and GC content in particular, can be used as a complementary information to further detect possible contaminants. In this respect, BlobTools [54] can be very useful. While it is unlikely that contaminant contigs will have both the same coverage and the same GC content expected for hemizygous regions of the target species, we recommend extra caution to avoid including suspect sequences in the pangenome assembly. Therefore, further controls, such as the use of Kraken 2 [55] or BLASTn-based filtering against the complete genome assembly of known contaminants could be applied.

    We also recommend including in the collection of pangenomic contigs only those exceeding a length of 1 kb, as those shorter than this threshold may correspond to local intergenic or intronic regions characterized by high heterozygosity.

    The obtained pangenomic contigs may be then subject to gene annotation, making sure to apply the same annotation pipeline used for the reference genome, providing a list of *dispensable* genes absent from the reference assembly, but present in one or more resequenced genomes of the same species.

17. Considering that GO terms tend to be strongly biased toward model species [56] and that orphan, taxonomically restricted genes without detectable primary sequence homology are highly abundant in the genomes of marine invertebrates [57],

we suggest to use Hidden Markov Model (HMM)-based conserved protein domain annotations to improve the inference of functional enrichment in nonmodel species. Several alternative resources, such as PFAM [45], InterPro [46, 47], and others, may be used for this purpose. Due to the BLAST-independent nature of HMM searches, these allow to improve the annotation rate of genes lacking significant primary sequence homology with entries deposited in public repositories.

References

1. Feuk L, Marshall CR, Wintle RF et al (2006) Structural variants: changing the landscape of chromosomes and design of disease studies. Hum Mol Genet 15:R57–R66
2. Marroni F, Pinosio S, Morgante M (2014) Structural variation and genome complexity: is dispensable really dispensable? Curr Opin Plant Biol 18:31–36
3. Read BA, Emiliania huxleyi Annotation Consortium, Kegel J et al (2013) Pan genome of the phytoplankton Emiliania underpins its global distribution. Nature 499(7457): 209–213. https://doi.org/10.1038/nature12221
4. McInerney JO, McNally A, O'Connell MJ (2017) Why prokaryotes have pangenomes. Nat Microbiol 2:17040. https://doi.org/10.1038/nmicrobiol.2017.40
5. Medini D, Donati C, Tettelin H et al (2005) The microbial pan-genome. Curr Opin Genet Dev 15:589–594
6. Vernikos G, Medini D, Riley DR et al (2015) Ten years of pan-genome analyses. Curr Opin Microbiol 23:148–154
7. Aherfi S, Andreani J, Baptiste E et al (2018) A Large Open Pangenome and a Small Core Genome for Giant Pandoraviruses. Front Microbiol 9:1486. https://doi.org/10.3389/fmicb.2018.01486
8. Song J-M, Guan Z, Hu J et al (2020) Eight high-quality genomes reveal pan-genome architecture and ecotype differentiation of Brassica napus. Nat Plants 6:34–45
9. Alonge M, Wang X, Benoit M et al (2020) Major impacts of widespread structural variation on gene expression and crop improvement in tomato. Cell 182:145–161.e23
10. Golicz AA, Bayer PE, Bhalla PL et al (2020) Pangenomics comes of age: from bacteria to plant and animal applications. Trends Genet 36:132–145
11. McCarthy CGP, Fitzpatrick DA (2019) Pan-genome analyses of model fungal species. Microb Genom 5:e000243
12. Sherman RM, Forman J, Antonescu V et al (2019) Assembly of a pan-genome from deep sequencing of 910 humans of African descent. Nat Genet 51:30–35
13. Tian X, Li R, Fu W et al (2020) Building a sequence map of the pig pan-genome from multiple de novo assemblies and Hi-C data. Sci China Life Sci 63:750–763
14. Li R, Li Y, Zheng H et al (2010) Building the sequence map of the human pan-genome. Nat Biotechnol 28:57–63
15. Rosa RD, Alonso P, Santini A et al (2015) High polymorphism in big defensin gene expression reveals presence–absence gene variability (PAV) in the oyster Crassostrea gigas. Dev Comp Immunol 49(2):231–238. https://doi.org/10.1016/j.dci.2014.12.002
16. Gerdol M, Moreira R, Cruz F et al (2020) Massive gene presence-absence variation shapes an open pan-genome in the Mediterranean mussel. Genome Biol 21:275
17. Vos M, Eyre-Walker A (2017) Are pangenomes adaptive or not? Nat Microbiol 2:1576–1576
18. Calcino AD, Kenny NJ, Gerdol M (2021) Single individual structural variant detection uncovers widespread hemizygosity in molluscs. Philos Trans R Soc Lond Ser B Biol Sci 376: 20200153
19. Martinez AS, Willoughby JR, Christie MR (2018) Genetic diversity in fishes is influenced by habitat type and life-history variation. Ecol Evol 8:12022–12031
20. Olsen KC, Ryan WH, Winn AA et al (2020) Inbreeding shapes the evolution of marine invertebrates. Evolution 74:871–882
21. Seppey M, Manni M, Zdobnov EM (2019) BUSCO: assessing genome assembly and

annotation completeness. Methods Mol Biol 1962:227–245

22. Zdobnov EM, Tegenfeldt F, Kuznetsov D et al (2017) OrthoDB v9.1: cataloging evolutionary and functional annotations for animal, fungal, plant, archaeal, bacterial and viral orthologs. Nucleic Acids Res 45:D744–D749
23. Bushnell B. et al. (2014) BBMap: A Fast, Accurate, Splice-Aware Aligner. No. LBNL-7065E. Ernest Orlando Lawrence Berkeley National Laboratory, Berkeley, CA.
24. Neph S, Kuehn MS, Reynolds AP et al (2012) BEDOPS: high-performance genomic feature operations. Bioinformatics 28:1919–1920
25. Quinlan AR, Hall IM (2010) BEDTools: a flexible suite of utilities for comparing genomic features. Bioinformatics 26:841–842
26. Li H, Durbin R (2009) Fast and accurate short read alignment with burrows-wheeler transform. Bioinformatics 25:1754–1760
27. Li H (2013). Aligning sequence reads, clone sequences and assembly contigs with BWA-MEM. http://github.com/lh3/bwa
28. fastp, Github. https://github.com/OpenGene/fastp
29. Andrews S FastQC, Github. https://github.com/s-andrews/FastQC
30. Marçais G, Kingsford C (2011) A fast, lock-free approach for efficient parallel counting of occurrences of k-mers. Bioinformatics 27: 764–770
31. Pedersen BS, Quinlan AR (2018) Mosdepth: quick coverage calculation for genomes and exomes. Bioinformatics 34:867–868
32. Harris CR, Millman KJ, van der Walt SJ et al (2020) Array programming with NumPy. Nature 585:357–362
33. McKinney W (2010) Data Structures for Statistical Computing in Python. Proceedings of The 9th Python in Science Conference, pp. 51-56. https://doi.org/10.25080/majora-92bf1922-00a
34. Pacific Biosciences (2017) pbmm2, Github. https://github.com/PacificBiosciences/pbmm2
35. Pacific Biosciences (2017) pbsv, Github. https://github.com/PacificBiosciences/pbsv
36. Li H, Handsaker B, Wysoker A et al (2009) The sequence alignment/map format and SAMtools. Bioinformatics 25:2078–2079
37. Virtanen P, Gommers R, Oliphant TE et al (2020) Author correction: SciPy 1.0: fundamental algorithms for scientific computing in python. Nat Methods 17:352
38. Benson G (1999) Tandem repeats finder: a program to analyze DNA sequences. Nucleic Acids Res 27:573–580
39. Li H (2018) Minimap2: pairwise alignment for nucleotide sequences. Bioinformatics 34: 3094–3100
40. Wingett SW, Andrews S (2018) FastQ screen: a tool for multi-genome mapping and quality control. F1000Res 7:1338
41. Danecek P, Bonfield JK, Liddle J et al (2021) Twelve years of SAMtools and BCFtools. Gigascience 10:giab008
42. Falcon S, Gentleman R (2008) Hypergeometric testing used for gene set enrichment. Analysis:207–220. https://doi.org/10.1007/978-0-387-77240-0_14
43. Ashburner M, Ball CA, Blake JA et al (2000) Gene ontology: tool for the unification of biology. The Gene Ontology Consortium. Nat Genet 25:25–29
44. Gene Ontology Consortium (2021) The gene ontology resource: enriching a GOld mine. Nucleic Acids Res 49:D325–D334
45. Mistry J, Chuguransky S, Williams L et al (2021) Pfam: the protein families database in 2021. Nucleic Acids Res 49:D412–D419
46. Jones P, Binns D, Chang H-Y et al (2014) InterProScan 5: genome-scale protein function classification. Bioinformatics 30:1236–1240
47. Blum M, Chang H-Y, Chuguransky S et al (2021) The InterPro protein families and domains database: 20 years on. Nucleic Acids Res 49:D344–D354
48. Stancu MC, van Roosmalen MJ, Renkens I et al (2017) Mapping and phasing of structural variation in patient genomes using nanopore sequencing. Nat Commun 8:1–13
49. Heller D, Vingron M (2019) SVIM: structural variant identification using mapped long reads. Bioinformatics 35:2907–2915
50. Jiang T, Liu Y, Jiang Y et al (2020) Long-read-based human genomic structural variation detection with cuteSV. Genome Biol 21:189
51. Rhie A, Walenz BP, Koren S et al (2020) Merqury: reference-free quality, completeness, and phasing assessment for genome assemblies. Genome Biol 21:245
52. Bemm F, Weiß CL, Schultz J et al (2016) Genome of a tardigrade: horizontal gene transfer or bacterial contamination? Proc Natl Acad Sci U S A 113(22):E3054–E3056
53. Espinas NA, Tu LN, Furci L et al (2020) Transcriptional regulation of genes bearing intronic heterochromatin in the rice genome. PLoS Genet 16:e1008637

54. Laetsch DR, Blaxter ML (2017) BlobTools: interrogation of genome assemblies. F1000Res 6:1287
55. Wood DE, Lu J, Langmead B (2019) Improved metagenomic analysis with kraken 2. Genome Biol 20:257
56. Gaudet P, Dessimoz C (2017) Gene ontology: pitfalls, biases, and remedies. Methods Mol Biol 1446:189–205
57. Khalturin K, Hemmrich G, Fraune S et al (2009) More than just orphans: are taxonomically-restricted genes important in evolution? Trends Genet 25:404–413

Chapter 5

From Sequences to Enzymes: Comparative Genomics to Study Evolutionarily Conserved Protein Functions in Marine Microbes

Angelina Cordone, Alessandro Coppola, Angelica Severino, Monica Correggia, Matteo Selci, Antonio Cascone, Costantino Vetriani, and Donato Giovannelli

Abstract

Comparative genomics is a research field that allows comparison between genomes of different life forms providing information on the organization of the compared genomes, both in terms of structure and encoded functions. Moreover, this approach provides a powerful tool to study and understand the evolutionary changes and adaptation among organisms. Comparative genomics can be used to compare phylogenetically close marine organisms showing different vital strategies and lifestyles and obtain information regarding specific adaptations and/or their evolutionary history. Here we report a basic comparative genomics protocol to extrapolate evolutionary information about a protein of interest conserved across diverse marine microbes. The outlined approach can be used in a number of different settings and might help to gain new insights into the evolution and adaptation of marine microorganisms.

Key words Comparative genomics, Ancestral protein, Evolution of function, Adaptation

1 Introduction

Comparative genomics is a field of biological research in which the genome composition of different organisms is compared, both in terms of structure and encoded functions [1]. Recent advances in sequencing technologies, modelling and molecular techniques have improved and accelerated the complete or near-complete genome sequencing of many organisms and microorganisms [2]. These advances, together with the ability to recover near-complete Metagenome Assembled Genomes (MAGs) from the environment [3] have facilitated large scale comparative genomics and phylogenomics analysis. Comparative genomics aims to identify portions of DNA, RNA, or proteins conserved or divergent between

Cinzia Verde and Daniela Giordano (eds.), *Marine Genomics: Methods and Protocols*,
Methods in Molecular Biology, vol. 2498, https://doi.org/10.1007/978-1-0716-2313-8_5,

different species, thus providing an important tool to study how evolution shaped the genomes of organisms (e.g., to identify ancestral and evolved functional traits) [4, 5].

In this contest, comparative genomics can be used to compare phylogenetically close organisms showing different vital strategies and lifestyles to analyze their adaptations through evolution [6, 7]. For example, in marine microbiology, this approach has been used to compare gene functions between environmental and clinical relevant bacteria belonging to the *Vibrio* genus, thus identifying the unique and the common core functions of the pathogenic strains [8, 9], or to identify conserved functions such as nitrate respiration or quorum sensing among deep-sea Epsilonproteobacteria (recently proposed to be reclassified as phylum Campylobacterota) [10] and their pathogenic relatives [11, 12].

Here, we show how to apply comparative genomics methods to study and analyze the evolution of proteins conserved across marine microbes adapted to different lifestyles (free-living, epibionts, and obligate pathogens). Specifically, we showcase how the in silico identification of an unknown conserved protein and the study of its characteristics can be used to assess its evolutionary trajectory within defined groups of marine microorganisms. The results of this approach can be used downstream to inform heterologous expression experiments allowing comparative biochemical and functional characterization of the protein(s) (*see* Chapter *[From sequences to enzymes: heterologous expression of genes from marine microbes, this volume]*). The outlined approach can be used in a number of different settings and might help to gain new insight into the evolution and adaptation of marine microorganisms.

2 Materials

1. A computer with enough RAM to handle 3D visualization and phylogenetic tree construction might be required. A Linux-based operating system (such as Ubuntu, Mint, or Fedora) is preferred. Most programs used also work on Windows or Mac OS operating systems, but have not been tested with this protocol. An internet browser (Google Chrome, Firefox, etc.), and a reliable internet connection are also necessary. Plots and statistical tests have been performed using the free software environment for statistical computing and graphics R (https://www.r-project.org/), and figure editing carried out using the free and open-source vector graphics software Inkscape (https://inkscape.org/). Other software might be used in place of the ones described in this chapter and the ultimate selection rests upon the researcher.

2. Access to public genome repositories such as the National Genetic Sequence Database (GenBank, https://www.ncbi.nlm.nih.gov/search/) and the European Molecular Biology Laboratory European Bioinformatics Institute (EMBL-EBI, https://www.ebi.ac.uk/).
3. A local install of the NCBI Basic Local Alignment Search Tool (BLAST) or familiarity with the use of the online interface available at https://blast.ncbi.nlm.nih.gov/Blast.cgi.
4. A program for multiple sequence alignment. We used SeaView 4.0 (http://doua.prabi.fr/software/seaview) but several multi-platform alternatives are available [13].
5. A phylogenetic reconstruction software to compute and manipulate phylogenetic trees. We used PhyLM 3.0, embedded in SeaView 4.0 [14] but several other well-known options are available.
6. An open-source molecular visualization system for high quality three dimensional (3D) visualization and structure analysis of proteins. We used PyMOL for the analysis [15], but a number of other options are available.
7. A freely available program able to detect pockets and cavities of proteins of known 3D structure and to predict putative substrate-binding sites. In the present protocol we have used *POcket-CAvity Search Application* (POCASA) available online at http://g6altair.sci.hokudai.ac.jp/g6/service/pocasa/ [16].
8. A working knowledge of a comparative modeling server for 3D protein structures. For the present protocol we used the web based version of SWISS-MODEL available at https://swissmodel.expasy.org/ [17].

3 Methods

This protocol assumes that the protein of interest has been already identified through other means, for example large scale comparative genomic and pangenomic approaches [18], metagenomic analysis [3] or metabolic pathway reconstruction [6, 7, 11]. The identification of key proteins of interest is beyond the scope of this chapter.

3.1 Retrieving Closely Related Amino Acid Sequences from the Database

1. Using the Basic Local Alignment protein Search Tool (BLASTp) against the NCBI (or EMBL-EBI) database [19], manually search amino acid sequences closely related to the protein of interest from several representative strains belonging to the same taxonomic group or ecological group (*see* **Note 1**).

2. Download the amino acid FASTA files of all the examined proteins together with the relative accession numbers (WP_) (*see* **Notes 2** and **3**).

3.2 Retrieving 16S rRNA Gene Sequences for the Selected Organisms from the Database

1. Inspect the obtained amino acid sequences and identify the hosting organisms. The operation can be also done directly on the NCBI database by clicking on the organism name while on the GenBank view of the selected sequence.
2. Obtain and download the nucleotide FASTA files of the 16S rRNA genes from all the identified species, along with their respective accession numbers (NR_) (*see* **Note 4**).

3.3 Amino Acid Sequence Alignment and Evaluation of Conserved Regions

1. Align all the download amino acid sequences using the sequence alignment software and the algorithm of choice. We suggest the ClustalO alignment algorithm [20], available within SeaView [13]. Several alternatives are available.
2. Manually inspect the obtained alignment and refine it if necessary.
3. Conserved amino acids sequence regions and motifs can be identified when aligning the sequence against a known template protein sequence (Fig. 1).
4. The conserved regions can be annotated using BOXSHADE (version 3.21) available at https://embnet.vital-it.ch/software/BOX_form.html.

3.4 Nucleotide Sequences Alignment

1. Align all the downloaded nucleotide sequences using the sequence alignment software and the algorithm of choice. We suggest the ClustalW alignment algorithm [20], available within SeaView [13]. Several alternatives are available.
2. Manually inspect the obtained alignment and refine it if necessary.

3.5 Phylogenetic Trees Construction

1. Identify the best fitting amino acid or nucleotide substitution to use for tree generation from the alignment. We suggest ProtTest3 [21] for the amino acid sequences and JModelTest [22] for the nucleotide sequences.
2. Once the best substitution model has been identified, construct the maximum-likelihood phylogenetic tree for both the protein sequence and the 16S rRNA gene sequences using the identified parameters. We use PhyML for this purpose [14], selecting alternative likelihood ratio tests (a-LRT) for evaluating branch support.
3. Root the obtained tree according to a preselected strategy. Appropriate rooting is required to infer robust evolutionary information (*see* **Note 5**). The obtained trees can be efficiently

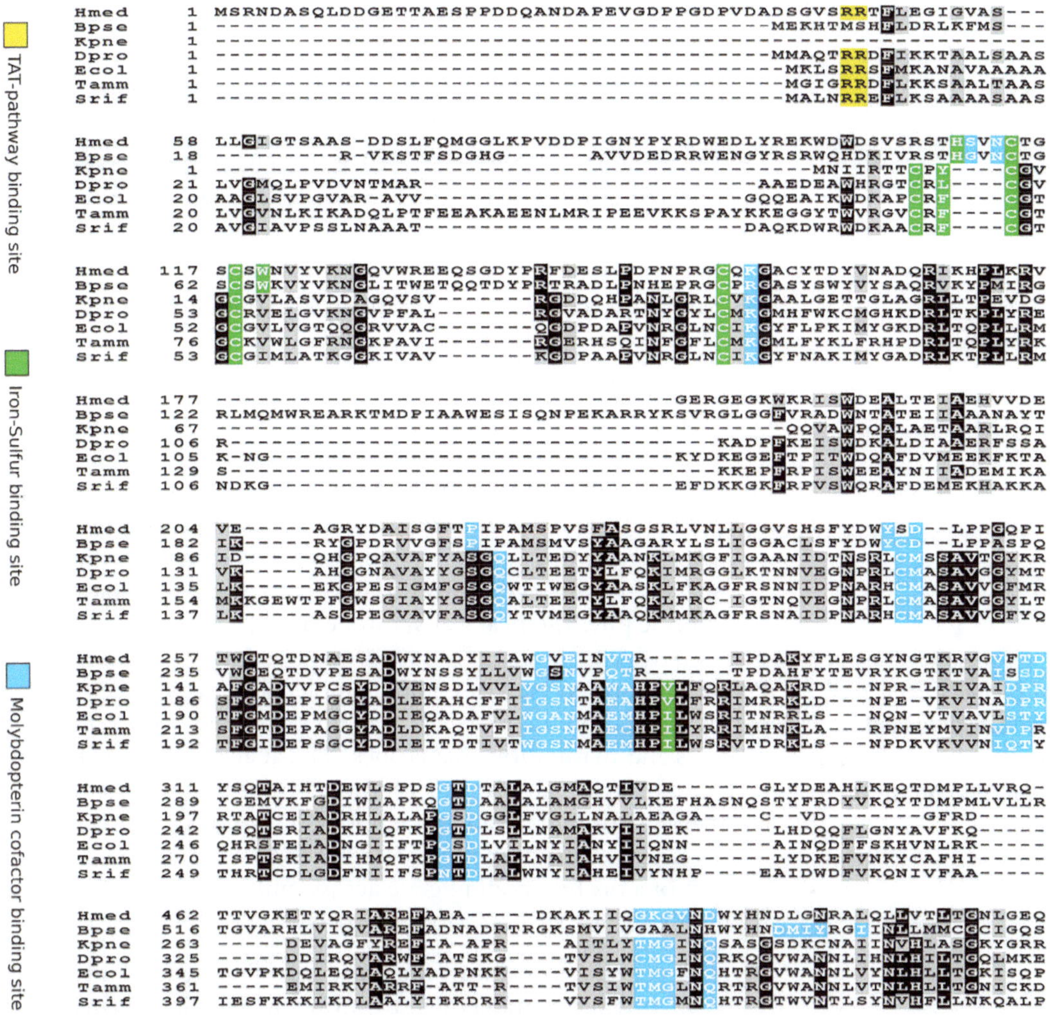

Fig. 1 A multiple sequence alignment of amino acid sequences encoding for the catalytic subunit of different nitrate reductases. Hmed—*Haloferax mediterranei* pNarG; Bpse—*Burkholderia pseudomallei* NarG; Kpne—*Klebsiella pneumoniae* NasA; Dpro—*Desulfobulbus propionicus* monomeric NapA (NapC-independent); Ecol—*Escherichia coli* dimeric NapA (NapC-dependent); Tamm—*Thermovibrio ammonificans* monomeric NapA (NapC-dependent); Srif—*Sulfurovum riftiae* dimeric NapA (NapC-independent). The conserved regions are highlighted on the basis of the information available in the literature

visualized and manipulated using tools like FigTree [23] and the Interactive Tree Of Life tool (ITOL) [24].

3.6 Phylogenetic Trees Reconciliation (Optional Step)

The phylogenetic trees reconstructed as described in the previous section can be directly reconciled to obtain more reliable information regarding potential horizontal gene transfer events and gene shuffling across taxa. Tree reconciliation can be done in a number of different ways. We take advantage of the "Reconcile tree" function of PhyML (*see* **Note 6**).

1. Open the functional protein and the 16S rRNA gene tree with PhyML in the SeaView program.
2. Organize the 16S rRNA gene tree as preferred, being sure to root the tree accordingly.
3. Once the 16S rRNA gene tree has been organized according to selected criteria, select the "reconcile tree" function from the PhyML menu, and reconcile the protein tree to the 16S rRNA tree.
4. Save both trees in a vector format. We routinely use svg or .eps to save the visualization of the trees.
5. The obtained vector files can be imported into a vector graphic software such as Inkscape. The tree can now be arranged side by side.
6. To highlight changes in the branching of the organismal and protein trees, segments connecting genes originating from the same organisms can be drawn. Also, coloring according to a specific factor, such as phylogeny, ecological adaptations, or specific physiological traits, can be used (Fig. 2).

3.7 Identification of the Genomic Context for the Protein-Coding Gene of Interest

Proteins are often part of complex molecular machineries created by multiple subunits that can also be associated with regulatory proteins and maturation chaperones. The analysis of the gene context of a protein can provide additional information regarding its putative function, its regulations and its evolutionary history. Several alternative approaches can be used to investigate the genomic context of a protein-coding gene across multiple organisms. Here, we highlight one such possible approach.

1. The genomic context of the protein-coding gene of interest of each organism under investigation can be directly inspected using the Genome portal of the NCBI database.
2. The operonic structure can be investigated for each genome using either FGenesB (http://www.softberry.com/berry.phtml?topic=fgenesb&group=programs&subgroup=gfindb), searching for the organisms of interest in the precomputed ProOpDB [25], through Patric [26], BioCyc.org [27] or the CoGe GenomeEvolution web service [28] (Fig. 3) (*see* **Note** 7).
3. The obtained genome context information can be used to directly compare operonic structures and genes upstream and downstream from the gene of interest, providing information on the evolutionary history of the entire molecular machinery involved in the process under investigation.

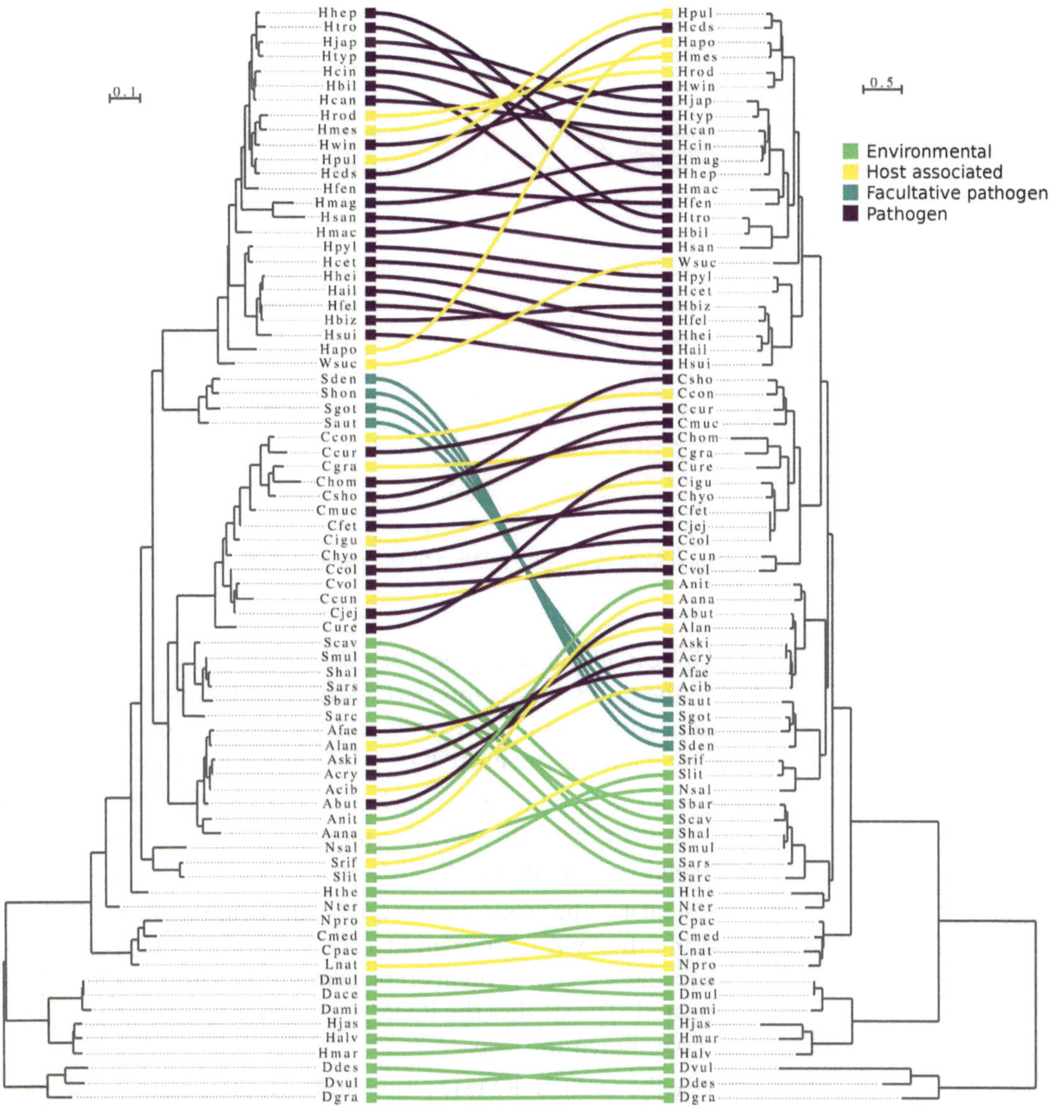

Fig. 2 Example of reconciled tree between the 16S rRNA gene and the PrtC-like gene from 75 representative species within the class *Epsilonproteobacteria* (recently proposed to be reclassified as the phylum *Campylobacterota*) [10]. The PrtC-like protein is a putative protease conserved in all *Epsilonproteobacteria*, including those found at deep-sea hydrothermal vents, and important for the virulence of the human pathogen *Helicobacter pylori* [32]

3.8 Protein 3D Structure Homology Modeling

Homology modeling of the protein of interest is carried out through a computational analysis on Swiss-Model [17]. Prior to starting homology modelling, verify how many 3D structures have been resolved for closely related proteins. The modelling results will be more accurate if the number of protein structures available is higher. It is possible to access all known protein structures through the RCSB Protein Data Bank database (https://www.rcsb.org/).

Fig. 3 Example of screenshot obtained from Patric and relative to the gene context of the periplasmic nitrate reductase (*napA*) gene

1. Upload all the amino acid sequences for the selected proteins to the Swiss-Model interface (https://swissmodel.expasy.org/interactive), which performs an homology search of the structure database (*see* **Note 8**).
2. Once the best match for each protein to be modelled has been identified, proceed with the modelling and download all the protein 3D models in PDB format.
3. Calculate the root-mean-square deviation values (RMSD) using the PyMol program [15] between the protein of interest and the selected proteins representing the cluster being investigated. The smaller the RMSD, the closer the two structures in three-dimensional space.
4. Use the ColorByRMSD plugin in PyMol to obtain visual information regarding the distribution of the RMSD between two protein structures according to the structure atoms deviation (Fig. 4) (*see* **Note 9**).
5. Additional measurements can be done on the obtained models, and compared across the organisms under investigation. For example, in the case of enzymes with an active pocket, the protein ligand binding sites and the volume of the pocket can be predicted using the POCASA program [16]. Molecular docking experiments can be carried out using the ligand of interest and the UCSF Chimera program [29] with the extension AutoDock-Vina [30] (*see* **Note 10**).

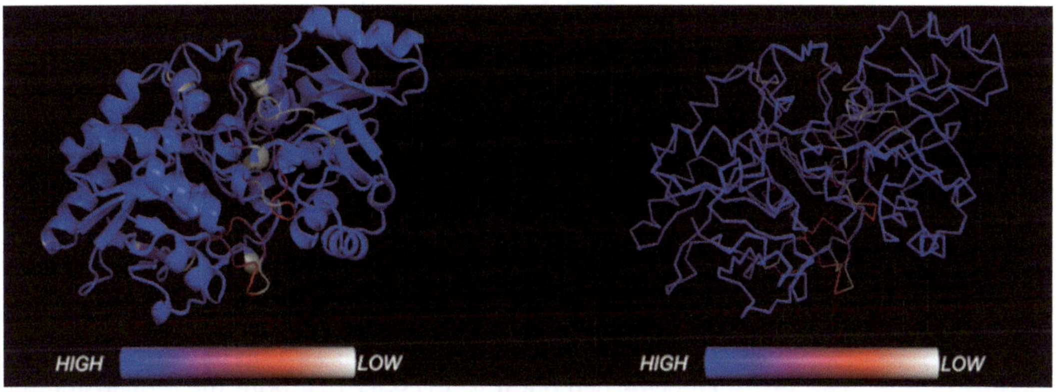

Fig. 4 Superimposition of the PrtC-like protein of *Sulfurovum riftiae* (an epibionts of the deep-sea hydrothermal vent tubeworm *Riftia pachyptila*) [33] and *Helicobacter pylori* (an obligate human pathogen) colored by root-mean-square deviation values

4 Notes

1. The boundaries of the group of interest can be defined either based on taxonomic, phenotypic, or ecological characteristics. The way the different organisms are grouped can and need to be changed according to the research question. For example, taxonomically related groups of organisms can be used to investigate relationships between vertical inheritance and adaptations, while groups based on similar lifestyle and phenotype can be used to study the protein adaptations to specific ecological niches.
2. We suggest that you rename the obtained sequences with a meaningful name that is human readable. We have found it useful to use the first letter of the genus name followed by the first three letters of the species name, for example *Ddes* for *Desulfovibrio desulfuricans,* to uniquely identify the organism from which each sequence originated. Other solutions are possible and might be required when comparing strains of the same species or in similar cases.
3. Keeping a table with all the organisms information, accession numbers and bibliographic references will help in the downstream processing of the results. The same table can also be annotated with taxonomic, physiologic, metabolic, and ecological information to help interpret the results.
4. While it is possible to obtain information about the evolutionary history of the protein of interest with respect to the organisms comparing the functional protein and the organismal tree, to be able to perform a couple reconstruction it is necessary that for each functional protein the corresponding 16S rRNA gene is also downloaded.

5. Rooting phylogenetic trees might be of primary importance depending on the type of hypothesis being tested in the study. Trees can be rooted using a number of different approaches [31]. We have found that robust inferences might be obtained when selecting a rooting strategy based on established outgroups for the organismal (16S rRNA) tree, and using the corresponding proteins from those organisms in the protein tree. This strategy might, however, lead to misleading results if the functional protein has undergone lateral gene transfer in the outgroup only, or if it is not conserved in the outgroup.
6. Tree reconciliation in PhyML is based on the sequence name. To be carried out it is necessary that both the protein and the 16S rRNA sequences have the exact same name.
7. While all the outlined software can provide information regarding the gene context, manual inspection is often required together with multiple BLAST searches to confirm the findings. Additionally, the best quality figures are obtained using a vector based graphic software such as Inkscape and drawing the genes to scale.
8. The Swiss-Model interface will return multiple targets ranked from best to worst. Before selecting the best target for modeling, be sure to inspect the model parameters for the goodness of fit, such as the coverage and the identity. The higher these parameters the better the computed model.
9. RMSD if computed pairwise for all the available proteins can be used to generate a RMSD matrix to be used as input distance matrix in the R statistical software to obtain a distance based tree from the protein structures. This can be then reconciled to the phylogenetic tree obtained from the amino acid sequences. The patristic distances obtained from the two trees can be plotted as a scatter plot to visualize discontinuities in the relationship between sequence evolution and structural changes.
10. The obtained information like pocket volume, docking score, and so on can all be used to relate information regarding the protein 3D structure to the phylogenetic analysis providing insight into the evolutionary history of the protein.

Acknowledgments

This work was partially supported by the European Research Council (ERC) under the European Union's Horizon 2020 research and innovation programme (grant agreement No. 948972) and by NSF grants MCB 15-17567 and IOS 19-51690.

References

1. Tripathi A (2019) Comparative genomics a new arena of biological research: a review. Genet Mol Biol Res 3(1):10
2. Mardis ER (2017) DNA sequencing technologies: 2006-2016. Nat Protoc 12:213–218
3. Malmstrom RR, Eloe-Fadrosh EA (2019) Advancing Genome-Resolved Metagenomics beyond the Shotgun. mSystems 4: e00118–e00119
4. Ellegren H (2008) Comparative genomics and the study of evolution by natural selection. Mol Ecol 17:4586–4596
5. Hardison RC (2003) Comparative genomics. PLOS Biol 1:e58
6. Giovannelli D, Vetriani C (2017) From extreme environments to human pathogens: an evolutionary journey. Biochemist 39:4–9
7. Giovannelli D, Sievert SM, Hügler M et al (2017) Insight into the evolution of microbial metabolism from the deep-branching bacterium, *Thermovibrio ammonificans*. eLife 6: e18990
8. Hasan NA, Grim CJ, Haley BJ et al (2010) Comparative genomics of clinical and environmental *Vibrio mimicus*. Proc Natl Acad Sci U S A 107:21134–21139
9. Le Roux F, Wegner KM, Baker-Austin C et al (2015) The emergence of vibrio pathogens in Europe: ecology, evolution, and pathogenesis (Paris, 11-12th march 2015). Front Microbiol 6:830
10. Waite DW, Chuvochina MS, Hugenholtz P (2019) Road map of the phylum Campylobacterota. In: Bergey's manual of systematics of archaea and bacteria. American Cancer Society, Atlanta, Georgia, pp 1–11
11. Vetriani C, Voordeckers JW, Crespo-Medina M et al (2014) Deep-sea hydrothermal vent Epsilonproteobacteria encode a conserved and widespread nitrate reduction pathway (nap). ISME J 8:1510–1521
12. Pérez-Rodríguez I, Bolognini M, Ricci J et al (2015) From deep-sea volcanoes to human pathogens: a conserved quorum-sensing signal in Epsilonproteobacteria. ISME J 9: 1222–1234
13. Gouy M, Guindon S, Gascuel O (2010) SeaView version 4: a multiplatform graphical user Interface for sequence alignment and phylogenetic tree building. Mol Biol Evol 27:221–224
14. Guindon S, Dufayard J-F, Lefort V et al (2010) New algorithms and methods to estimate maximum-likelihood phylogenies: assessing the performance of PhyML 3.0. Syst Biol 59: 307–321
15. Schrodinger LLC (2010) The PyMOL molecular graphics system, version 1.5. 0.4. Schrodinger LLC
16. Yu J, Zhou Y, Tanaka I et al (2010) Roll: a new algorithm for the detection of protein pockets and cavities with a rolling probe sphere. Bioinformatics 26:46–52
17. Waterhouse A, Bertoni M, Bienert S et al (2018) SWISS-MODEL: homology modelling of protein structures and complexes. Nucleic Acids Res 46:W296–W303
18. Delmont TO, Eren AM (2018) Linking pangenomes and metagenomes: the Prochlorococcus metapangenome. PeerJ 6:e4320
19. Altschul SF, Madden TL, Schäffer AA et al (1997) Gapped BLAST and PSI-BLAST: a new generation of protein database search programs. Nucleic Acids Res 25:3389–3402
20. Sievers F, Wilm A, Dineen D et al (2011) Fast, scalable generation of high-quality protein multiple sequence alignments using Clustal omega. Mol Syst Biol 7:539
21. Darriba D, Taboada GL, Doallo R et al (2011) ProtTest 3: fast selection of best-fit models of protein evolution. Bioinformatics 27: 1164–1165
22. Posada D, Crandall KA (1998) Modeltest: testing the model of DNA substitution. Bioinformatics 14:817–818
23. Rambaut A (2009) FigTree v1.3.1, https://tree.bio.ed.ac.uk/software/figtree/. Accessed 11 November 2020
24. Tung M-C, Hsu B-M, Tao C-W et al (2013) Identification and significance of Naegleria fowleri isolated from the hot spring which related to the first primary amebic meningoencephalitis (PAM) patient in Taiwan. Int J Parasitol 43:691–696
25. Taboada B, Ciria R, Martinez-Guerrero CE et al (2012) ProOpDB: prokaryotic operon DataBase. Nucleic Acids Res 40:D627–D631
26. Wattam AR, Abraham D, Dalay O et al (2014) PATRIC, the bacterial bioinformatics database and analysis resource. Nucleic Acids Res 42: D581–D591
27. Karp PD, Billington R, Caspi R et al (2019) The BioCyc collection of microbial genomes and metabolic pathways. Brief Bioinform 20: 1085–1093
28. Lyons E, Pedersen B, Kane J et al (2008) Finding and comparing Syntenic regions among Arabidopsis and the outgroups papaya, poplar, and grape: CoGe with Rosids. Plant Physiol 148:1772–1781

29. Pettersen EF, Goddard TD, Huang CC et al (2004) UCSF chimera—a visualization system for exploratory research and analysis. J Comput Chem 25:1605–1612
30. Trott O, Olson AJ (2010) AutoDock Vina: improving the speed and accuracy of docking with a new scoring function, efficient optimization and multithreading. J Comput Chem 31: 455–461
31. Huelsenbeck JP, Bollback JP, Levine AM (2002) Inferring the root of a phylogenetic tree. Syst Biol 51:32–43
32. Kavermann H, Burns BP, Angermüller K et al (2003) Identification and characterization of helicobacter pylori genes essential for gastric colonization. J Exp Med 197:813–822
33. Giovannelli D, Chung M, Staley J et al (2016) *Sulfurovum riftiae sp.* nov., a mesophilic, thiosulfate-oxidizing, nitrate-reducing chemolithotrophic Epsilonproteobacterium isolated from the tube of *Riftia pachyptila*, a deep-sea hydrothermal vent tube worm. Int J Syst Evol Microbiol 66(7):2697–2701. https://doi.org/10.1099/ijsem.0.001106

Chapter 6

VenomFlow: An Automated Bioinformatic Pipeline for Identification of Disulfide-Rich Peptides from Venom Arsenals

Eleonora Achrak, Jennifer Ferd, Jessica Schulman, Trami Dang, Konstantinos Krampis, and Mande Holford

Abstract

Animal venoms are among the most complex natural secretions known, comprising a mixture of bioactive compounds often referred to as toxins. Venom arsenals are predominately made up of cysteine-rich peptide toxins that manipulate molecular targets, such as ion channels and receptors, making these venom peptides attractive candidates for the development of therapeutics to benefit human health. With the rise of omic strategies that utilize transcriptomic, proteomic, and bioinformatic methods, we are able to identify more venom proteins and peptides than ever before. However, identification and characterization of bioactive venom peptides remains a significant challenge due to the unique chemical structure and enormous number of peptides found in each venom arsenal (upward of 200 per organism). Here, we introduce a rapid and user-friendly in silico bioinformatic pipeline for the de novo identification and characterization of raw RNAseq reads from venom glands to elucidate cysteine-rich peptides from the arsenal of venomous organisms.

Implementation: This project develops a user-friendly automated bioinformatics pipeline via a Galaxy workflow to identify novel venom peptides from raw RNAseq reads of terebrid snails. While designed for venomous terebrid snails, with minor adjustments, this pipeline can be made universal to identify secreted disulfide-rich peptide toxins from any venomous organism.

Key words Venom, Bioinformatics, Transcriptome, Galaxy, RNA-Seq, Next-generation sequencing, Disulfide-rich peptide

1 Introduction

Recent advancements in omics technologies and informatics have allowed scientists to generate and process large and complex biological data sets which are stored and shared on repository sites like GigaDB (https://www.re3data.org/repository/r3d100010478) and generate connective maps such as the human proteome [1]. Genomics, transcriptomics, and proteomics

Cinzia Verde and Daniela Giordano (eds.), *Marine Genomics: Methods and Protocols*,
Methods in Molecular Biology, vol. 2498, https://doi.org/10.1007/978-1-0716-2313-8_6,

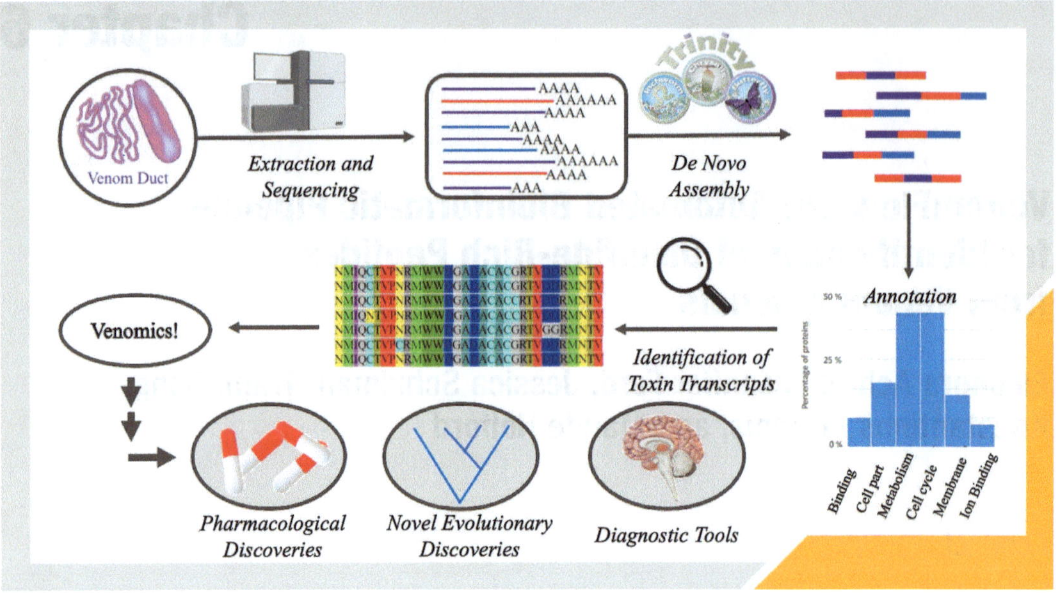

Fig. 1 Rise of omic technologies has increased identification and application of peptides from animal venoms. Pipeline to identify disulfide-rich venom peptides from Next Generation Sequencing of extracted mRNA from venom glands to identification of peptide toxins that can be assayed for discoveries pertaining to pharmacological applications, novel evolutionary knowledge, and diagnostic tools

methods promoted a paradigm shift in biological study, allowing for systematic analyses of the molecular components (genes, mRNA and proteins) that drive complex biological processes [2]. The rise in omics techniques also helped to expand venom research because it allowed for the study of animals that produce small amount of venoms (Fig. 1). Historically, venom research was focused on large venom producing species, such as snakes and scorpions, as they were more readily available to collect or obtained from purchase.

Next generation sequencing (NGS) allows for detection and quantification of different RNA transcripts; millions of transcripts can be detected even in a small sample size (Fig. 1). With this technological advancement, more robust ways of filtering and characterizing data are needed. As each snail venom arsenal can contain upward of 250 unique peptides, in addition to proteins and small molecules, multiple bioinformatic tools are needed to assemble, annotate, align and identify putative venom peptides using NGS.

We study marine snail venoms to better understand how venoms evolved over time and for its potential to modulate malfunctioning cellular signals in humans [3]. The efficiency with which venom toxins manipulate their molecular target has led to new treatments of various human diseases and disorders [4, 5] (Table 1). Captopril® and Prialt® (ziconotide) are two of the

Table 1
Snail venom peptides and associated therapeutic targets

Conus species	Peptide	Molecular receptor	Therapeutic target
C. purpurascens	k-Conotoxin PVIIA	Potassium channels	Cardioprotection
C. catus	w-Conotoxin CVID	N-type calcium channels	Cancer pain
C. magus	w-Conotoxin MVIIA		Neuropathic pain
C. marmoreus	x-Conotoxin MrIA	Noradrenaline receptors	Neuropathic pain
C. victoriae	a-Conotoxin Vc1.1	Nicotinic ach receptors	Neuropathic pain
C. regius	a-Conotoxin RgIA		Neuropathic pain
C. Geographus	Contulakin-G	Neurotensin receptors	Neuropathic pain
	Conantokin-G	NDMA receptors	Epilepsy, neuropathic pain
C. tulipa	Conantokin-T		Neuropathic pain

breakthrough FDA-certified treatments derived from venom toxins. Captopril®, derived from the Brazilian pit viper *Bothrops jararaca*, started a new class of drugs, angiotensin converting enzyme (ACE) inhibitors, commonly used to treat hypertension and heart failure. Similarly, ziconotide, from the *Conus magus*, is a novel nonaddictive nonopioid peptide analgesic used to treat chronic pain in patients with HIV and cancer [6]. The success of Captopril® and ziconotide has led to an increase in demand for the discovery and characterization of venom peptides as agents of change and innovation in the search for groundbreaking therapeutic candidates [7–9].

Venomous snails are an unsuspecting ally in the search for medicinal treatments, however as ziconotide demonstrated, these mollusks pack a powerful punch in their toxin arsenals. There are ~400 known species of terebrid snails, if each produced approximately 250 peptide toxins (teretoxins), that would result in a library of approximately 100,000 peptides to be characterized [10, 11]. Using methods previously described from our lab [12], we have analyzed the venom gland transcriptome of 25 terebrid snails and have successfully identified ~3000 novel teretoxins and teretoxin gene families. We created VenomFlow, an open-sourced resource for the venom community to standardize and speed up bioinformatic analyses for identifying disulfide-rich peptides from a venom arsenal (Fig. 2).

VenomFlow utilizes the Galaxy platform [13] to provide an open source graphical user interface (GUI) for creating, executing, and sharing bioinformatic workflows and freely available applications, tools, and scripts (Fig. 3). VenomFlow automates our in-house bioinformatic pipeline to overcome the challenge of assembling and characterizing de novo nonmodel organism

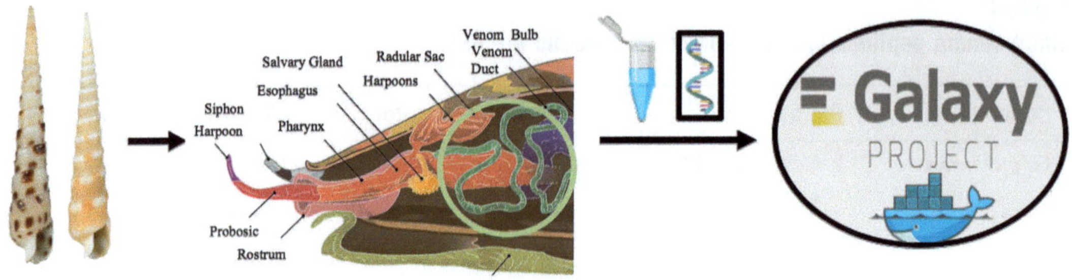

Fig. 2 VenomFlow is used to characterize mRNA extracted from terebrid snail venom glands. The mRNAs are sequenced and the raw transcriptome data is analyzed in a Galaxy pipeline

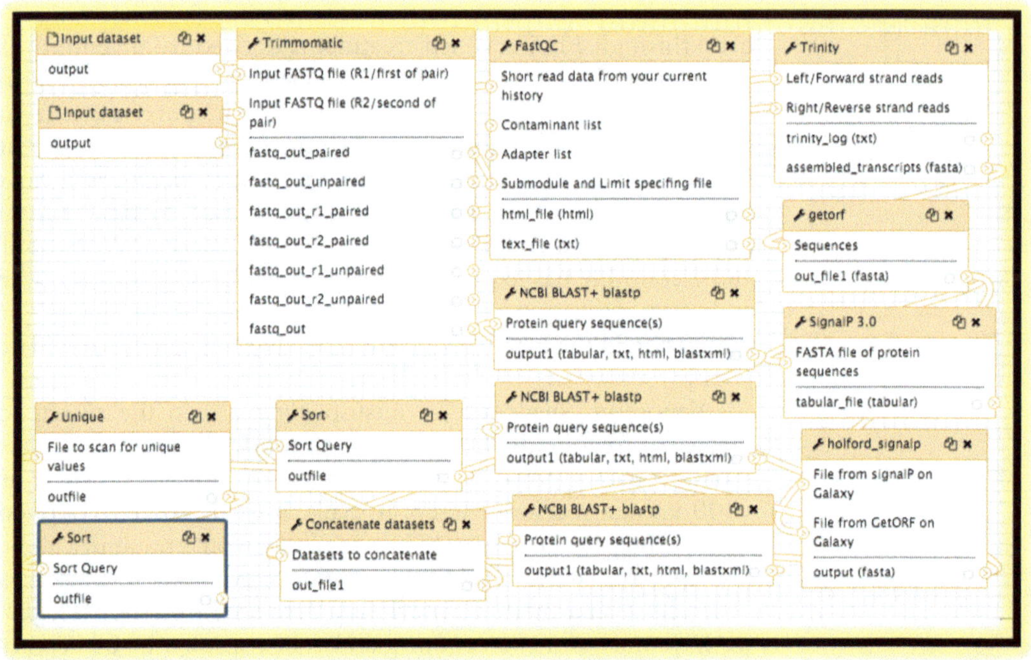

Fig. 3 VenomFlow Galaxy pipeline outlining steps and programs used to analyze de novo assemblies of RNASeq venom gland transcriptomes

transcriptomes. Tools are customized in Galaxy utilizing a "wrapper" with eXtensible Markup Language (XML) files describing the user interface, executables, input and output data types. We used a Docker wrapper, a method for wrapping up a tool along with its dependencies into portable and lightweight containers of packaged code [14]. Most venom peptides can be identified by looking for the following in the transcriptome: (1) venom peptides typically have a signal sequence and pro region that enable secretion from the cell [15]. (2) signal and pro regions are followed by a dibasic cleavage site. (3) the carboxy terminus of the gene contains mature cysteine-rich peptide toxins. Venom peptides identified by the VenomFlow pipeline are further validated by comparative analyses with proteomic and experimental data. This pipeline has been used routinely to identify putative venom peptide toxins in the Holford lab to build our in-house teretoxin library [11, 12].

2 Implementation

VenomFlow begins with installation of Docker version 18.06.0 for the Linux operating system (see GitHub link for step by step details: https://venomflow.github.io/Venom_Flow/). A Galaxy Docker image (https://github.com/bgruening/docker-galaxy-stable) with Galaxy installation is processed as a Docker container. Each updated image is executed using *Docker run* to access the internal Galaxy platform. Required software and tools are installed using the Galaxy Toolshed or manual integration with Python's hash algorithm. Implementation of the appropriate tools enables users to input files and receive accurate output data without manual bioinformatic tools. Our pipeline begins by uploading raw RNAseq data into the docker container. The data is then evaluated for potential sequencing errors using FastQC version 0.11.5 with standard parameters [16] and any remaining Illumina adapters are removed using Trimmomatic 0.36. At this point users have the option to add an additional quality control step to ensure sample quality before reads are de novo assembled with Trinity version 2.4.0 [17]. Assembled reads are passed through EMBOSS GetORF to identify and tag open reading frames (ORFs) and likely protein-coding regions with a minimum and maximum size of 45 to 500 respectively. Sequences with open reading frames are fed into SignalP to identify secreted venom proteins by looking for signal peptide cleavage sites in the amino acid sequence. Using a custom Python script, transcripts that survive both filtering steps (GetORF, SignalP) are pulled into a new FASTA file used as the query file for BLAST filtering. Transcripts are searched against SwissProt/Uniprot database (https://www.uniprot.org/downloads), NCBI venom peptide database (https://www.ncbi.nlm.nih.gov/), and our local in-house teretoxin database with a cut off e-value of 0.01. Results from all three BLAST searches are concatenated into a singular output and duplicates are removed using the Unique tool. Sequences with hits to the venom databases are considered putative teretoxins and are further investigated. The final steps in the pipeline filtered our sequences to 1105 reads, while our started material had over 80 million reads (Table 2).

Table 2
Example of identified transcripts after each filtering step for our test data set using RNAseq data from *Terebra*

Sample ID	Trimmomatic	Trinity	GetORF	SignalP	SignalP extraction	BLAST; in house database	BLATS NCBI database	BLAST Uniprot database	BLAST alignment; final results
T. jimiculata	86,146,978	217,633	1,157,528	1,100,000	189,278	4405	2698	67,196	1105

3 Notes

As stated in the implementation section, VenomFlow was designed and executed on Linux operating system, but with minor modification, it can be used by any other operating system. Firstly, the installation of Docker, as mentioned in the implementation section, needs to be installed for the appropriate operating system with help of Docker documentation (https://docs.docker.com/get-docker/). When following the GitHub step by step protocol, each step may need to be altered for operating systems different from what we used. If possible, Virtual Machine can be installed with Linux operating system, which will adhere to our instructions and provide access to the pipeline in the same manner we used it.

Referencing our introduction, VenomFlow was designed to speed up the processes of venom peptide discovery, but it took our team a substantial amount of time to develop and test each step. To minimize the time, we used multiple clusters to run each step in the pipeline. If you need to troubleshoot the pipeline, we recommend you also use multiple clusters if available. Furthermore, before installing and creating tools in Galaxy GUI, we uploaded them to the cluster to measure the spread and accuracy of the results. Through Galaxy, each step of the VenomFlow can be modified and altered for specific venoms of choice, and there is an option to update each tool to the most current version. If one chooses to update one tool, all other tools should be updated as well for consistent results. For example, we used SignalP version 3.0 in VenomFlow, and the current version is 5.0, to update this without losing accuracy of the results, GetORF and BLAST would need to be adjusted with different parameters.

4 Discussion

Less than 0.01% of venom toxins have been identified and even less have been characterized [18]. VenomFlow automates the Holford Lab's integrated venomic strategy to enable fast and accurate identification of teretoxins across species [1, 7]. Using a Docker image provides us with a functionally reproducible workflow available to run on a virtual machine via a graphical user interface. While our lab focuses on the characterization of venoms from terebrid snails, VenomFlow is equipped to characterize transcriptome data from any organism's venom arsenal. VenomFlow is a fast and comprehensive method for venom scientists with little to no experience with command line coding to efficiently increase the pace of discovery and characterization of venom peptides.

Acknowledgments

This work was supported by research grants from the Camille & Henry Dreyfus Foundation Teacher Scholar Award and the National Institutes of Health (NIH-NIMHD grant 8-G-12-MD007599). Student support was provided by the Hunter College MARC program, Grant #GM007823, and NYU Tandon School of Engineering. Original test data and assistance was provided by Holford Lab member Juliette Gorson.

References

1. Fagerberg L, Hallström BM, Uhlén M et al (2015) Proteomics tissue-based map of the human proteome. Science 347:1260419
2. Charpentier E, Doudna JA (2014) Genome editing. The new frontier of genome engineering with CRISPR-Cas9. Science 346: 1258096
3. Anand P, Gorson J, Jannetti S, Kelly P et al (2016) From mollusks to medicine: a Venomics approach for the discovery and characterization of therapeutics from Terebridae peptide toxins. Toxins (Basel) 8(4):117. https://doi.org/10.3390/toxins8040117
4. Chen N, Xu S, Zhang Y et al (2018) Animal protein toxins: origins and therapeutic applications. Biophys Rep 4(5):233–242. https://doi.org/10.1007/s41048-018-0067-x
5. Bryan J (2009) From snake venom to ACE inhibitor—the discovery and rise of captopril. Pharm J. https://doi.org/10.1211/PJ.2021.1.69712
6. Brogan SE, Olivera BM, Safavi-Hemami H et al (2019) Pain therapeutics from cone snail venoms: from Ziconotide to novel non-opioid pathways. J Proteome 190:12–20. https://doi.org/10.1016/j.jprot.2018.05.009
7. Holford M, Daly M, King G et al (2018) Venoms to the rescue. Science 361(6405): 842–844. https://doi.org/10.1126/science.aau7761
8. Angell Y, Holford M, Moos WH (2018) Building on success: a bright future for peptide therapeutics. Protein Pept Lett 25(12): 1044–1050. https://doi.org/10.2174/0929866525666181114155542
9. Holford M, Napolitano T (2018) Breakthroughs in venom peptide screening methods to advance future drug discovery. Protein Pept Lett 25(12):1137–1148. https://doi.org/10.2174/0929866525666181101103047
10. Olivera BM, Watkins M, Bandyopadhyay P et al (2012) Adaptive radiation of venomous marine snail lineages and the accelerated evolution of venom peptide genes. Ann N Y Acad Sci 1267: 61–70. https://doi.org/10.1111/j.1749-6632.2012.06603.x
11. Anand P, Grigoryan A, Bhuiyan MH et al (2014) Sample limited characterization of a novel disulfide-rich venom peptide toxin from Terebrid marine snail *Terebra variegata*. PLoS One 9(4):e94122. https://doi.org/10.1371/journal.pone.0094122
12. Ramrattan G, Verdes A, Wright EM et al (2015) Molecular diversity and gene evolution of the venom arsenal of Terebridae predatory marine snails. Genome Biol Evol 7(6): 1761–1778. https://doi.org/10.1093/gbe/evv104
13. Baker D, Batut B, Van Den Beek M et al (2018) The Galaxy platform for accessible, reproducible and collaborative biomedical analyses: 2018 update. Nucleic Acids Res 46(W1): W537–W544. https://doi.org/10.1093/nar/gky379
14. Di Tommaso P (2015) The impact of Docker containers on the performance of genomic pipelines. PeerJ 3:e1273. https://doi.org/10.7717/peerj.1273
15. Fassio G (2019) Venom diversity and evolution in the Most divergent cone snail genus Profundiconus. Toxins 11:623. https://doi.org/10.3390/toxins11110623
16. Haas BJ, Yassour M, Levin JZ et al (2011) Full-length transcriptome assembly from RNA-Seq data without a reference genome. Nat Biotechnol 29(7):644–652. https://doi.org/10.1038/nbt.1883
17. Lohse M, Usadel B, Bolger AM et al (2014) Trimmomatic: a flexible trimmer for Illumina sequence data. Bioinformatics 30(15):

2114–2120. https://doi.org/10.1093/bioinformatics/btu170

18. Liu ZH, Qian W, Li J et al (2009) Biochemical and pharmacological study of venom of the wolf spider Lycosa singoriensis. J Venom Anim Toxins Incl Trop Dis 15: 79–92. https://doi.org/10.1590/S1678-91992009000100008

Chapter 7

Population Genomics Analysis with RAD, Reprised: Stacks 2

Angel G. Rivera-Colón and Julian Catchen

Abstract

Restriction enzymes have been one of the primary tools in the population genetics toolkit for 50 years, being coupled with each new generation of technology to provide a more detailed view into the genetics of natural populations. Restriction site-Associated DNA protocols, which joined enzymes with short-read sequencing technology, have democratized the field of population genomics, providing a means to assay the underlying alleles in scores of populations. More than 10 years on, the technique has been widely applied across the tree of life and served as the basis for many different analysis techniques. Here, we provide a detailed protocol to conduct a RAD analysis from experimental design to de novo analysis—including parameter optimization—as well as reference-based analysis, all in Stacks version 2, which is designed to work with paired-end reads to assemble RAD loci up to 1000 nucleotides in length. The protocol focuses on major points of friction in the molecular approaches and downstream analysis, with special attention given to validating experimental analyses. Finally, the protocol provides several points of departure for further analysis.

Key words Linkage disequilibrium, Population genetics, Population structure, RAD analysis, Simulation

1 Introduction

Type II restriction enzymes have been powering significant aspects of population genetics since their discovery in the early 1970s [1, 2]. Each new generation of genome technology has been combined with restriction enzymes (RE) over the successive 50 years resulting in a continuously improving view of the distribution of genetic variation in natural populations. From assaying single loci with restriction fragment length polymorphisms [3], combining REs with electrophoresis in polyacrylamide gels, to anonymously assaying whole genome sets of loci with amplified fragment length polymorphisms [4], combining REs with PCR amplification, to labeling all genomic loci with Restriction site-Associated DNA (RAD, [5]), combining REs with DNA microarrays. The trend accelerated with the advent of second-generation sequencing

Cinzia Verde and Daniela Giordano (eds.), *Marine Genomics: Methods and Protocols*,
Methods in Molecular Biology, vol. 2498, https://doi.org/10.1007/978-1-0716-2313-8_7,

technologies, such as Illumina short-read platforms, that could sequence millions of small DNA fragments at low cost. The family of restriction site-Associated DNA sequencing (RADseq) protocols [6, 7] has democratized the field of population genetics by enabling large-scale studies of natural variation [8, 9], demographics [10], population structure [11], genetic maps [12, 13], landscape genetics [14, 15], methylation [16, 17], and phylogenetics [18, 19]. As RADseq has matured over the last decade, several underlying processes affecting the application of the protocol and the resulting data have become better understood, including issues involving DNA quality and allelic dropout, and with that knowledge, the protocol has become a reliable instrument in the biologist's toolkit.

In this chapter, we will present a full protocol for the analytical components of a RADseq experiment. After broadly introducing the set of RAD protocols (Subheading 1.1), we will discuss several important points, with respect to the molecular protocols, that often separate successful from failed experiments (Subheading 1.2). We then introduce the Stacks version 2 software (Subheading 1.3) and we present a full analytical protocol consisting of several major parts: how to set up your experiment and choose your enzymes via *RADinitio* simulation (Subheading 2.3), then, once the molecular library is created and sequenced (n.b., beyond our scope), we walk through cleaning and demultiplexing the data (Subheading 2.4), assembling RAD loci without a reference genome (Subheading 2.5), including optimizing de novo assembly parameters, assembling with a reference genome (Subheading 2.6), as well as a hybrid of the two methods (Subheading 2.7). We also demonstrate how to apply a population genetics frame to the data via a population-level analysis (Subheading 2.8). At the end of each analysis section, we focus on how to assay the success or failure of the underlying molecular library and analytical results. Finally, we provide four examples of exporting data outside of Stacks to further pursue a population genetics analysis, including assessing SNP and haplotype-based population structure, genetic divergence, and linkage disequilibrium.

1.1 The RAD Family of Protocols

A number of RADseq protocols have been introduced over the past decade (*see* [6] for an introduction and [7] for an update), but they can be generalized into two broad categories, both of which have been successfully applied in many experimental contexts. For single-digest protocols (sdRADseq), DNA is extracted from an individual and is digested with a single restriction enzyme; a barcoded adaptor is attached, the DNA molecules are sheared to a uniform size, and the sample is then multiplexed into a library of dozens to hundreds of samples. Finally, the multiplexed library is PCR amplified to increase the fraction of RE-associated DNA from a small to a large fraction of the overall DNA in the library. We highlight a variant of this protocol, BestRAD [20] as it improves yield significantly by physically isolating the RE-associated DNA from the larger pool of DNA prior to sequencing.

Double-digest restriction enzyme protocols (ddRADseq, [21]) define the second class. This group of protocols uses two restriction enzymes to isolate DNA loci from the genome—typically a rare cutting enzyme combined with a frequent cutting second enzyme—replacing the shearing step from sdRADseq. ddRADseq has a simpler implementation at the bench; however, it introduces some additional analytical challenges. The choice of enzyme combinations can be useful when working with very large genomes to select the right subset of DNA loci to work with, while imprecise size selection, in conjunction with dual enzymes, can result in loci being variably present across different individuals. In addition, PCR duplicates cannot be identified in a ddRADseq analysis unless molecules are labeled with a random oligo sequence as part of the adaptor [22].

1.2 Top Molecular Impediments to Good Results

Bioinformatic data analysis is itself rarely fruitful unless it is preceded by the successful execution of a molecular protocol at the bench. Our experience in designing and implementing Stacks and subsequently in RADseq data analysis, has highlighted three aspects of molecular library construction that are critical to achieving a successful experimental analysis. First, a sufficient amount of DNA must be available from each sample to be included in the molecular library. In the hands of an experienced technician, and with a thorough understanding of the tradeoffs, a successful RAD library can be prepared using small amounts of DNA. More frequently, however, limited amounts of DNA will cascade negatively into later parts of the protocol, increasing the likelihood of library construction failure—for a fraction of the samples, or for the entire library. For this reason, RADseq experimental design should emphasize maximizing DNA input.

Second, and related to the DNA input, are the effects of PCR amplification on a library. In any sequencing library, the goal is to obtain many copies of the host DNA—by extracting it from many cells in the organism—this will provide a high amount of information with respect to which alleles are present to enable genotyping. Nearly all RAD protocols include an amplification step to increase the prevalence of RE-associated DNA above the remaining genomic DNA (which will not be sequenced). PCR is very effective at increasing copies of template DNA molecules; however, if too few templates are present from the host genome, then the PCR process will instead amplify copies of the limited templates. These PCR duplicates do not add information with respect to the underlying alleles in the host genome—if an allele was not initially sampled it will remain missing after PCR amplification—and they will definitely consume sequencing capacity. Thus, amplification can present the illusion of high information content. Sequencing a library rich in PCR duplicates yields a reduced amount of informative DNA sequence from which to genotype. Sequencing such a library to very high depths will slightly increase the amount of informative sequence but ultimately, the library cannot provide the information it never contained due to a lack of template DNA. And while it may be tempting to associate these affects with the number of PCR cycles used (and to therefore adjust cycles down), ultimately it will not solve the problem. Thus, the amount of DNA present should be assayed independently of amplification.

Third, and still the most common problem we see in the Stacks analysis help forum is a lack of sequencing coverage. Fortunately, this is a problem that is eminently preventable with the proper planning. But for a successful analysis, a RAD practitioner should aim for high coverage with the ability to accept lower coverage in case of molecular difficulties. Therefore, the RAD practitioner should not try to overeconomize a library, which can end up producing coverage too low to reliably call polymorphisms and yielding a homozygous bias in those polymorphisms that are called. A practitioner ideally should plan for a minimum of 20–30× coverage per individual in the sequenced library.

1.3 Stacks Version 2: De Novo and Reference-Based RAD Analysis

Stacks was one of the first analysis pipelines released to work specifically with RAD data—initially for the de novo creation of genetic map markers [12, 23], and then expanded to work with both de novo and reference-aligned data more generally for any population genetics analysis [24]. More recently, a major update (version 2) was made expanding the software to work specifically with paired-end reads, using the single- and paired-end reads to construct individual RAD loci up to 1 kilobasepair (kbp) in size [25].

Stacks is architected to execute in three major phases: first data are demultiplexed, breaking the molecular library back down into individual samples, the existence of the RE cutsite remnant is verified, and the data are cleaned (low quality reads are discarded). Second, the main pipeline executes. If data are to be assembled de novo—without a reference genome—then the single-end reads are assembled into loci in each sample independently (`ustacks`). These loci are merged together, matching homologous loci across samples, in the construction of a metapopulation catalog (`cstacks`), and samples are then independently matched to that catalog (`sstacks`). Next, data are reorganized by the `tsv2bam` program so that loci from all samples are in the same order (loci start off in an arbitrary order among samples) to increase analytical efficiency. Finally, `gstacks` executes, assembling a paired-end contig in the metapopulation for each locus. It does this for each locus by taking the single-end reads used to assemble that locus and identifying the corresponding set of paired-end reads. These reads are collected across the metapopulation, down-sampled to a reasonable number (~1000 by default) and fed into a *de bruijn* graph assembly algorithm to form the paired-end contig. The paired-end contig is then merged with the single-end locus, using a suffix tree algorithm, and all reads from all samples in the metapopulation are then aligned to the resulting locus. Single nucleotide polymorphisms (SNPs) are then called, using a Bayesian model-based approach. First, the presence of a polymorphism is detected, using all reads in the metapopulation. This likelihood then serves as the Bayesian prior to genotype each sample individually at that locus. In the case of reference-aligned data, the user aligns their processed reads to a genome (with e.g., BWA [26]) and the main pipeline starts directly at the `gstacks` program.

A third, hybrid mode to run the main Stacks pipeline exists: one can take loci assembled de novo, align them to an arbitrary genome after assembly, and then inject those genomic coordinates back into the dataset (we show this ***integration*** method in Subheading 2.7).

Finally, given the output of `gstacks`, the final component of the pipeline to run is the `populations` program, which will consider the resulting `gstacks` genotypes in a particular biological context. Given a ***population map*** (a list of sample names and the populations they belong to) the `populations` program will calculate summary statistics (e.g., π, F_{IS}, loci that are in Hardy–Weinberg Equilibrium), divergence statistics (e.g., D_{XY}, and F_{ST}), as well as produce a number of exports for further analysis (e.g., VCF, STRUCTURE, and/or GenePop to name just a few). The `populations` program is designed to be executed multiple times on the same assembled and genotyped catalog, adjusting filters each time. Many filters are available, including a minor allele frequency filter, a population frequency filter, and so on. The biological frame can also be modified (without changing the core genotypes calculated previously) by running `populations` with a different population map to calculate statistics in different ways (e.g., dividing the individuals into populations by geography, sex, or phenotype).

1.4 Stacks Haplotypes Versus SNPs

One of the strongest aspects of RAD-derived data is that a RAD locus can easily be phased. That is, the alleles from the underlying SNPs can be matched up into haplotypes. This was one of the drivers behind the construction of paired-end contigs with Stacks version 2, and it enables haplotypes of up to 1kbp to be reconstructed from each RAD locus in the metapopulation (locus length is dependent on the size selection step of the underlying molecular protocol). These haplotypes are useful for estimates of nucleotide diversity (π), for input into phylogenetics models, the estimation of divergence statistics, such as D_{XY}, and F_{ST}, and the inference of coalescent genealogies. The `populations` program produces haplotype-level statistics alongside the SNP-based statistics (e.g., Φ_{ST}) and it also provide several haplotype-specific filters (e.g., `--filter-haplotype-wise` will filter rare SNPs from haplotypes to decrease the amount of missing data).

2 Methods

2.1 Presequencing: RADseq Experimental Design

Before sequencing, experimenters should seek to understand at least the base differences between protocols. Each protocol has its own molecular and experimental biases and can produce different magnitude and quality of data in different experimental systems. This means that, first, not all RADseq or RAD-like protocols are interchangeable and, second, that the success of an experiment using the technology is strongly dependent on the selection and implementation of a molecular protocol. Moreover, the problem is exacerbated by similar (and often uninformative) names of some of the protocols [27], for example, "Genotype-by-sequencing" or GBS is both a specific protocol and a generic reference to RADseq-like experiments.

Before protocol selection, researchers should determine what genomic resources are available for their system of study, including the most closely related species with a reference genome (the NCBI Assembly organism browser is a good resource, https://www.ncbi.nlm.nih.gov/assembly/organism/). This reference can be used for the direct estimation of RADseq loci, in addition to being a resource for reference-based analyses. Researchers should also check the literature for any available genomic studies in the system (or in closely related organisms) to determine if RADseq has been previously applied to the system, verifying the specific protocol used and its implementation. The results of these previous publications can be used as guidelines from which to design a new experiment.

To begin planning, a researcher should focus on the tradeoffs between the number of samples they want to include in the experiments, the number of RAD loci (and variant markers) they want to produce from those samples, and the required sequencing capacity to achieve those goals. The number of loci generated varies from species to species and from protocol to protocol. The right number of markers to obtain depends on the biological questions being asked. For example, an assessment of population structure might be performed with a few thousand independent sites of the genome and might allow for the sequencing of fewer sites across more individuals. In contrast, a genome scan might need tens of thousands of makers evenly distributed along the genome and might require a marker-dense RAD protocol. Sequencing requirements are also tightly linked to the estimation of RAD loci. Choosing a protocol which produces a few markers in a library of a few individuals might result in oversequencing the selected markers, while a very marker-dense library might result in undersequencing if too many samples are added in the library and sequencing capacity is not scaled accordingly.

Below we show examples of how to calculate the expected number of RAD loci from a given reference genome, followed by an example of how to calculate sequencing capacity (in the form of number of samples and coverage) from these loci estimates.

2.2 Data Set Used in this Protocol

In this protocol, we will be using a RADseq dataset from two populations of the Patagonian blenny, *Eleginops maclovinus*. This fish species is native to the coastal regions of southern South America and is an outgroup to the cold-specialized radiation of Antarctic notothenioid fish [28]. Sixty individuals from two populations in southern Chile (Valdivia $n = 37$, Puerto Natales $n = 23$) were processed into a single-digest RAD library [29, 30] generated using the RE *SbfI*. The library was sequenced (2×150 bp) on a single lane of an Illumina NovaSeq6000 at the Roy J. Carver Biotechnology Center sequencing facility at the University of Illinois, Urbana-Champaign.

A chromosome-level genome assembly was additionally generated for the species using long molecules from a PacBio Sequel 2 sequencer, which were then scaffolded using Hi-C. The assembly contains a total of 628 sequences, including 24 chromosome-level contigs and smaller unplaced scaffolds, for a total size of 613 Mb and a scaffold and contig N50 of 26.7 Mb and 7.6 Mb, respectively.

All data analyzed in this protocol, as well as shell scripts to run the analysis and R scripts to generate summary statistics and plots, are available online at https://bitbucket.org/CatchenLab/mimb-stacks2-protocol.

2.3 Simulations

In designing an experiment, we need to understand the architecture of the genome of our species of interest, how many RAD loci particular enzymes will yield, and how much sequencing will be required to assay those loci in the populations we wish to study. The first step is to determine the effects of RE and RAD protocol of choice.

2.3.1 Calculating an Estimated Tally of RAD Loci

To estimate the number of RAD loci we will tally the number of recognition sequences of one or more restriction enzymes in a reference genome. For sdRADseq, each enzyme recognition sequence will yield two tags, one from each DNA strand, while double-digest protocols require the recognition sequences for both enzymes to be within a certain insert-size window. The number and distribution of these RE recognition sequences are dependent on the size of the genome, its overall GC content, and the distribution of di- and trinucleotides in the sequence [31]. However, the number of RAD loci experimentally recovered can deviate from the number of observed cut sites. For example cut sites in close proximity may interfere with one another, reducing coverage [32].

There are several software packages designed around this initial estimation of RAD loci, including *SIMRLLS* [18], *simRAD* [33], *ddRADseqtools* [34], *DDRAGE* [35], *PredRAD* [31], and *RADinitio* [36]. All these programs base their estimation on counting restriction enzyme cut sites on a reference sequence but vary in the way they account for the variation introduced by the presence of repetitive loci, natural polymorphism, and technical error. For our estimates, we will be using the software *RADinitio*, as it can model the effects of genetic variation, protocol and enzyme choice, PCR duplicates, and sequencing coverage affecting the RADseq data generated.

Since there is a reference assembly available for *E. maclovinus*, we can perform the estimation of RAD loci directly. First, *RADintio* requires the reference genome in FASTA format, as well as a list of the sequences to simulate. The list of sequences can be obtained directly from the FASTA file by first extracting all the sequence ID lines and removing the ID identifier '>' alongside any comments after the FASTA ID, saving them to a new file:

```
$ cate_maclovinus.genome.fasta | \
  grep '^>' | \
  sed -E 's/^>(\S+)\s?.?/\1/' > chromosome_list
```

Note! As we will be presenting shell commands in this protocol, we will symbolize the command line prompt with the '$' character. This character is not part of the command but signifies the prompt at which the reader will type the command. To make our commands more readable in this limited space, we will divide them across lines using the backslash character, '\'. If the reader is typing commands all on one line, '\' is not part of the command itself and should be omitted. Otherwise, '\' tells the shell to continue the command onto the next line, instead of executing it.

The file `chromosome_list` will now contain a list of all the chromosomes, one per line, which in the case of the *E. maclovinus* assembly are named as `HiC_scaffold_X`. We can look at the first 10 lines of the file using the `head` command.

```
$ head chromosome_list
HiC_scaffold_1
HiC_scaffold_2
HiC_scaffold_3
HiC_scaffold_4
HiC_scaffold_5
HiC_scaffold_6
HiC_scaffold_7
HiC_scaffold_8
HiC_scaffold_9
HiC_scaffold_10
```

Since we want a tally of all possible RAD loci in the genome, we will pass all the available sequences to *RADinitio*. However, this chromosome list file can be edited if simulating over a subset of specific sequences is needed.

Now that we have the list of chromosomes, we can run `radinitio --tally-rad-loci`. The software requires the path to the reference genome in FASTA format (`--genome`), the path to the text file containing the list of chromosomes (`--chromosomes`), the path to an output directory to write the results (`--out-dir`), the type of library type to simulate (`--library-type`), and the RE (`--enz`). In this instance, we are simulating a single-digest RAD library using the enzyme *SbfI*. Additional options to simulate double-digest libraries, additional enzymes, and controlling insert sizes can be found in the *RADinitio* manual (http://catchenlab.life.illinois.edu/radinitio/manual/).

First, we will create a new output directory, which we will name according to the type of library being simulated. This will make it easier to identify the contents of the directory once we simulate libraries across several parameters.

```
$ mkdir ri_libs_sdRAD_SbfI
```

Next, we will run *RADinitio* using the parameters specified above. When reading this command remember that the '\' character is a special escape character that allows us to continue the command on the next line, making longer commands easier to read. Additionally, the '#' character is used to specify comments, meaning the text after it will not be interpreted as part of the command. These comments are a useful way to annotate your code, helping describe the purpose of specific options.

```
$ radinitio \
    --tally-rad-loci \                          # tally rad loci mode
    --genome ./e_maclovinus.genome.fasta \ # reference genome
    --chromosomes ./chromosome_list \       # list of chromosomes
    --out-dir ./ri_libs_sdRAD_SbfI \        # output directory
    --library-type sdRAD \                  # library type
    --enz SbfI                              # restriction enzyme
```

If we were to type this same command in a single line without using any comments it would look like this:

```
radinitio --tally-rad-loci --genome ./e_maclovinus.genome.fasta
--chromosomes ./chromosome_list --out-dir ./ri_libs_sdRAD_SbfI
--library-type sdRAD --enz SbfI
```

Executing the command simulates the *SbfI* single-digest library on the *E. maclovinus* reference genome. The produced tally of loci can be found in the `radinitio.log` file. We can observe that *RADinitio* found 52,644 RAD loci.

```
$ cat radinitio.log
RADinitio version 1.1.1
radinitio.tally_rad_loci started on 2021-06-28 14:11:22.

Simulating for 628 chromosomes.

RAD Library Options:
    library type : sdRAD
    restriction enzyme : SbfI (CC/TGCAGG)
    insert size mean : 350 bp
    insert size std dev : 37 bp
    read length : 150 bp
    sequencing coverage : 20 X

Loading the genome...

Extracting RAD loci...
    Extracted 52644 loci.

radinitio.tally_rad_loci completed on 2021-06-28 14:17:55.
```

Note that the `--tally-rad-loci` command only accounts for variation in the proximity of loci and the effects of insert size when performing the tally of loci. More complex simulations, such as those introducing the effects of population polymorphism and sequencing can be performed with other *RADinitio* options (such as `--simulate-all)`.

When no reference genome is available, a reference from a closely related species, for example one in the same genus or family, can substitute to estimate the number of RAD loci. These heterospecific genomes can also serve as an anchor for loci generated de novo using an *integrated* approach (Subheading 2.7). In the case that no single close reference is available, a good strategy is to perform simulation on several references to bound the expected results and ameliorate the effects of the evolutionary distance between the organism of study and the available references. Note that, when making these estimates, it is less important to have an exact count of the loci in the genome—after all any reference genome is bound to be different from the sequenced individuals—and more important to have an estimate of the magnitude of data expected.

To elaborate on the process described above, we performed several simulations and estimates of recovered RAD loci using various library protocols on seven different taxa (Table 1). We simulated both single- and double-digest loci using a variety of enzymes with varying GC content in their recognition sequences. For sdRADseq we used the GC-rich 8 bp cutter *SbfI* (CC/TGCAGG), the GC-rich 6 bp cutter *PstI* (C/TGCAG), and the AT-rich 6 bp cutter *EcoRI* (G/AATTC). For double-digest RAD, we used these three main cutters (*SbfI*, *PstI*, and *EcoRI*) in combination with the AT-rich 4 bp *MseI* (T/TAA), the GC-rich 4 bp *MspI* (C/CGG), and the 4 bp *AluI* (AG/CT) which contains 50% GC in its recognition sequence.

We observe several important patterns. First, a given enzyme (or combination of enzymes) does not produce the same magnitude of loci across all genomes. Notice how *EcoRI* yields 2.5 million loci in the Tasmanian devil (*Sarcophilus harrisii*, genome size 3.08 Gbp), but only 136 k in the bush monkeyflower (*Mimulus aurantiacus*, genome size 207 Mbp). This is expected, as the frequency of cut sites is dependent on the size of the genome and the properties of the sequence (e.g., GC content). Thus, the larger the genome of interest, the more RAD loci should be expected on average. Second, for a given taxon, not all RAD protocols and enzyme combinations yield the same number of loci. Enzymes with shorter recognition sequence will cut more often and yield more loci. For example, *PstI* and *SbfI* have similar GC-rich recognition sequences. However, the shorter recognition sequence in *PstI* results in a yield of loci an order of magnitude higher than *SbfI* across all genomes. GC content has similar effects. In the AT-rich genome of the *Heliconius melpomene* butterfly (GC: 32.8%), an AT-rich cutter such as *EcoRI* yields far more loci than the GC-rich *PstI*, even when their recognition sequences are both 6-bp long. Last, for a given main cutter, double-digest libraries result in fewer loci than single-digest libraries. This happens because single-digest RAD loci are defined just by the presence for the recognition sequence of the main cutter. In other words, each cut site found will always yield two loci. In contrast, in ddRADseq, the loci are defined by the presence of both enzymes. A locus will only be recovered if the recognition sequences of the primary and secondary enzymes are found within the desired insert size.

2.3.2 Estimating Sequencing Coverage

The simulated number of RAD loci can then be used to estimate the sequencing capacity required for the experiment. For a fixed amount of sequencing (i.e., a single Illumina sequencing lane), the sequencing capacity establishes the relationship between the number of samples that can be multiplexed and the depth of coverage that will result:

$$\textit{Number of samples} = \frac{\textit{Read throughput}}{(\textit{Number of loci} \times \textit{Depth of coverage})}$$

Table 1
Tally of RAD loci based on RADinitio simulations

Protocol	Enzyme	Heliconius butterfly	Mimulus Monkeyflower	Swainsons Thrush	Threespine Stickleback	Patagonian Blenny	S. Georgia Icefish	Tasmanian Devil
sdRAD	*SbfI*	2050	2328	134,310	38,666	52,644	70,432	67,676
sdRAD	*PstI*	53,452	77,168	1,894,498	549,448	868,478	1,074,684	1,154,034
sdRAD	*EcoRI*	216,792	135,520	690,630	144,204	209,284	287,342	2,458,596
ddRAD	*SbfI-MspI*	318	337	17,844	10,059	7939	14,853	10,403
ddRAD	*SbfI-MseI*	1558	1398	58,227	20,377	31,135	40,080	37,403
ddRAD	*SbfI-AluI*	753	916	78,856	17,949	25,495	32,548	32,858
ddRAD	*PstI-MspI*	6474	10,745	171,384	135,973	123,178	208,906	116,396
ddRAD	*PstI-MseI*	42,833	47,505	960,386	297,370	519,804	629,062	696,153
ddRAD	*PstI-AluI*	18,184	32,178	1,103,549	254,298	419,391	493,069	540,275
ddRAD	*EcoRI-MspI*	20,589	17,746	49,055	34,058	27,914	54,182	82,368
ddRAD	*EcoRI-MseI*	179,728	90,052	388,664	80,593	129,999	171,451	1,681,957
ddRAD	*EcoRI-AluI*	73,855	49,573	373,678	65,770	100,383	129,502	1,029,578

Table 2
Estimates of sequencing coverage

Enzyme	Loci	Number of samples at 30×	Coverage for 50 samples
SbfI	52,644	221.6	133.0
PstI	868,478	13.4	8.1
EcoRI	209,284	55.7	33.4
SbfI-MspI	7939	1469.5	881.7
SbfI-MseI	31,135	374.7	224.8

Using the number of loci simulated above for *E. maclovinus*, we can then calculate the sequencing capacity of the experiment (Table 2). Ideally, we want to sequence each sample at a 30x coverage, meaning that each individual locus is observed in 30 independent reads. On a single Illumina HiSeq 4000 lane, which yields roughly 350 million read pairs, we would be able to sequence about 222 individuals at a depth of coverage of 30x. If for example, we only had 50 individuals to sequence and we prepare a library with *SbfI*, we would obtain a coverage of 133x per sample. In this example, this high coverage would indicate that we are likely oversequencing each sample and that, in the absence of more individuals, we could afford to use a library preparation that yields more loci.

2.4 Cleaning and Demultiplexing Data

After library construction and sequencing, the user will receive one or more FASTQ files (or pairs of files, if working with paired-end data) containing the multiplexed data for all individuals sequenced in the library. Before assembling loci, in most cases the raw data needs to be demultiplexed (i.e., split between the different individuals), and it should always be cleaned. We encourage users to accept only raw data from the sequencing core, as opposed to length-trimmed or quality-filtered reads. We will create an output directory to hold our unmodified, raw data we received from the sequencing core, our barcodes file, which we will create, and our cleaned/demultiplexed data that will be output by `process_radtags`:

```
$ mkdir -p ./01_process_radtags/raw          # Unmodified, raw files stored here
$ mkdir -p ./01_process_radtags/info         # Barcodes archived here
$ mkdir -p ./01_process_radtags/samples      # Cleaned/demultiplexed data here
```

When preparing the molecular library, each individual sample was marked with a barcode, a unique molecular identifier that is present in the P1 adapter that gets ligated to the cut site. This barcode gets sequenced alongside the rest of the DNA molecule, meaning that we can then use it to assign reads to their individual of origin. We specify these identifiers (which should have been recorded when the molecular protocol was performed) in a file that we then pass to the `process_radtags` program in Stacks, containing both the sequence of the barcode and the name of the sample. In the case of our *E. maclovinus* library, after creating the file with any text-editing program (e.g., `nano`, `vi`, or `emacs`) the barcode file looks like this:

```
$ head barcodes.tsv
ACTAGGA<tab>emac_va_01
CCGATAG     emac_va_02
TACATTG     emac_va_03
GGCGGAG     emac_va_04
ACTACTT     emac_va_05
CGTGTGA     emac_va_06
TCCGCCA     emac_va_07
GTAGCTT     emac_va_09
CAACGCA     emac_va_10
TAGTACA     emac_va_11
```

Note! The '`<tab>`' text in the barcodes file is meant to signify a tab character as the file is a tab-separated file, do not include '`<tab>`' but rather, press the Tab key.

The file contains two columns separated by tabs. The first column specifies the barcode sequence and the second the sample ID. Each row corresponds to a single sample.

Different RAD library preparations can have different barcode configurations, including combinatorial barcodes on the single- and paired-end reads and/or incorporating the Illumina index barcodes. More information can be found in the Stacks manual.

In addition to demultiplexing, `process_radtags` can also be used to clean the data, removing reads with uncalled or low-quality bases, as well as discarding those without a recognizable barcode (since they cannot be assigned to an individual) or an intact RE remnant cut site sequence (as they might not originate from a RAD locus in the genome).

For our *E. maclovinus* dataset, we tell `process_radtags` the path to our raw data files (`--path`), as well as the path to the output directory where all the individual sample files will be saved (`--outpath`), and the location of the barcode file shown above (`--barcodes`). We also specify that our input data contains paired reads (`--paired`). To clean the data, we want to make sure that all reads with uncalled bases (`--clean`) or low-quality scores (`--quality`) are removed. Both barcodes and restriction sites will be *rescued* (`--rescue`), meaning that they will be corrected if a small number of mismatches from the expected sequence are detected. To check for the RE remnant sequence, we also specify that the library was prepared with a single RE, in this case *SbfI* (`--renz-1 sbfI`).

Our complete `process_radtags` command looks like this.

```
$ process_radtags \
    --path ./01_process_radtags/raw \                        # Path to raw FASTQ
    --outpath ./01_process_radtags/samples \                 # Path to output files
    --barcodes ./01_process_radtags/info/barcodes.tsv \      # Path to barcode file
    --paired \                                               # Input is paired reads
    --clean \                                                # Remove uncalled bases
    --quality \                                              # Remove low quality
    --rescue \                                               # Rescue barcode/cutsite
    --renz-1 sbfI                                            # Single enzyme SbfI
```

2.4.1 Assaying the Processing and Cleaning of the Raw Data

Once this command is complete, a detailed breakdown of all the filtering and demultiplexing can be found in the `process_radtags.raw.log` file. From this log file, we can use the `stacks-dist-extract` utility to extract specific distributions from the different Stacks log files (Stacks log files are text files that can be viewed with any text-viewing program, using `stacks-dist-extract` is a convenience). First, we want to look at the library-wide statistics—how well did the library perform overall—examining the `total_raw_read_counts` distribution:

```
$ stacks-dist-extract process_radtags.raw.log total_raw_read_counts
Total Sequences        601046598
Reads containing adapter sequence        17306290
Barcode Not Found     31864524
Low Quality    30109741
RAD Cutsite Not Found        4540000
Retained Reads        517226042
```

We had 601 million reads in our library, of which we retained 517 million (86%). Notice that the majority of reads lost were due to the lack of an intact barcode (31.8 million) or due to low sequencing quality (30 million). Even after these losses, we still retain a significant proportion of the sequencing and can confidently proceed with further analysis. Normally, we expect to retain between 80% and 90% of the total sequenced reads in the library. In scenarios were the proportion of retained reads is low, the user should assess the distribution above. Are the majority of reads lost because due to missing barcodes? If so, verify that the information in the barcode file is correct. Do many reads lack a RAD cutsite? Perhaps the enzyme passed to `process_radtags` does not match the library preparation, or the specific RAD P1 adapter used in the protocol may have included extra nucleotides in the barcode/RE cut site region. Lastly, are many reads lost due to low quality? Here, the user should contact the sequencing facility to assess any problems with the sequencing lane.

Once the library-wide assessment is performed, the next step is to verify the proportion of reads kept between the different samples. We can again use the `stacks-dist-extract` program to extract the `per_barcode_raw_read_counts` distribution from the log.

Looking at the first ten lines of the distribution we can see that it counts the number of total reads in the library for each barcode (sample), as well as the number of reads retained and lost due to the lack of a restriction enzyme cut site or low sequencing quality.

```
$ stacks-dist-extract process_radtags.raw.log \
  per_barcode_raw_read_counts | head
Barcode  Filename    Total    Rad Cutsite Not Found  Low Quality  Retained
ACTAGGA  emac_va_01  4931342  25416                  262001       4591725
CCGATAG  emac_va_02  7199608  29237                  380085       6710605
TACATTG  emac_va_03  6515476  41815                  343706       6056243
GGCGGAG  emac_va_04  2181034  25823                  115770       2013546
ACTACTT  emac_va_05  800126   48240                  41209        695349
CGTGTGA  emac_va_06  4027288  35938                  215757       3728410
TCCGCCA  emac_va_07  1412670  19049                  74505        1301546
GTAGCTT  emac_va_09  7922452  157646                 416670       7251495
CAACGCA  emac_va_10  2478420  64198                  129562       2248435
```

The user can capture this data into a file to load into R [37] or to another statistical software and calculate detailed distributions of the proportion of reads kept and lost for each sample. Regardless of the method used, the user should check for at least three things: (1) the proportion of each sample in the total library, as well as (2) the number and (3) proportion of reads retained in each sample. The Appendix links to an R script providing an example on how to perform these calculations.

The proportion of reads per sample (Fig. 1a) roughly represents the proportion at which individuals were multiplexed in the library. For example, here we can observe that some samples composed over 3.5% of the library, while others were present in less than 0.04%. Such differences are expected given a combination of effects during library preparation (e.g., human error, DNA quality, equipment sensitivity). However, the user should monitor for larger effects, such as biases between populations (Fig. 1b) or sequencing lanes, or for cases in which one or a few samples dominate the sequencing as this may imply that library contamination occurred.

Next, the user should check the number of reads retained per sample, as it is an early metric to assess sequencing coverage (Fig. 1c). In our library, we can observe variation between samples, ranging from 130 k to 19 million reads. This variation is expected given the differences in per-sample library composition described above. Similarly, while there are some small differences in the expected coverage between populations (Fig. 1d), they do not indicate any large-scale biases in library preparation. Note that, given the very low number of retained reads for some samples, it is a good practice to remove these individuals from further analyses. In the case of this library, we will remove the two samples retaining less than one million reads (`emac_va_05` and `emac_va_12`).

> Note! This threshold of 1 million reads used to remove samples showing early low coverage is not an absolute one. It is used here based on the properties of the dataset and could vary depending on the number of markers recovered from the genome and the overall number of reads per sample.

The last check is the percentage of reads retained per sample. While we originally had observed that we had retained 86% of all reads in the library, this original assessment did not account for variation between samples. As we can observe from Fig. 1e, a subset of the samples retained only a small fraction of the total reads and should be assessed in more detail.

2.5 De Novo Analysis

When a reference genome is not available, is of low or questionable quality (e.g., low N50, low contiguity, certain regions missing), or when a researcher desires to conduct an analysis independent from a particular assembly (e.g., the construction of a genetic map), then a de novo analysis is the recommended approach. When a high quality, complete reference is available, it will typically yield superior results, but the option to assemble de novo provides the researcher with a lot of flexibility and the ability to test hypotheses independent of the genome. Here we will show the major steps involved, starting with parameter optimization, followed by the execution of the core pipeline, and following up with a set of methods to evaluate the success of the assembly.

2.5.1 Parameter Optimization

When analyzing RADseq data without a reference genome, it is important to optimize the parameters that will define the de novo assembly of loci. There are two main parameters in `ustacks` that control the assembly of alleles into loci within individuals and the matching of these loci, in a catalog, between individuals. The initial assembly of loci within each sample happens in `ustacks`, where the parameter `-M` specifies the number of mismatches allowed when matching stacks, that is, putative alleles, into loci. Setting this parameter too low results in undermerging—leaving two alleles as

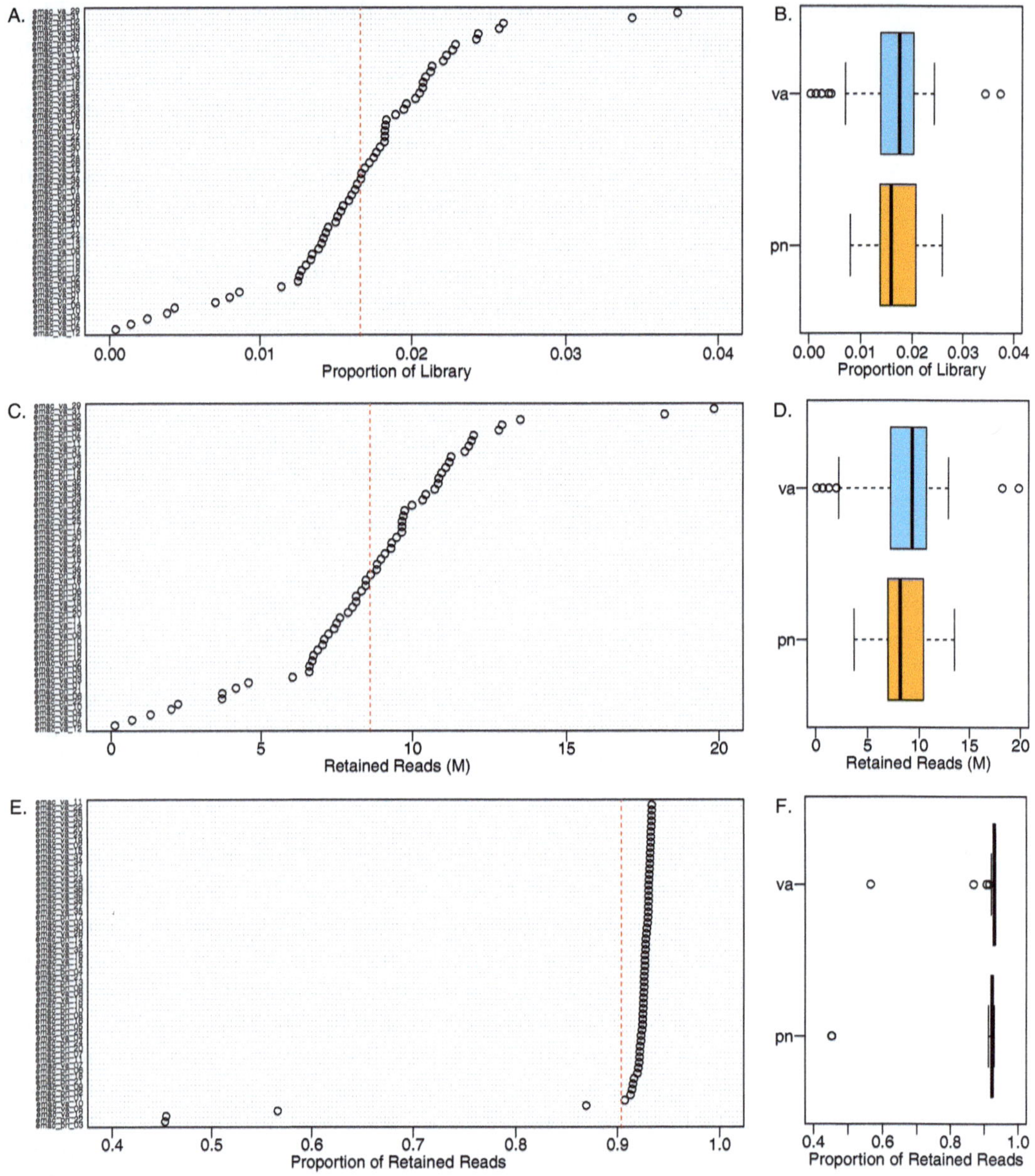

Fig. 1 Distribution of reads per sample after `process_radtags`. (**a**) Proportion of raw sample reads in the total library (sum of raw reads across all samples). Red dashed line shows the mean proportion of samples reads in the total library. (**b**) Proportion of total library between Valdivia (`va`, blue) and Puerto Natales (`pn`, yellow) samples. (**c**) Number of retained reads per sample, in millions of reads. (**d**) Number of retained reads between populations. (**e**) Proportion of retained reads per sample. (**f**) Proportion of retained reads between populations

separate, homozygous loci. In contrast, if `-M` is too high alleles can be overmerged, collapsing alleles of different loci into a single locus. The value is `-M` is roughly equivalent to the level of biological polymorphism present in an individual given a particular read length (e.g., `-M` should be set lower when using very short reads).

Once loci are assembled within samples, they are then merged by `cstacks` into a catalog of loci, representing the metapopulation. The `-n` parameter specifies the number of mismatches between the loci of individuals and the catalog, playing the same role as `-M` in `ustacks`. For example, a variable locus may have been assembled as a homozygous locus in one individual, and as a homozygous locus with the alternative allele in a second individual. The `-n` parameter allows `cstacks` to merge these loci into a proper heterozygous locus for the population with the same side effects if `-n` is set too low (undermerging) or too high (overmerging). The value of `-n` is roughly equivalent to the level of biological polymorphism present across the population. If working with several populations, `-n` should also account the genetic distance between them (i.e., `-n` might need to be larger if the populations are divergent or are separate species).

Several protocols have been published regarding the optimization of assembly parameters for de novo analysis [38–41]; McCartney-Melstad et al. (2019) used a method based on the relation of sequence similarity and the fraction of sequences flagged paralogous, while Heller et al. (2021) performed a de novo catalog assembly with default parameters on a subset of individuals, after which reads for all samples are aligned to the catalog consensus sequences and analyzed as whole-genome resequencing data. In this tutorial, we will be focusing on the R80 method first proposed by Paris et al. (2017) and described further by Rochette and Catchen (2017). Briefly, this method looks for the assembly parameters that maximize the number of R80 loci in the dataset (i.e., the number of polymorphic loci present in at least 80% of the samples). Values of `-M` and `-n` outside of this optimum result in the over- and undermerging of loci, leading to an increased proportion of missing data across the dataset.

To perform the parameter optimization using the R80 method, we will run the core pipeline several times on a subset of individuals, iterating over increasing values of `-M` and `-n` in each run. In the first run, both `ustacks -M` and `cstacks -n` will be set to "`1`" — allowing for one mismatch between alleles in both the individuals and the population. These values will then be increased until a maximum of 10–12 mismatches is reached. Once the iterations are complete, the user will calculate the change in R80 loci between subsequent runs (i.e., did the number of R80 loci increase or decrease after the assembly values were increased by one). When optimum assembly parameters are found, the change in the number of R80 loci will approach zero and remain positive. Exceeding the optimal value will result in a negative number, indicating a loss of R80 loci.

Before we can optimize, we must select a subset of samples on which to run the analysis. These samples should be representative of the whole library, so it is ideal to include samples across all studied populations or groups. Similarly, it is best to include samples with average amounts of input data (which can be assessed from the method described in Subheading 2.4.1). For the *E. maclovinus* library, we selected 20 individuals for parameter optimization, 10 each from the Valdivia and Puerto Natales populations. Each sample contained between 7.6 to 10.9 million reads (average of 8.9 million). We will record the prefix sample names (i.e., the filename for the sample without any suffixes such as "`0.1.fq.gz`") of these samples in a ***population map*** text file (`popmap.tsv`), a list associating individuals with a population grouping. Here we have a single "population", and we arbitrarily refer to it as "`opt`", to signify *optimization*.

```
$ cat popmap.tsv
emac_pn_01<tab>opt
emac_pn_06      opt
emac_pn_08      opt
emac_pn_11      opt
emac_pn_14      opt
...
emac_va_26      opt
emac_va_27      opt
emac_va_32      opt
emac_va_35      opt
emac_va_38      opt
```

After creating the population map, the next step is to create an output directory for the current run of the pipeline. For example, here we make a new directory using the current value of `-M` as part of the name.

```
$ mkdir denovo.M1
```

With the population map and output directory, we can now prepare the Stacks `denovo_map.pl` program (`denovo_map.pl` is a Perl wrapper script that will execute the core pipeline for us). First, we specify the path to our input reads (`--samples`), the path to the population map file (`--popmap`), and the path to our output directory (`--out-path`). In the case of the *E. maclovinus* library, we have paired-end reads and thus want them included in the analysis (`--paired`). We also want to remove PCR duplicate reads (`--rm-pcr-duplicates`), as well as only output R80 loci

(`--min-samples-per-pop 0.8`). Last, we set our values if `-M` and `-n` to our current iteration of the pipeline, in this case "`1`."

```
$ denovo_map.pl \
    --samples ./01_process_radtags/samples \
    --popmap ./02_param_opt/info/popmap.tsv \
    --out-path ./02_param_opt/denovo.M1 \
    --paired \
    --min-samples-per-pop 0.8 \
    --rm-pcr-duplicates \
    -M 1 \
    -n 1
```

The code above runs the pipeline once on our 20 samples, using the specified assembly parameter values. The iteration across all desired values can be automated using a shell loop, and we provide an example script, listed in the Appendix.

Once Stacks is complete, the number of polymorphic R80 loci can be extracted from the output of `populations`. In this case, we are using the `population.sumstats.tsv` file, which contains the summary statistics for each variant site per population in the dataset (sample counts, allele frequencies, observed and expected heterozygosity, etc.). The first column of this file contains the locus ID from which each variant site originates, and performing a tally of these unique values would provide us with the number of R80 polymorphic loci:

```
$ cat populations.sumstats.tsv | \
    grep -v '^#' | \                # Remove Comment line
    cut -f 1 | \                    # Cut first column (Locus ID)
    sort -n -u | \                  # Sort numerically and unique values
    wc -l                           # Count the number of lines
36721
```

For this dataset, allowing for 1 mismatch between the alleles of individuals and across the population provides 36,721 R80 loci. This value is similar to that reported in the `population.log` file, except that the value in the log reports the total number of polymorphic and invariant R80 loci (37,576).

```
$ cat populations.log | grep -A 1 'Removed'
Removed 93314 loci that did not pass sample/population constraints from 130890
loci.
Kept 37576 loci, composed of 23691256 sites; 49964 of those sites were filtered,
277665 variant sites remained.
```

Table 3
Change in R80 loci

M/n	Number of R80 Loci	Change in R80 Loci
1	36,721	–
2	37,647	926
3	37,768	121
4	37,728	−40
5	37,589	−139
6	37,518	−71
7	37,449	−69
8	37,430	−19
9	37,301	−129
10	37,244	−57
11	37,190	−54
12	37,066	−124

Once the core de novo pipeline has been run for all `-M/-n` values, we can proceed to calculate the change in R80 loci between runs (Table 3). To do this, we can subtract the number of R80 loci found in a run, with the number of R80 loci found in the previous run (i.e., `M = 1` to `M = 2`, `M = 2` to `M = 3`, etc.). Plotting these results, we look for the value in which the change in R80 loci approaches zero, while remaining positive (Fig. 2). Here, increasing `-M/-n` from 2 to 3 results in a gain of 121 R80 loci, while increasing from 3 to 4 results in a loss of 40 loci. This indicates that an `-M/-n` of 3 is the optimal value for de novo loci assembly for the *E. maclovinus* dataset.

Note! The optimization procedure will not always converge for all datasets. It is possible that there are several, equivalent values for `-M` and `-n` in a dataset, as well as scenarios in which `-M` and `-n` are not equal. Additionally, when working with several divergent populations, the optimal parameters might even be different for each population. Other factors can also affect parameter optimization, such as the presence of repetitive loci in the genome and highly polymorphic indels in the metapopulation. We encourage users to explore these parameters, but we advise against infinitely iterating over all possible combinations in order to find perfect parameters. The maximization of R80 loci is just the metric of the optimization process, not the underlying biological signal present in the data.

2.5.2 Core De Novo Pipeline Execution

With the completion of parameter optimization, we are ready to employ those parameters to assemble our RAD loci and genotype all samples. Recall that R80 loci are simply a target we employed for

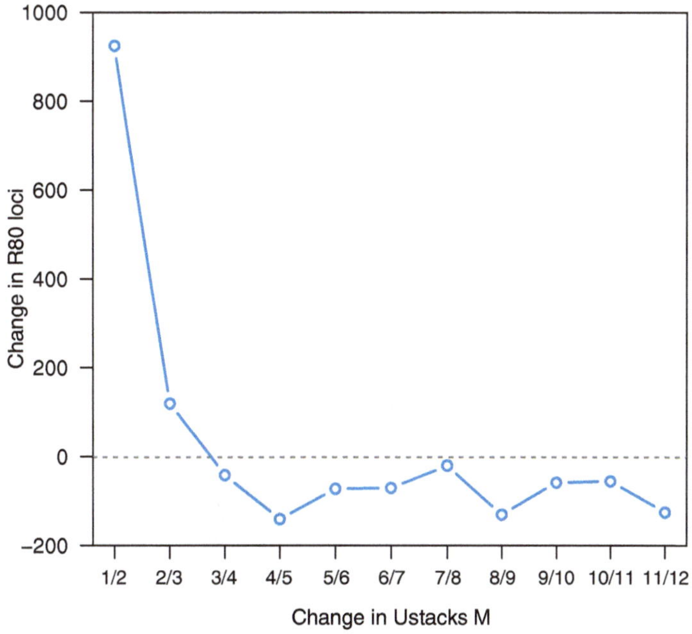

Fig. 2 Change in R80 loci during de novo parameter optimization. Blue line shows the change in loci kept in 80% of individuals with the successive increase of `ustacks -M`. The optimal assembly parameter value for the *E. maclovinus* dataset is an `-M` of 3

optimization. Now, we will deal with the full dataset, which of course includes loci present in a range of sample numbers, from as few as one to all samples. We will have the option to filter these data using `populations` (Subheading 2.8), once the core pipeline is complete.

We will prepare a new population map that includes all samples and now lists our biologically meaningful population associations (`va` and `pn`, here based on geography). Note that based on the analysis of our `process_radtags` results, we have removed the two samples with less than 1 M reads, as including them will skew our filtering results:

```
$ cat popmap.tsv
emac_va_01<tab>va
emac_va_02      va
emac_va_03      va
emac_va_04      va
emac_va_06      va
...
emac_pn_20      pn
emac_pn_21      pn
emac_pn_22      pn
```

```
emac_pn_24      pn
emac_pn_25      pn
```

We make a new output directory:

```
$ mkdir denovo.M3.rm_pcr_dups
```

And execute the denovo_map.pl wrapper, which will run Stacks de novo component programs, removing the PCR duplicates found from our paired-end reads:

```
$ denovo_map.pl \
    --samples ./01_process_radtags/samples \        # Path to samples
    --popmap ./03_denovo/info/popmap.tsv \          # Popmap
    --out-path ./03_denovo/denovo.M3.rm_pcr_dups \ # Output directory
    --paired \                                     # Include paired reads
    -M 3 \                                         # ustacks M
    -n 3 \                                         # cstacks n
    --rm-pcr-duplicates                            # Remove PCR duplicates
```

2.5.3 Assaying the Effectiveness of De Novo Assembly

Once the pipeline is finished, we will focus on three aspects of the analysis to assay our results: sequencing coverage, number of PCR duplicates, and phasing of SNPs into haplotypes within each fully assembled (single- and paired-end) locus. Sequencing coverage can be assessed at different stages of a RAD analysis (n.b., we initially assessed coverage after demultiplexing the raw, unassembled reads in Subheading 2.4.1). The first stage during the core pipeline run is following the execution of ustacks, in which the single-end reads from each sample are processed independently and prior to the removal of PCR duplicates. We can check these results by extracting the per-sample coverage distribution from the denovo_map.log file using the stacks-dist-extract utility.

```
$ stacks-dist-extract denovo_map.log cov_per_sample \
  | grep -v '^#'
samples      depth of cov   max cov   n_reads incorporated   %_reads incorporated
emac_va_01   38.89          496       1843431                86.2
emac_va_02   53.05          697       2714362                86.8
emac_va_03   48.89          650       2467227                87.6
emac_va_04   19.39          187       805017                 86.4
emac_va_06   32.9           327       1495664                86.5
...
emac_pn_20   59.67          500       3258540                89.2
emac_pn_21   33.3           271       1711132                88.7
```

```
emac_pn_22    4.77         32         882                  5
emac_pn_24    64.98        618        3562075              89.1
emac_pn_25    62.15        532        3349985              88.6
```

The distribution demonstrates a high variation in coverage between samples, which matches the distribution of reads observed after library demultiplexing. While average coverage is very high, a few samples are outliers. For example, `emac_pn_22` contained only a small number of forward reads (~17k) of which only 882 (5%) were incorporated into loci by `ustacks`.

While the final assessment of coverage, accounting for PCR duplicates, will take place after running `gstacks`, this early assessment is important as it might highlight problems with some of the samples. Trends like these, in which only a small fraction of reads were used in `ustacks`, can imply problems in library preparation, such as poor DNA quality and even the presence of contamination.

After `gstacks` executes, building the paired-end contig and incorporating it into the single-end locus, we can verify coverage once again, as well as checking the levels of PCR duplicates and rates of phasing. A summary of these metrics for the whole library can be found in the `gstacks.log` file, as well as information on the locus assembly and genotyping.

```
$ cat gstacks.log | grep -B 2 -A 3 '^Genotyped'
Removed 90954688 read pairs whose insert length had already been seen in the
same sample as putative PCR duplicates (55.5%); kept 73010865 read pairs.

Genotyped 167698 loci:
  effective per-sample coverage: mean=31.7x, stdev=11.6x, min=2.4x, max=64.0x
  mean number of sites per locus: 389.3
  a consistent phasing was found for 882805 of out 912101 (96.8%) diploid loci
needing phasing
```

We first see that the final coverage is 31.7×, almost half of the coverage observed in `ustacks`. This is because this final coverage is calculated after the removal of PCR duplicate reads, which account for 55.5% of reads in this library.

> Note! For single-digest RAD libraries, PCR duplicates are determined by looking at each pair of reads in a particular locus and if the paired-end aligns to the same position as another pair of reads, it is assumed to be a duplicate produced during PCR amplification.

Our post-duplicate levels of coverage are still robust and will yield confident genotypes. The high number of PCR duplicates in this library likely comes as a function of the very high sequence coverage. At this point, if coverage has dropped significantly below 10x, the researcher should consider that genotype calls may be of low confidence and biased toward homozygotes, and data may be widely missing from loci throughout the population (*see* Subheading 2.9.3).

Last, we can check how well individual genotypes could be phased into two distinct haplotypes (diploid alleles) in each individual at each locus. The `gstacks` program can phase SNPs into haplotypes by linking alleles together that occur in the same set of paired (and linked) reads. In *E. maclovinus*, 96.8% of all loci could be phased. The proportion of unphased loci often represent the presence of repetitive and paralogous sequences in the genome, though typically low phasing demonstrates problems in the assembly of loci (e.g., excessive overmerging).

A further breakdown of the results from `gstacks` can be extracted from the `gstacks.log.distribs` file using the `stacks-dist-extract` utility. It provides information on the number of raw loci present in each sample, as well as coverage, and PCR duplicate rates. Coverage is reported both as general average coverage (`mean_cov`) and the average coverage adjusted by the number of samples present at a locus (`mean_cov_ns`). We will use the adjusted mean to assess the coverage across samples in this section.

```
$ stacks-dist-extract gstacks.log.distribs effective_coverages_per_sample \
  | grep -v '^#' \
  | cut -f 1-2,4-5,8
sample       n_loci   mean_cov   mean_cov_ns   pcr_dupl_rate
emac_pn_01   44838    26.745     29.761        0.555
emac_pn_02   44568    41.911     46.344        0.566
emac_pn_03   262      2.508      2.577         0.455
emac_pn_04   43242    37.022     39.807        0.561
emac_pn_05   42860    40.023     42.657        0.559
...
emac_va_35   43054    36.571     39.266        0.549
emac_va_36   43314    36.708     39.643        0.560
emac_va_37   43366    38.790     41.952        0.559
emac_va_38   42192    30.821     32.492        0.551
emac_va_39   43792    41.274     45.032        0.565
```

Note! We are using the `cut` command to show only some of the columns in this file for readability purposes.

If we save this distribution to a file, we can use R to calculate further statistics on the coverage and PCR duplicate data. In particular, we want to observe the distribution of coverage across the samples (Fig. 3a), so we can determine if any additional samples need to be removed from the analysis (*See* the Appendix for a link to

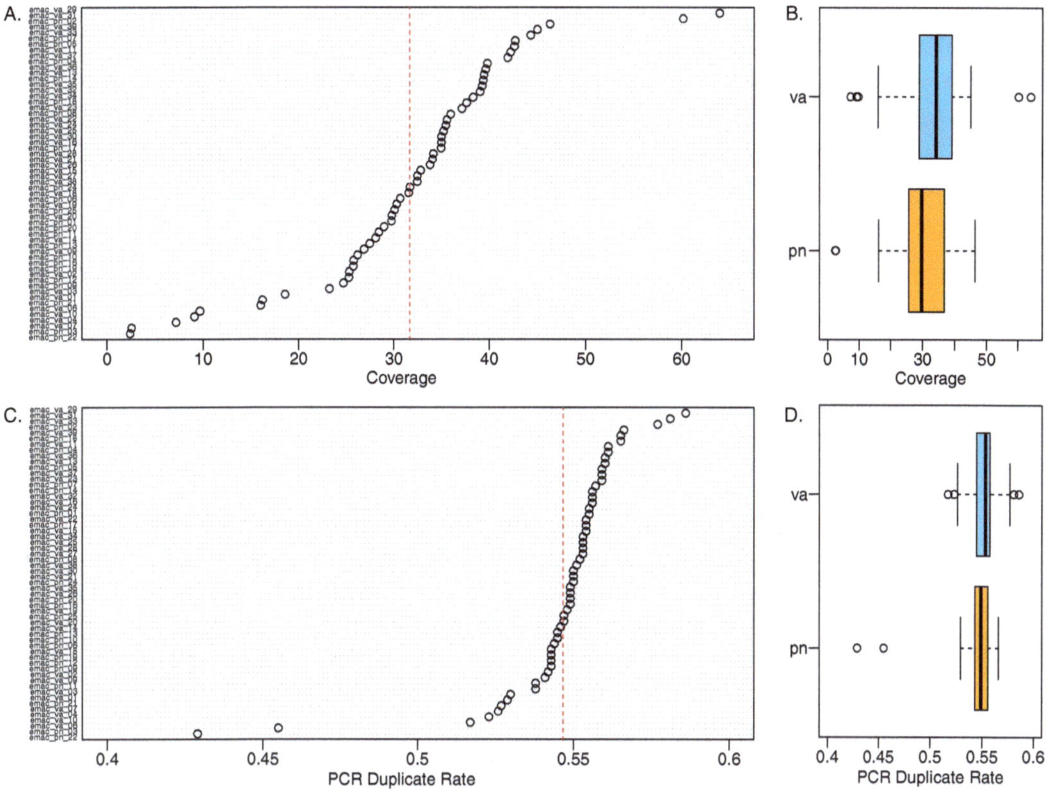

Fig. 3 Sequencing coverage and PCR duplicate statistics. (**a**) Nonredundant depth of coverage per sample. Red dashed line shows the mean coverage across all samples. (**b**) Nonredundant depth of coverage between Valdivia (`va`, blue) and Puerto Natales (`pn`, yellow) populations. (**c**). PCR duplicate rate, proportion of reads identified as duplicates, per sample. (**d**). PCR duplicate rate between populations

the R code). In accordance with the results from `ustacks`, we observe a subset of samples (e.g., `emac_pn_03` and `emac_pn_22`) with very low coverage (<5×), which will be removed from downstream analyses. The same samples also display a reduced proportion of PCR duplicates (Fig. 3c), but this is likely a function of their reduced coverage. In addition, we can also verify if there are any biases between the populations of groups in the library. While we, again, observe differences between the `pn` and `va` populations, we do not see any large-scale biases in coverage (Fig. 3b) or PCR duplicate rate (Fig. 3d).

We can obtain additional information by extracting per-sample phasing statistics. From this distribution we can look at both the number of genotypes called per individual, as well as the proportion of loci that could not be phased. These metrics can also be used to identify any problematic samples that should be removed from the analysis. In this case the same two samples with low coverage (`emac_pn_03` and `emac_pn_22`) also display outlier patterns of phasing and number of genotypes.

```
$ stacks-dist-extract gstacks.log.distribs phasing_rates_per_sample
sample      n_gts  n_multisnp_hets  n_phased  misphasing_rate
emac_pn_01  40407  16617            16131     0.029
emac_pn_02  40510  17596            16845     0.043
emac_pn_03  219    22               22        0.000
emac_pn_04  40467  17454            16835     0.035
emac_pn_05  40522  17881            17239     0.036
...
emac_pn_05  40522  17881            17239     0.036
emac_pn_06  40521  16988            16478     0.030
emac_pn_07  40566  17560            16924     0.036
emac_pn_08  40497  17123            16544     0.034
emac_pn_09  40252  16312            15849     0.028
```

From here, the core pipeline run is complete. While the outlier samples we identified have contributed some data to the metapopulation catalog, we can prevent them from being included in our downstream analysis simply by removing them from the population map. If we want to ensure that no data at all was included from the outliers, we would reexecute `denovo_map.pl` after adjusting the population map.

We will next show how to complete an analogous core analysis using a reference genome, followed by a hybrid de novo/reference assembly method called *integration*. After that, the three methods converge when we will apply a population genetics frame to the processed data using the `populations` program.

2.6 Reference-Aligned Analysis

When a reference genome is available for the species of interest, Stacks can be run in reference mode. A local assembly of loci is still performed, using the position of the aligned reads in the genome to identify the reads belonging to a locus, in contrast of the clustering based on sequencing similarity used by the de novo approach. Thus, no parameter optimization is required. However, note that locus assembly is performed independent of the reference genome, as the reference is not used as a template or as a source of alleles for SNP calling. A reference-aligned analysis allows us to place the genotypes and the population-level metrics in the context of the chromosomes, enabling analyses such as genome scans of genetic divergence [42], measuring linkage disequilibrium and identifying structural variants [13, 28], and fine-scale coalescent reconstruction along chromosomes [43].

Before we can run Stacks, we need to align the reads generated by `process_radtags` (Subheading 2.4.1) to the reference of interest. This can be done using any standard short-read aligner, we will be using BWA [26] here. Alignments will then be sorted using SAMtools [44]. We will create a set of directories to hold the reference genome and the aligned samples.

```
$ mkdir -p ./04_refmap/genome/
$ mdkir -p ./04_refmap/aligned_samples
```

Before we can align, we need to generate a searchable database of the reference genome that can be accessed by the aligner. For our *E. maclovinus* reference, we will generate this database using the `index` mode of `bwa`, and we will name the resulting database as `emac_db`:

```
$ bwa index \
    ./04_refmap/genome/e_maclovinus.genome.fasta \ # reference genome
    -p ./04_refmap/genome/emac_db                  # output database
```

Once the database is generated, we will align each sample independently using `bwa` in `mem` mode. The output of `bwa` will then be directly piped into `samtools view`, to compress the alignments (`-b`) while preserving the headers (`-h`), and these data are likewise piped into `samtools sort`. Aside from sorting, no further processing or filtering is done.

```
$ bwa mem ./04_refmap/genome/emac_db \                      # BWA database
    ./01_process_radtags/samples/emac_va_01.1.fq.gz \      # forward reads
    ./01_process_radtags/samples/emac_va_01.2.fq.gz | \ # paired reads
    samtools view -b -h | \                                # convert to BAM
    samtools sort \                                        # sort alignments
    -o ./04_refmap/aligned_samples/emac_va_01.bam          # BAM output
```

The alignment of all samples can be automated using a shell script, an example of which is linked in the Appendix, or they could be run as independent jobs on a cluster, to speed up execution.

Next, we will create the output directory for the Stacks reference run:

```
$ mkdir refmap_rm_pcr_dups
```

We can now prepare the command for the `ref_map.pl` program. Similar to the de novo wrapper, `ref_map.pl` is a Perl wrapper script used to execute the core reference-based pipeline. We will specify the path to our processed reads (`--samples`), the path to our new output directory (`--out-path`), and the path to our population map (`--popmap`). This population map file is the same used in Subheading 2.5.2, which classifies the samples between the va and pn populations and excludes the two individuals with low read counts. Last, we tell gstacks to remove PCR duplicates (--rm-pcr-duplicates).

```
$ ref_map.pl \
    --samples ./04_refmap/aligned_samples \      # Aligned samples
    --popmap ./04_refmap/info/popmap.tsv \       # Popmap
    --out-path ./04_refmap/refmap_rm_pcr_dups \ # Output directory
    --rm-pcr-duplicates                         # Remove PCR duplicates
```

2.6.1 Assaying the Effectiveness of a Reference-Based Assembly

With the completion of pipeline execution, we first want to assess our input data, particularly the quality of the alignments. The gstacks.log file contains information about the input alignments.

```
$ cat gstacks.log | grep -A 5 'BAM records'
Read 474544871 BAM records:
  kept 421697658 primary alignments (89.7%), of which 204681793 reverse reads
  skipped 26444533 primary alignments with insufficient mapping qualities (5.6%)
  skipped 17370711 excessively soft-clipped primary alignments (3.7%)
  skipped 4712448 unmapped reads (1.0%)
  skipped some suboptimal (secondary/supplementary) alignment records
```

Users should verify the proportion of retained alignments and examine the main reasons alignments were discarded. While the proportion of alignments kept in this dataset is high (89.7%), users should check for the fractions of alignments discarded due to low mapping quality or excessive soft-clipping, as they might imply problems with the underlying reference, presence of nontarget DNA in the library, or, in the case of using a heterospecific reference, a large evolutionary distance between the reference the species sequenced.

A quick way of visually assessing the alignments is by looking at the CIGAR string [44] present in column six of the SAM/BAM file format. We can extract some of these CIGAR strings using the following command.

```
$ samtools view emac_pn_22.bam | cut -f 6 | head
143M
150M
76S67M
29M114S
60S22M68S
150M
```

Note! Remember that BAMs are binary and need to be processed and translated to text using the `samtools view` command.

The CIGAR strings report information on the structure of the alignment. If the reads are 150 bp long, a CIGAR string of 150 M would present a one-to-one alignment of 150 "matches" between the reference and query sequence (n.b., *matches* implies two nucleotides are *aligned*, not necessarily identical). The presence of indels in an alignment are recorded with "`I`" and "`D`" designations (e.g., `20M1I129M` represents an alignment with 20 matches, a 1-bp insertion in the subject, and then 129 additional matches). CIGAR strings can also be used to represent partial alignments, in which a large portion of the subject read does not match the reference, represented by the "`S`" and "`H`" operations, in which part of the subject is excluded from the alignment (e.g., given `76S67M`, the first 76 bases of the reads do not align to the reference and are soft-clipped). The presence of these partial, soft-clipped alignments is expected; however, excessive soft-clipping might be a sign of a poor-quality reference, too much genetic distance between the subject species and reference, and/or improper processing of the raw reads. Moreover, the presence of soft-clipped alignments is not accounted for by the mapping quality (`MAPQ`) calculation present in the BAM file, meaning that filtering for quality alone might not remove all the partially aligned reads.

For reference-aligned data, a per-sample distribution of alignment statistics can be obtained from the `gstacks.log.distribs` file using the `stacks-dist-extract` utility.

```
$ stacks-dist-extract gstacks.log.distribs bam_stats_per_sample \
  | cut -f 2-4,6-7
sample        primary_kept  kept_frac  primary_disc_mapq  primary_disc_sclip
emac_pn_01    6956086       0.882      437268             323095
emac_pn_02    11273163      0.901      593824             424064
emac_pn_03    38264         0.471      3987               6005
emac_pn_04    9357193       0.891      587694             388855
emac_pn_05    9950393       0.894      606845             395133
emac_pn_06    6902795       0.889      426681             282369
emac_pn_07    9945776       0.891      610522             401058
emac_pn_08    8309644       0.889      517207             346634
emac_pn_09    5494906       0.888      348349             227379
```

It is useful to verify two at least two alignment statistics including, (1) the number of primary records (alignments) kept per sample, and (2) the fraction of records kept. We observed that an average of 7.3 M records were kept per sample, although variation is seen between the samples (Fig. 4a)—which is expected given the

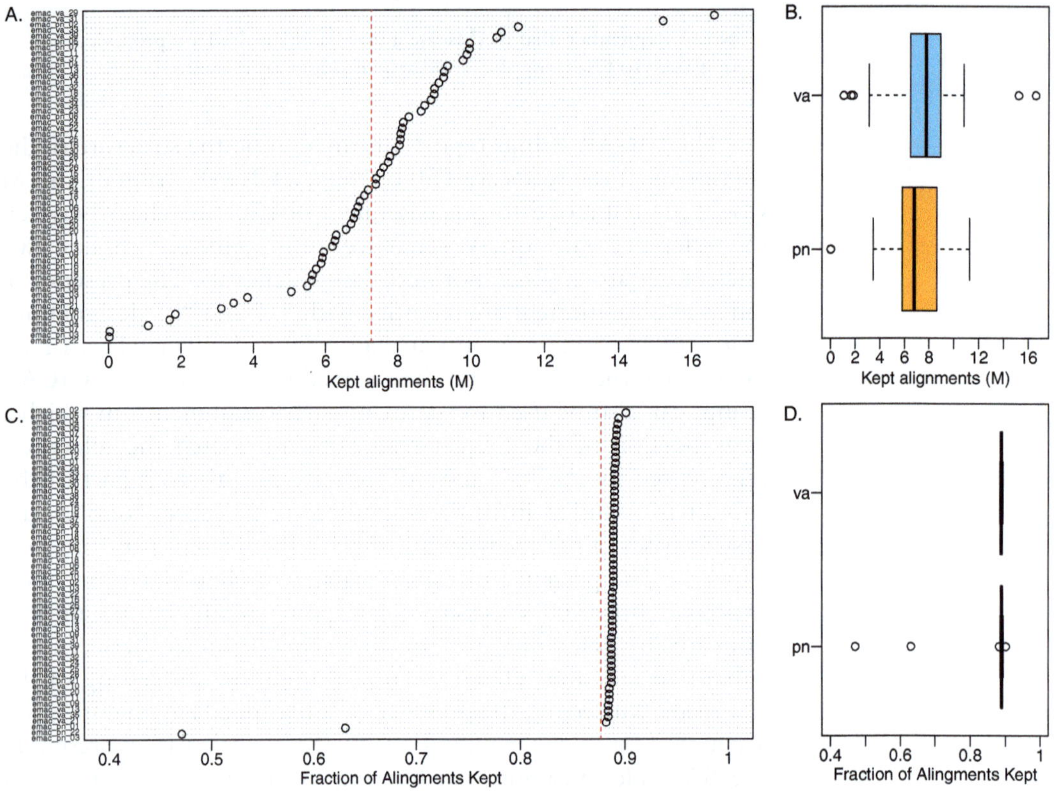

Fig. 4 Reference-based aligned statistics. (**a**). Number of alignments per-sample kept by `gstacks`, in millions of alignments. Red dashed line shows the mean number of kept alignments per sample. (**b**) Number of kept alignments between the Valdivia (`va`, blue) and Puerto Natales (`pn`, yellow) populations. (**c**). Fraction of alignments kept per sample, out of the total number of reads. (**d**) Fraction of kept alignments between populations

differences in the number of input reads and resulting coverage. When we compare the number of records kept between the two populations, we observe that, while the `va` population keeps more records overall, no large biases are observed between the two groups (Fig. 4b).

While the library has a high average proportion of kept alignments, there are some samples displaying outlier proportions of kept reads (Fig. 4c). These samples (e.g., `emac_pn_03` and `emac_pn_22`) also display very reduced coverage and can be removed from the population map for subsequent analyses. Aside from this section regarding alignments, the output of `gstacks` is otherwise identical to the one obtained in de novo mode. Therefore, coverage, PCR duplicates, and phasing can be assessed as described in Subheading 2.5.3.

2.7 Integrated Analysis

A high-quality reference genome (with a large N50 and chromosome-level contiguity) provides the best foundation for a RAD-based population genetics analysis. However, a de novo analysis can still reveal a lot about the underlying genetics of a natural population, particularly with respect to population structure, levels of heterozygosity, and coalescent and phylogenetic inferences. Stacks provides a third option: conducting a de novo assembly, followed by alignment of those loci to a heterospecific reference genome. This *integrated* method allows a researcher to assemble loci without a distantly related or low-quality reference affecting the results, but still providing information with respect to the colocalization of loci along the chromosomes (e.g., to visualize runs of divergence). The `stacks-integrate-alignments` program provides the ability to inject locus alignment coordinates into the Stacks catalog and related files, while providing filters to eliminate spurious alignments.

The procedure starts by assembling loci de novo (Subheading 2.5.2), and then aligning the loci in the catalog (i.e., `catalog.fa.gz`) against a reference genome of interest, obtaining a BAM output file of those alignments. Next, the `stacks-integrate-alignments` Python program is run, feeding it the de novo output directory, the aligned loci BAM, and specifying any filters, which can be used to require a certain percentage of the length of a locus to be aligned, as well as a minimum percent identity for an underlying alignment. These filters prevent spurious matches between loci and, for example, repetitive elements in the heterospecific reference.

To demonstrate this functionality, we will align our *E. maclovinus* de novo loci against another non-Antarctic notothenioid, the Channel bull blenny, *Cottoperca gobio* (NCBI accession GCA_900634415.1). This species is in a different family from *E. maclovinus*, so we expect only a fraction of our de novo loci will be alignable, none the less, it is a high-quality, chromosome-level, assembly with an N50 > 25 Mbp.

We create a series of directories to hold the integrated data.

```
$ mkdir -p ./05_integrated/genome/                # Genome FASTA and BWA index
$ mkdir -p ./04_integrated/info/                  # Population map
$ mkdir -p ./05_integrated/integrate_alns.cgob/   # Output of integration
```

Next, we align the catalog generated in the de novo assembly to this *C. gobio* reference genome (code not shown) and then we can integrate the alignment positions into the de novo catalog, storing the results in `integrate_alns.cgob`.

```
$ stacks-integrate-alignments \
--in-path ./03_denovo/denovo.M3.rm_pcr_dups \                    # De novo Stacks path
--bam-path ./05_integrated/integrate_alns.cgob/catalog.cgob.bam \ # Catalog BAM
--out-path ./05_integrated/integrate_alns.cgob \                  # Output directory
--min_alncov 0.8 \                                # Min fraction of catalog aligned
--min_pctid 0.8                                 # Min percent identity of alignment
```

The two filters we employ, minimum alignment coverage (`--min_alncov`) and percent identity (`--min_pctid`), will make sure we avoid spurious alignments of de novo loci.

Once `stacks-integrate-alignments` is complete, we then execute the `populations` program on the new catalog in `integrate_alns.cgob` and we are free to proceed with standard filtering, calculating summary and divergence statistics, and exporting data for further analysis.

The strength of the *integrated* results will be related to the evolutionary distance and quality of the heterospecific reference used. Here, with *C. gobio*, many of our loci do not align; starting with 167,698 de novo loci, the majority did not align or were filtered, yielding ~8k loci for analysis. Of course, for demonstration purposes, we used very strict filters and a somewhat distant genome. If multiple high-quality genomes are available, a good approach would be to align the de novo catalog to several of them to determine which of the heterospecific assemblies retains the highest proportion of aligned loci.

2.8 Populations

In the final analysis stage, the `populations` program puts the data into a population genetics context; in the prior analysis stages, the Stacks pipeline has considered all individuals as a single metapopulation, but now Stacks will divide samples into groups, which may represent geographical populations, divisions according to phenotype, sex, or generation (e.g., F0 and F2) in the case of a genetic map, or species in the case of a phylogenetic analysis. The `populations` program will calculate the frequencies of the observed alleles, according to these divisions, alongside several measures of population diversity and divergence. Importantly, `populations` can apply filters to the data, such as removing alleles and genotypes not observed in the majority of the samples and can export the results for further analysis in many standard file formats (e.g., VCF, Genepop, and STRUCTURE, among others).

The `populations` program is designed to be applied multiple times, independently, over the catalog of loci and genotypes, which were generated by the core pipeline—regardless of if it was generated de novo or with a reference genome. In other words, loci are assembled and genotyped once, after which `populations` can be run several times to manipulate, subset, and export the data according to the needs of a given analysis.

We will show an example of a populations command to filter the data generated de novo in Subheading 2.5.2. After the core pipeline execution, the catalog contains all the loci in the metapopulation, including erroneous loci produced by artefacts from the molecular protocol and from noise in the underlying species genome (e.g., repeat elements). A common practice is to only include loci which are present in a fraction of the individuals—with the independent assembly of a locus in multiple individuals serving as evidence that a locus is real. Similarly, to differentiate between real polymorphisms in the population and variation produced by molecular and sequencing error, genotypes can be filtered to only include variant sites present in some fraction of individuals. To begin, we will apply a baseline set of filters to our data to assay how many loci and variant sites are retained and are therefore usable in other downstream analyses. We want to keep loci observed in both populations and present in 50% or more of individuals in each population, and only keep variant sites where the minor allele was observed at least three times in the metapopulation (which guarantees they are observed in a minimum of 2 individuals, excluding erroneous SNP calls made only in single individuals). In addition, we want to run this analysis excluding the four samples which we previously characterized as having low quality.

Within the directory of our de novo Stacks run we will create a new output directory for this specific `populations` run.

```
$ mkdir -p ./03_denovo/denovo.M3.rm_pcr_dups/pop_N56_r50_p2_mac3
```

We can then create our `populations` command. First, we specify the path to the base Stacks run we will use for the analysis (`--in-path`). Here, we are using a de novo Stacks run, but the analysis can be equivalently performed with reference-aligned or integrated data. Next, we specify the path to our new, filtered population map file (`--popmap`), and the path to our newly created output directory (`--out-path`). We then apply filters, processing only the loci present in at least 50% of samples in a population (`--min-samples-per-pop`), present in both populations (`--min-populations`), and only keeping variant sites with a minimum allele count of three (`--min-mac`). We also determine which loci and variant sites are in Hardy–Weinberg Equilibrium (`--hwe`). Several other options can be found in the Stacks documentation.

```
$ populations \
    --in-path ./03_denovo/denovo.M3.rm_pcr_dups \                    # Stacks run
    --popmap ./03_denovo/info/popmap.rm_low_cov.tsv \                # Popmap
    --out-path ./03_denovo/denovo.M3.rm_pcr_dups/pop_N56_r50_p2_mac3 \ # Output dir
    --min-samples-per-pop 0.50 \                                     # Loci in 50% samples
    --min-populations 2 \                                            # Loci in 2 pops
    --min-mac 3 \                                                    # Min allele count 3
    --hwe                                                            # Calculate loci in HWE
```

2.8.1 Assessing the Populations Results

Once the `populations` run is complete, we can evaluate our results by checking the `populations.log` file.

```
$ cat populations.log | grep -A 4 '^Removed'
Removed 127971 loci that did not pass sample/population constraints from 167698 loci.
Kept 39727 loci, composed of 25553775 sites; 283468 of those sites were filtered,
271068 variant sites remained.
Number of loci with PE contig: 39727.00 (100.0%);
  Mean length of loci: 633.23bp (stderr 0.35);
```

First, we want to check the number of kept loci. We see that, using these filtering parameters, we keep a total of 39,727 loci in this de novo *E. maclovinus* dataset. At first glance, this looks like a small fraction of the 167,698 loci present in the catalog. However, we must remember that the size of the catalog is not representative of the real data as it also contains error loci originating from all samples. Going back to our simulations, we identified around 52k *SbfI* RAD loci in the *E. maclovinus* genome. While this number is higher than the 40k loci seen here, recall that (1) not all loci in the genome will be properly assembled (i.e., those in repeats and recent paralogs), (2) the empirical data contains noise and technical error from the library preparation and sequencing, and (3) that 40k represent the fraction of loci kept after applying moderate filters to the data. The 40k loci kept at this level of filtering stringency should be sufficient for most evolutionary applications, including the 271,068 filtered variant sites (demonstrating that most RAD loci produced more than one SNP). The program also reports to us the mean length of assembled loci (~630 bp). As we increase our filtering stringency, we will typically see this mean length increase.

An additional source of information from the `populations` execution is the `populations.log.distribs` file. This file contains a number of diagnostic distributions, many of them presented both before and after filtering, demonstrating the effects of the chosen filtering parameters. We can view the distribution of the number of samples per locus as well as the number of SNPs per locus. Again, using `stacks-dist-extract` we can pull them out one at a time, first the number of samples per locus, before filtering.

```
$ stacks-dist-extract populations.log.distribs samples_per_loc_prefilters
# Distribution of valid samples matched to a catalog locus prior to filtering.
n_samples   n_loci
0           41
1           111582
2           6144
3           2251
...
53          2499
54          4231
55          8028
56          17396
```

And we can view the number of SNPs per locus, again before filtering.

```
$ stacks-dist-extract populations.log.distribs snps_per_loc_prefilters
# Distribution of the number of SNPs per catalog locus prior to filtering.
n_snps   n_loci
0        119076
1        2427
2        1948
3        1965
4        2129
...
```

After filtering, it is clear that we have removed loci that do not meet our 50% threshold, as there are no entries for less than 29/56 total samples (51% of our samples).

```
$ stacks-dist-extract populations.log.distribs samples_per_loc_postfilters
# Distribution of valid samples matched to a catalog locus after filtering.
n_samples   n_loci
29          32
30          36
31          71
...
54          4231
55          8028
56          17396
```

Finally, we can evaluate the amount of data per sample, including the number of missing loci per sample, or the number of missing variant sites per sample. Here, as an example, is the number loci per sample.

```
$ stacks-dist-extract populations.log.distribs loci_per_sample
# Number of loci per individual sample (after filtering).
Sample      n_loci  present_loci  missing_loci  frequency_missing
emac_va_01 39727   35562         4165          0.1048
emac_va_02 39727   37538         2189          0.0551
emac_va_03 39727   37542         2185          0.0550
emac_va_04 39727   31846         7881          0.1984
emac_va_06 39727   34181         5546          0.1396
...
```

These outputs allow us to see how different individuals in the analysis are performing (e.g., missing data per locus or SNP) as well as evaluating the data set as a whole (e.g., samples per locus) and the underlying biology of our system (SNPs per locus). Two more `populations` outputs provide additional, biologically related information for the system under study. The `populations.sumstats_summary.tsv` file contains a summary of the number of variant sites, heterozygosity, and related measures per population, while the `populations.sumstats.tsv` file contains the same detailed measures per SNP per population.

It is important to note that the filtering parameters shown here are only just a base example of the functionality of the `populations` program. At this point, the onus of interpretation moves from the effects of the molecular protocol and sequencing run to the underlying biology. It is up to the researcher to determine what level of SNPs or loci are appropriate and these values will differ widely, for example searching for population structure in closely related populations versus building phylogenetic trees among distantly related taxa will present very different requirements. Researchers should not forget that the filtering parameters are not the goal of the analysis. Instead, practitioners should shift their thinking to ask: What is required to answer the biological questions proposed in the experiment?

In the following sections, we provide several possible routes for further analysis, beyond Stacks itself.

2.8.2 Post-analysis Directions After Stacks

Given a set of RAD loci, and their endemic SNPs, nearly any population genetic analysis is possible. However, different assumptions apply to the underlying data in different analyses, and we can use the populations program to prepare different sets of loci and or SNPs for these analyses. We provide several examples below.

Note! In the examples below we apply specific filtering parameters for each of our analyses related to the quality and coverage of the E. maclovinus dataset and based on the requirements of the downstream application. Specific filters should be tested and verified to be appropriate for each independent dataset.

SNP-Based Population Structure

One of the most common population genetic analyses is to search for population structure, or to understand how many genetic groups are present in a set of individuals. This can be done via a PCA analysis, or by running the program STRUCTURE [45] or one of the newer derivatives. To determine population structure, we only need a small number of SNPs, and we do not need genomic coordinates for those SNPs (they can be from a de novo analysis). However, since we want to understand the underlying neutral relationships between individuals, we want to avoid outlier SNPs whose allele frequencies may be driven by selection—choosing SNPs in Hardy–Weinberg Equilibrium (HWE) can ameliorate this concern. Further, multiple SNPs from the same locus are by definition linked, so we only want to choose one representative SNP from each locus. The `populations` program will calculate how far out of HWE a locus is, providing a p-value, and this data is included for individual SNPs (`populations.sumstats.tsv`) and for loci (`populations.hapstats.tsv`). In addition, we can instruct `populations` to only include a single SNP per locus (chosen after filtering is complete) and we can supply a *whitelist* of loci—a specific set of loci to include, enabling us to randomly select, for example, 5000 loci for the STRUCTURE analysis.

First, we will create a whitelist file, based on the loci recovered in the previous `pop_N56_r50_p2_mac3` run. We will randomly select 5000 loci that are in HWE for both populations. Here we employ a shell pipe to combine Stacks output data to generate our list of loci.

```
$ cat populations.sumstats.tsv | grep -v '^#' \
  | awk '$20 > 0.05 {print $0}' \          # Choose loci in HWE (p-value > 0.05)
  | cut -f 1| \                            # Print the first column (Locus ID)
  | sort -n -u \                           # Sort and return unique values
  | shuf -n 5000 \                         # Shuffle 5000 random loci
  | sort -n > wl.txt                       # Sort and redirect to file
```

We will also create a new directory to output this new run.

```
$ mkdir pop_wl_structure
```

Once we have created our whitelist, we can form the populations command. Note we no longer need filters since the list of loci in our whitelist originated from an already-filtered set of loci.

```
$ populations \
    --in-path ./03_denovo/denovo.M3.rm_pcr_dups \
    --popmap ./03_denovo/info/popmap.rm_low_cov.tsv \
    --out-path ./03_denovo/denovo.M3.rm_pcr_dups/pop_wl_structure \
    --whitelist ./03_denovo/denovo.M3.rm_pcr_dups/pop_wl_structure/wl.txt \
    --write-single-snp \
    --structure
```

After the populations run is complete, we will find a `populations.structure` file in the output directory, which can then be loaded into STRUCTURE (after removing the Stacks timestamp comment on the first line and modifying the population identifiers). We can visualize the output for our *E. maclovinus* dataset for K = 3 (Fig. 5a), where we observe a substructuring of samples within the Valdivia population (`va`). Little to no substructuring is observed within the two geographic sites.

Haplotype-Based Population Structure

An alternative method for assessing population structure uses haplotypes instead of the single SNPs used by STRUCTURE. A haplotype approach maximizes the biological information encoded by each marker—single SNPs can only describe two alleles present in the metapopulation, while haplotypes can describe several alleles. This increase in information content by markers often allows for an increase in resolution both when identifying differentiation between populations and when assigning individuals to a population [46, 47].

Identifying population structure using RADseq-derived haplotypes can be performed using `RADpainter` and `fineRADstructure` [48], a pair of sister programs designed for the generation of a coancestry matrix derived from RADseq haplotypes, followed by the inference of populations among a group of individuals based on their coancestry. This analysis was specifically designed around RADseq data and thus can take full advantage of the haplotypes natively produced by Stacks.

Similar to STRUCTURE, Stacks can export data directly in the `RADpainter/fineRADstructure` format (`--radpainter`). Since the software is designed to work around a larger number of markers, it is not necessary to subsample loci using a whitelist. Instead, we can just apply our filters directly (`--min-samples-per-pop 0.75`, `--min-populations 2`). In addition, we tell populations to filter data by haplotypes (`--filter-haplotype-`

wise), which applies the missing data filters to the whole locus, instead of applying them to the individual variant sites. This approach will prune the variant sites with the highest proportion of missing data in the locus, in turn guaranteeing the least possible missing data within the kept haplotypes. This option is recommended for all applications based on the haplotype outputs from populations.

```
$ populations \
    --in-path ./03_denovo/denovo.M3.rm_pcr_dups \
    --popmap ./03_denovo/info/popmap.rm_low_cov.tsv \
    --out-path ./03_denovo/denovo.M3.rm_pcr_dups/pop_N56_r75_p2_mac3_haps \
    --min-samples-per-pop 0.75 \
    --min-populations 2 \
    --min-mac 3 \
    --filter-haplotype-wise \
    --radpainter
```

The generated populations.haps.radpainter file can be imported directly into RADpainter/fineRADstructure for analysis. Applying this method to our *E. maclovinus* dataset (Fig. 5b) we observe that there is a fraction of va samples clustering separately from all other individuals (*see* Subheading 2.9.3). In contrast to STRUCTURE, the extra information present in the haplotypes does allows for the sub-clustering of va and pn individuals.

Divergence Statistics

An alternative analysis route involves looking for divergent loci between populations. This can include traditional measures of F_{ST} and is again calculated at the SNP level. It is possible to calculate F_{ST} without a reference genome, typically plotting loci with high levels of heterozygosity against those loci with high values of F_{ST} [49]. However, once these loci have been identified, there are not a lot of further analytical options. Given a reference genome, however, populations can identify SNPs that are colinear on particular chromosomes indicating regions of divergence between populations. Below, we run populations on our reference-aligned dataset, and we employ several filters to export loci present in 80% of samples (--min-samples-per-pop 0.80) and present in two populations (--min-populations 2), and removing variant sites with a minimum allele frequency below 5% (--min-maf 0.05). We then ask for F statistics to be calculated (--fstats) and for the software to *smooth* the values of F_{ST}, giving us a moving average of divergence along each chromosome:

```
$ populations \
    --in-path ./04_refmap/refmap_rm_pcr_dups \
    --popmap ./04_refmap/info/popmap.rm_low_cov.tsv \
    --out-path ./04_refmap/refmap_rm_pcr_dups/pop_N56_r80_p2_maf05_fstats \
    --min-samples-per-pop 0.80 \
    --min-populations 2 \
    --min-maf 0.05 \
    --fstats \
    --smooth
```

Once `populations` is complete, an output file will be created for each pair of populations in the analysis, containing per-SNP divergence statistics. For the *E. maclovinus* dataset, this output file is named `populations.fst_va-pn.tsv` for the `va` and `pn` populations. In the case of genetic divergence, there are also several statistics that are applied at the locus/haplotype level, including absolute divergence, D_{XY}, and locus-level forms of F_{ST} including Φ_{ST} and F_{ST}'. These multiallelic statistics can provide a stronger signal, since the alleles can be distributed across multiple populations in more configurations than biallelic SNPs. D_{XY} also provides a more robust divergence statistic at more distant time scales as it is directly proportional to the inter-population coalescent time without being biased by allele frequency [50]. And may be better for the identification of islands of divergence, as well as providing evidence for additional models for divergence with ancient or nonexistent gene flow [51]. These files are output by `populations` in a similar way, including smoothed values, in `populations.phistats_va-pn.tsv`.

The output of these divergence analysis can then be used for several applications, one example is plotting genetic divergence along specific chromosomes. Here we can see the relative (F_{ST}) and absolute divergence (D_{XY}) between the `va` and `pn` *E. maclovinus* populations over the chromosome-level scaffold, `HiC_chromosome_4` (Fig. 5d).

Linkage Disequilibrium

When reference-aligned RADseq data are available, it can additionally be used to generate measures of linkage disequilibrium (LD) along the chromosomes to identify patterns of outlier recombination and linkage. Before we can perform any LD calculations, we first must statistically phase our variant sites into chromosome-level haplotypes. Since the phasing process imputes the genotypes for any missing sites, we will use very stringent filtering parameters (loci present in over 90% of individuals) to guarantee we export the least number of missing genotypes. Additionally, we will apply the `--ordered-export` flag to `populations` to ensure sites are

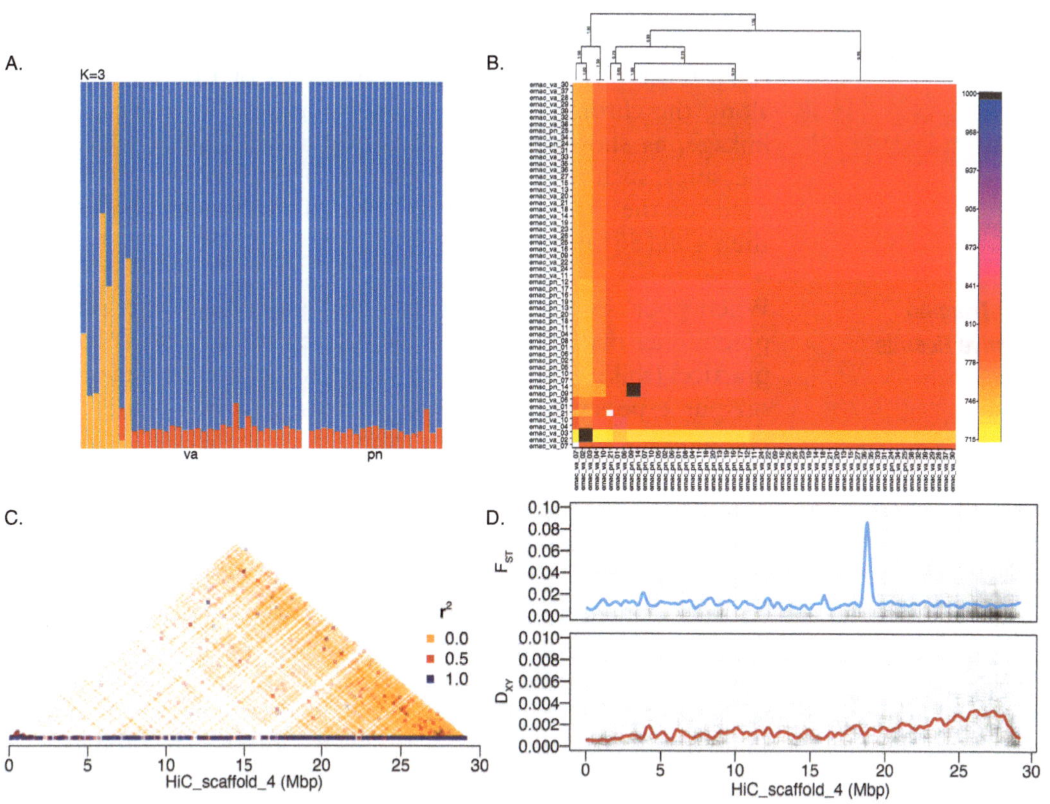

Fig. 5 Examples of downstream applications of the `populations` output. (**a**) SNP-based assessment of population structure using STRUCTURE, with a $K = 3$, for the Valdivia (`va`) and Puerto Natales (`pn`) *E. maclovinus* populations. (**b**) Haplotype-based assessment of population structure using `RADpainter/fineRADstructure`. Heatmap shows population clustering based on sample coancestry. (**c**) Pairwise linkage disequilibrium (measured by the correlation coefficient, r^2) in the *E. maclovinus* HiC_scaffold_4. This was calculated by phasing SNPs obtained via a reference-based analysis. (**d**) Genetics divergence along the *E. maclovinus* `HiC_scaffold_4`. Top plot shows relative divergence (F_{ST}). Black crosses show per-site F_{ST} values, while the blue line shows a smoothed average. Bottom plot shows absolute divergence (D_{XY}), with the black crosses showing per-locus D_{XY} values and red line showing a smoothed average

exported sorted by their genome coordinates, and in the case of overlapping RAD loci, only one representative is exported, guaranteeing compatibility with external software.

```
$ populations \
    --in-path ./04_refmap/refmap_rm_pcr_dups \
    --popmap ./04_refmap/info/popmap.rm_low_cov.tsv \
    --out-path ./04_refmap/refmap_rm_pcr_dups/pop_N56_r90_p2_mac3_vcf \
    --min-samples-per-pop 0.90 \
    --min-populations 2 \
    --min-mac 3 \
    --ordered-export \
    --vcf
```

The exported VCF can then be phased using `beagle` [52, 53], after which LD calculations can be performed using `vcftools` [54]. Once phasing is complete, measures of LD can be plotted along the chromosomes to identify regions displaying extended linkage, as shown in (Fig. 5c). Additionally, other LD-derived measures, such as extended haplotype homozygosity, can be computed on the chromosome-wide, phased results using, for example, the `rehh` [55] R package.

2.9 Triaging Negative Results

When poor results occur with a RADseq experiment, often the practitioner may try to *rescue* the data by applying *strict* analytical practices (this is inherently very subjective, hence the italics). Based on our experience in RADseq analysis, here we consider three common questions that result from a poor analysis.

2.9.1 Should I Trim Data?

A common procedure of many analysis pipelines for the processing of short-read Illumina data is the trimming of sequencing reads to remove fragments of the read with low quality and/or containing adapter sequences. While this produces datasets of uneven read length, the downstream effects of coverage are ameliorated by the random position of the shotgun reads along the genome (i.e., the trimmed portion of the reads will not always originate from the same position on the genome). This is, however, not the case for RADseq data—all reads from a single locus are anchored to the cutsite at the 5′-end of the locus. Trimming RADseq reads will therefore result in a localized decrease in coverage at the 3′ of the locus, which can have downstream effects on genotyping (since the calls are dependent on available coverage) and the assembly of contiguous loci (since the single- and paired-end contigs are merged based on their overlapping sequences). Counterintuitively, trimming low quality data may produce even worse results.

While not advised for the reasons described above, the Stacks pipeline can analyze trimmed data, as the gapped aligner in `ustacks` will handle uneven reads as if containing gaps. However, the software might throw out very short reads completely if they are unable to be matched to a putative locus. Sequencing a dataset to low coverage can exacerbate these behaviors, since the removal of reads results in a proportionally larger reduction of coverage across the whole locus, not just the 3′ end (Fig. 6a). The overall reduction in coverage subsequently results in a further increase of missing genotypes locus-wide (Fig. 6b).

Instead of trimming the datasets, the approach taken by Stacks during `process_radtags` is focused on the processing and discarding of whole reads instead of just partial fragments. As seen in Subheading 2.4, with proper initial sequencing coverage, whole reads can be filtered from the dataset without negative impacts on downstream coverage and genotypes. Reads with poor sequencing

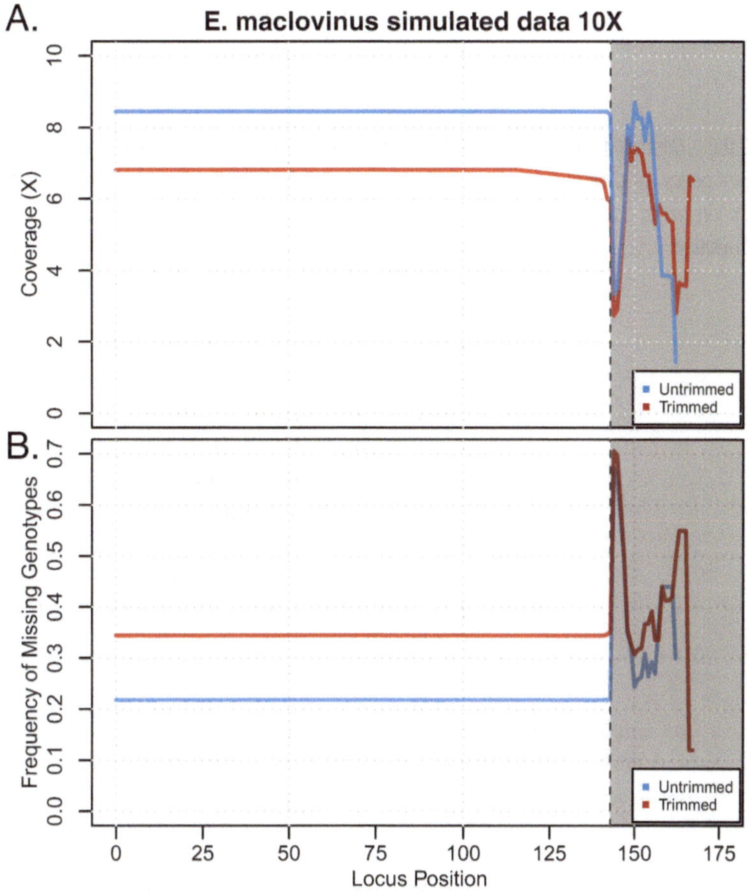

Fig. 6 Effects of read trimming on de novo locus assembly at low coverages. The results are shown for a simulated RADseq dataset, calculating values by averaging across all assembly loci. (**a**) Change in coverage along the single-end (forward read) region of the RAD locus. Grey dashed line shows the forward read length after barcode removal (143 bp). Grey box shows values for regions beyond the read length produced by indels between samples. The untrimmed dataset (blue) shows constant coverage across the locus, while the trimmed dataset (red) shows a decrease in coverage along the 3′ end of the locus. (**b**) Frequency of missing genotypes along the single-end region of the RAD locus. Grey dashed line shows the forward read length (143 bp). Grey box shows values for regions beyond the read length. The untrimmed dataset has a constant proportion of missing genotypes of around 20% along the locus, resulting from the low coverage. The trimmed dataset, while also having a constant value along the locus, shows an increase in missing genotypes, around 35%

quality are removed altogether with the `--quality` flag and checking for and discarding reads with adaptor fragment sequences can be done by using the `--adapter_1` and `--adapter_2` options. If coverage is low and adaptor is present, a better strategy may be to trim all reads to an even length, with `-t`. However, these options should only be applied after proper assessment of the quality of the input data to prevent the discarding of excess sequence.

2.9.2 Controlling the Stringency for Calling Genotypes at Low Coverage

RAD practitioners often want to apply a hard depth of coverage limit to assembled loci. Stacks does not provide this type of filter primarily because it results in arbitrary cutoffs—often removing loci that are statistically supported by their genotype calls, and/or not removing loci whose depth is inflated by PCR duplicate reads. Instead, if a user wants to have higher confidence in genotyping results (for which depth of coverage is used as a proxy) they should increase the level of evidence required for the genotyping model to make a call. Reducing the alpha value to the genotyping model (e.g., from 0.05 to 0.01) requires more evidence, for example, reads or depth of coverage, for a call to be made.

Returning to our original results from the de novo analysis (Subheading 2.5), after genotyping was performed in `gstacks` with the default parameters, and moderate filters were applied by the `populations` program (Subheading 2.8), the analysis produced:

```
$ cat populations.log | grep '^Kept'
Kept 39559 loci, composed of 25446496 sites; 284760 of those sites were filtered,
267285 variant sites remained.
```

We had 606,487 variant sites on this dataset by genotyping with default parameters. We then reran `gstacks` with `--gt-alpha 0.01`, to be more stringent in our genotype calls:

```
$ gstacks \
    -P ./03_denovo/denovo.M3.rm_pcr_dups \
    -M ./03_denovo/info/popmap.tsv \
    -O ./03_denovo/denovo.M3.rm_pcr_dups/gstacks.gt_alpha_01 \
    --rm-pcr-duplicates \
    --gt-alpha 0.01
```

We then rerun `populations`, applying the same filters as before (loci present in 50% of samples and two populations, variant sites observed three times in the catalog):

```
$ cat populations.log | grep '^Kept'
Kept 39559 loci, composed of 25446496 sites; 290505 of those sites were filtered,
233976 variant sites remained.
```

We retain the same number of loci (40k) and the same number of total sites (25M), but the original run retained 267k variant sites, while our new run with more stringent genotyping retains only 234k. This shows that we likely have 33k variant sites with low coverage which could not be genotyped after making `--gt-alpha` more stringent. However, this does indicate that the variant sites that are retained after this increase in genotype stringency are called with very high confidence.

These changes in genotyping stringency are a good approach for dealing with the effects of low coverage. Instead of applying an arbitrary coverage cutoff, users should simply let the genotyping algorithm assess the available evidence. The result will likely be a tradeoff between the magnitude and confidence of genotypes, with the final decision on stringency cutoff being dictated by the user according to the requirements of the experiment.

2.9.3 Removing Bad Apples

Cerca et al. (2021) described a method for the optimization on RADseq data based on the removal of samples with high missing data, or *bad apples* [56]. Briefly, a Stacks catalog is built (either de novo after parameter optimization or using a reference genome) after which the researcher identifies samples containing a high proportion of missing genotypes (described in Subheading 2.8.1). These samples drive a significant proportion of the catalog-wide fraction of missing data, and their removal improves overall retention of loci and genotypes after filtering, as well as minimizing biases derived from missing genotypes.

When viewing our *E. maclovinus* dataset in the light of bad apples, we do identify a fraction of individuals containing a higher proportion of missing data than the other samples in the analysis, even after the removal of some samples following raw read processing (Subheading 2.4.1) and catalog construction (Subheading 2.5.3). For example, when reviewing the `RADpainter/fineRADstructure` analysis performed previously in (Subheading 2.8.2), we notice that the proportion of `va` samples clustering separately from all other individuals are also the ones with the elevated proportion of genotypes.

```
$ stacks-dist-extract populations.log.distribs variant_sites_per_sample | \
    grep -v '^#' | \                   # Remove header
    grep -v '^sample\b' | \            # Remove column names row
    cut -f 1,5 | \                     # Extract sample and freq_miss missing column
    sort -k2 -n -r |                   # Sort numerically and reverse by 2nd column
    head -n 15                         # Print first 15 lines
emac_va_07      0.3816
emac_va_04      0.2250
emac_va_10      0.1932
emac_va_06      0.1412
```

```
emac_va_01      0.1011
emac_pn_21      0.0616
emac_va_03      0.0493
emac_va_02      0.0434
emac_va_09      0.0315
emac_pn_12      0.0280
emac_pn_16      0.0270
emac_pn_19      0.0266
emac_pn_10      0.0266
emac_pn_09      0.0255
emac_pn_13      0.0244
```

While the median proportion of missing data is less than 2% (median: 0.01845, mean: 0.03774), notice that some samples show outlier fraction of missing genotypes. This subset of samples can be classified as bad apples and removing them from subsequent analysis (by removing them from the population map file) can improve downstream analysis. For example, removing the apparent missing data bias in the results from `RADpainter`.

Appendix

Working example scripts referenced in this protocol can be found online at BitBucket in a Git repository: https://bitbucket.org/CatchenLab/mimb-stacks2-protocol. We list the available scripts below.

Plot Library Demultiplexing and Processing

The file `01_process_radtags/process_radtags_stats.R` uses the log reported by `process_radtags` and calculates several summary statistics on the proportion of samples in the library, number of total reads, and percentage of total reads kept per-sample. It also generates several plots showing the generated distributions. The input file can be obtained using the following.

```
stacks-dist-extract process_radtags.raw.log per_barcode_raw_read_counts >
per_sample_counts.tsv
```

De Novo Parameter Optimization Shell Loop

The file `02_param_opt/param_opt_loop.sh` loops over the values of 1 through 12 and uses the number in each iteration as the corresponding value for `ustacks -M` and `cstacks -n` for the Stacks de novo pipeline.

Gstacks Coverage R Script

The file `03_denovo/gstacks_stats.R` uses the log reported by `gstacks` and calculates several summary statistics on sample coverage and PCR duplicates. It also generates several plots showing the generated distributions. The input file can be obtained using the following.

```
stacks-dist-extract gstacks.log.distribs effective_coverages_per_sample | grep
-v '^#' > effective_coverages_per_sample.tsv
```

Bwa Alignment Shell Loop

The file `04_refmap/bwa_samples_loop.sh` loops over each sample in the provided population map file, and for each sample aligns the paired reads with `bwa mem` and processed alignments using `samtools view/sort`. For each iteration of the loop, the sample name present in the popmap is used as the prefix from which the path the input reads and output BAM files are constructed.

References

1. Smith HO, Welcox KW (1970) A restriction enzyme from Hemophilus influenzae. J Mol Biol 51:379–391
2. Kelly TJ, Smith HO (1970) A restriction enzyme from Hemophilus influenzae. J Mol Biol 51:393–409
3. Botstein D, White RL, Skolnick M et al (1980) Construction of a genetic linkage map in man using restriction fragment length polymorphisms. Am J Hum Genet 32:314–331
4. Vos P, Hogers R, Bleeker M et al (1995) AFLP: a new technique for DNA fingerprinting. Nucl Acids Res 23:4407–4414
5. Miller MR, Dunham JP, Amores A et al (2007) Rapid and cost-effective polymorphism identification and genotyping using restriction site associated DNA (RAD) markers. Genome Res 17:240–248
6. Davey JW, Hohenlohe PA, Etter PD et al (2011) Genome-wide genetic marker discovery and genotyping using next-generation sequencing. Nat Rev Genet 12:499–510
7. Andrews KR, Good JM, Miller MR et al (2016) Harnessing the power of RADseq for ecological and evolutionary genomics. Nat Rev Genet 17:81–92
8. Benestan L, Gosselin T, Perrier C et al (2015) RAD genotyping reveals fine-scale genetic structuring and provides powerful population assignment in a widely distributed marine species, the American lobster (*Homarus americanus*). Mol Ecol 24:3299–3315
9. Frugone MJ, López ME, Segovia NI et al (2019) More than the eye can see: genomic insights into the drivers of genetic differentiation in royal/macaroni penguins across the

Southern Ocean. Mol Phylogenet Evol 139: 106563

10. Marandel F, Charrier G, Lamy J et al (2020) Estimating effective population size using RADseq: effects of SNP selection and sample size. Ecol Evol 10:1929–1937
11. Carlen E, Munshi-South J (2021) Widespread genetic connectivity of feral pigeons across the northeastern megacity. Evol Appl 14:150–162
12. Amores A, Catchen J, Ferrara A et al (2011) Genome evolution and meiotic maps by massively parallel DNA sequencing: spotted gar, an outgroup for the teleost genome duplication. Genetics 188:799–808
13. Mérot C, Berdan E, Cayuela H et al (2021) Locally adaptive inversions modulate genetic variation at different geographic scales in a seaweed fly. Mol Biol Evol 38:3953–3971
14. Bay RA, Karp DS, Saracco JF et al (2021) Genetic variation reveals individual-level climate tracking across the annual cycle of a migratory bird. Ecol Lett 24:819–828
15. Dudaniec RY, Yong CJ, Lancaster LT et al (2018) Signatures of local adaptation along environmental gradients in a range-expanding damselfly (*Ischnura elegans*). Mol Ecol 27: 2576–2593
16. Schield DR, Walsh MR, Card DC et al (2016) Epi RAD seq: scalable analysis of genomewide patterns of methylation using next-generation sequencing. Methods Ecol Evol 7:60–69
17. Trucchi E, Mazzarella AB, Gilfillan GD et al (2016) BsRADseq: screening DNA methylation in natural populations of non-model species. Mol Ecol 25:1697–1713
18. Eaton DAR, Spriggs EL, Park B et al (2016) Misconceptions on missing data in RAD-seq Phylogenetics with a deep-scale example from flowering plants. Syst Biol 66:399–412
19. Near TJ, MacGuigan DJ, Parker E et al (2018) Phylogenetic analysis of Antarctic notothenioids illuminates the utility of RADseq for resolving Cenozoic adaptive radiations. Mol Phylogenet Evol 129:268–279
20. Ali OA, O'Rourke SM, Amish SJ et al (2016) RAD capture (rapture): flexible and efficient sequence-based genotyping. Genetics 202: 389–400
21. Peterson BK, Weber JN, Kay EH et al (2012) Double digest RADseq: an inexpensive method for De novo SNP discovery and genotyping in model and non-model species. PLoS One 7:e37135
22. Hoffberg SL, Kieran TJ, Catchen JM et al (2016) RADcap: sequence capture of dual-digest RADseq libraries with identifiable duplicates and reduced missing data. Mol Ecol Resour 16:1264–1278
23. Catchen JM, Amores A, Hohenlohe P et al (2011) *Stacks*: Building and Genotyping Loci *De Novo* From Short-Read Sequences. G3 (Bethesda) 1:171–182
24. Catchen J, Hohenlohe PA, Bassham S et al (2013) Stacks: an analysis tool set for population genomics. Mol Ecol 22:3124–3140
25. Rochette NC, Rivera-Colón AG, Catchen JM (2019) Stacks 2: analytical methods for paired-end sequencing improve RADseq-based population genomics. Mol Ecol 28:4737–4754
26. Li H (2013) Aligning sequence reads, clone sequences and assembly contigs with BWA-MEM. arXiv:1303.3997 [q-bio.GN]
27. Campbell EO, Brunet BMT, Dupuis JR et al (2018) Would an RRS by any other name sound as RAD? Methods Ecol Evol 9: 1920–1927
28. Chen L, Lu Y, Li W et al (2019) The genomic basis for colonizing the freezing Southern Ocean revealed by Antarctic toothfish and Patagonian robalo genomes. GigaScience 8(4):1–16
29. Baird NA, Etter PD, Atwood TS et al (2008) Rapid SNP discovery and genetic mapping using sequenced RAD markers. PLoS One 3: e3376
30. Etter PD, Bassham S, Hohenlohe PA et al (2012) SNP discovery and genotyping for evolutionary genetics using RAD sequencing. In: Orgogozo V, Rockman MV (eds) Molecular methods for evolutionary genetics. Humana Press, Totowa, NJ, pp 157–178
31. Herrera S, Reyes-Herrera PH, Shank TM (2015) Predicting RAD-seq marker numbers across the eukaryotic tree of life. Genome Biol Evol 7:3207–3225
32. Davey JW, Cezard T, Fuentes-Utrilla P et al (2013) Special features of RAD sequencing data: implications for genotyping. Mol Ecol 22:3151–3164
33. Lepais O, Weir JT (2014) SimRAD: an R package for simulation-based prediction of the number of loci expected in RADseq and similar genotyping by sequencing approaches. Mol Ecol Resour 14:1314–1321
34. Mora-Márquez F, García-Olivares V, Emerson BC et al (2017) DDRADSEQTOOLS: a software package for in silico simulation and testing of double-digest RADseq experiments. Mol Ecol Resour 17:230–246
35. Timm H, Weigand H, Weiss M et al (2018) DDRAGE: a data set generator to evaluate ddRADseq analysis software. Mol Ecol Resour 18:681–690

36. Rivera-Colón AG, Rochette NC, Catchen JM (2021) Simulation with RADinitio improves RADseq experimental design and sheds light on sources of missing data. Mol Ecol Resour 21:363–378
37. R Core Team (2021) R: a language and environment for statistical computing, R Foundation for Statistical Computing, Vienna, Austria
38. Paris JR, Stevens JR, Catchen JM (2017) Lost in parameter space: a road map for STACKS. Methods Ecol Evol 8:1360–1373
39. Rochette NC, Catchen JM (2017) Deriving genotypes from RAD-seq short-read data using stacks. Nat Protoc 12:2640–2659
40. McCartney-Melstad E, Gidiş M, Shaffer HB (2019) An empirical pipeline for choosing the optimal clustering threshold in RADseq studies. Mol Ecol Resour 19:1195–1204
41. Heller R, Nursyifa C, Garcia-Erill G et al (2021) A reference-free approach to analyse RADseq data using standard next generation sequencing toolkits. Mol Ecol Resour 21: 1085–1097
42. Bassham S, Catchen J, Lescak E et al (2018) Repeated selection of alternatively adapted haplotypes creates sweeping genomic remodeling in stickleback. Genetics 209:921–939
43. Nelson TC, Cresko WA (2018) Ancient genomic variation underlies repeated ecological adaptation in young stickleback populations. Evolution Lett 2:9–21
44. Li H, Handsaker B, Wysoker A et al (2009) The sequence alignment/map format and SAMtools. Bioinformatics 25:2078–2079
45. Pritchard JK, Stephens M, Donnelly P (2000) Inference of population structure using multilocus genotype data. Genetics 155:945
46. Leitwein M, Duranton M, Rougemont Q et al (2020) Using haplotype information for conservation genomics. Trends Ecol Evol 35: 245–258
47. Bootsma ML, Miller L, Sass GG et al (2021) The ghosts of propagation past: haplotype information clarifies the relative influence of stocking history and phylogeographic processes on contemporary population structure of walleye (*Sander vitreus*). Evol Appl 14: 1124–1144
48. Malinsky M, Trucchi E, Lawson DJ et al (2018) RADpainter and fineRADstructure: population inference from RADseq data. Mol Biol Evol 35:1284–1290
49. Allendorf FW, Hohenlohe PA, Luikart G (2010) Genomics and the future of conservation genetics. Nat Rev Genet 11:697–709
50. Cruickshank TE, Hahn MW (2014) Reanalysis suggests that genomic islands of speciation are due to reduced diversity, not reduced gene flow. Mol Ecol 23:3133–3157
51. Irwin DE, Milá B, Toews DPL et al (2018) A comparison of genomic islands of differentiation across three young avian species pairs. Mol Ecol 27:4839–4855
52. Browning SR, Browning BL (2007) Rapid and accurate haplotype phasing and missing-data inference for whole-genome association studies by use of localized haplotype clustering. Am J Hum Genet 81:1084–1097
53. Browning BL, Zhou Y, Browning SR (2018) A one-penny imputed genome from next-generation reference panels. Am J Hum Genet 103:338–348
54. Danecek P, Auton A, Abecasis G et al (2011) The variant call format and VCFtools. Bioinformatics 27:2156–2158
55. Gautier M, Klassmann A, Vitalis R (2017) REHH 2.0: a reimplementation of the R package REHH to detect positive selection from haplotype structure. Mol Ecol Resour 17:78–90
56. Cerca J, Maurstad MF, Rochette NC et al (2021) Removing the bad apples: a simple bioinformatic method to improve loci-recovery in de novo RADseq data for non-model organisms. Methods Ecol Evol 12:805–817

Chapter 8

A Metabarcoding Protocol to Analyze Coastal Planktic Communities Collected by Desalination Plant Filters: From Sampling to Bioinformatic Exploratory Analyses

Matteo Cecchetto, Andrea Di Cesare, Ester Eckert, Isabella Moro, Diego Fontaneto, and Stefano Schiaparelli

Abstract

The temporal dynamics of coastal planktic communities can be disclosed through DNA metabarcoding on the filters of reverse-osmosis desalination plants. Here, we describe the steps that are necessary to process the filters in order to create the subsamples used for DNA extraction and the bioinformatic pipeline to perform the first exploratory analyses on this kind of dataset.

Key words Coastal plankton, Monitoring, Desalination plant, DNA metabarcoding, Bioinformatic protocol

1 Introduction

One of the most important requirements for a thorough biodiversity monitoring program is the high sampling frequency, especially when investigating planktic communities [1]. This condition is particularly difficult to be achieved, especially considering the lengthy processing time and conspicuous resources that are usually required to analyze high amounts of data. High-throughput sequencing (HTS) technologies partially overcame these difficulties, by providing a fast, reproducible and cost effective methodology that increased both the amount of processed samples and the taxonomic spectrum of organisms that can be studied in respect to more traditional methods such as morphology-based ones [2]. Nonetheless, biodiversity monitoring research is still particularly difficult to be performed in remote and isolated regions such as Antarctica, mainly due to the harsh weather conditions and the high cost of personnel, thus limiting the extent of the sampling activity that can be carried out. A possible solution to overcome this

Cinzia Verde and Daniela Giordano (eds.), *Marine Genomics: Methods and Protocols*,
Methods in Molecular Biology, vol. 2498, https://doi.org/10.1007/978-1-0716-2313-8_8,

issue has been identified in the application of HTS technologies to the filters of reverse-osmosis desalination plants (e.g., [3–5]). These facilities provide clean, potable water to the research bases operating in Antarctica and start with a "prefiltration" process, usually employing disposable cartridge and/or bag filters that can be replaced daily and that are used to remove debris and organisms from the seawater. These steps are especially important during phytoplankton bloom events [5], when the whole system risks to be clogged. These disposable filters provide the high sampling frequency needed by many biodiversity monitoring programs and hence can be used to this aim. In addition, the reduced amount of time dedicated to the sampling process and the inexpensive technology make this methodology particularly useful in remote locations such as Antarctica.

Here, we describe the sampling strategy that can be adopted to process the entire disposable bag (25 μm mesh size) and cartridge (5 μ mesh size) filters of a desalination plant in order to obtain the subsamples necessary for the DNA extraction with a reproducible technique.

Then, we propose a concise bioinformatic pipeline that can be adopted to analyze a metabarcoding dataset of this kind and perform the basic, exploratory analyses typical of this methodology. The description of DNA amplification and sequencing is omitted from the protocol, as we believe that many lab protocols are already available in the literature and, in general, these steps are specific to the research questions and targeted organisms.

2 Materials

2.1 Subsampling Strategy

1. Nitrile gloves.
2. Scalpel with disposable blades.
3. Stainless-steel cork borer (approximately 26 mm in diameter).
4. Stainless-steel tweezers (or forceps).
5. Stainless-steel scissors.
6. Stainless-steel tweezers with smooth tips.
7. Autoclavable mortar and pestle.
8. Sterilized quartz sand.
9. Sterile plastic bag.
10. 50 mL falcon tubes.

2.2 Bioinformatic Analyses

2.2.1 Unix Environment

1. Linux kernel release "5.4.0-73-generic" and version "#82~18.04.1-Ubuntu SMP Fri Apr 16 15:10:02 UTC 2021".
2. FastQC—v0.11.5 [6].

3. Vsearch—v2.13.3_linux_x86_64 [7].
4. Usearch—v11.0.667_i86linux32 (32-bit version) [8].
5. Cutadapt—v2.3 [9].
6. Mothur—v1.44.1 [10].
7. Rscript—v4.1.0 (but the older version 3.6.3 can be used as well) [11].

2.2.2 R Environment (www.r-project.org)

1. R—v4.1.0 (but the older version 3.6.3 can be used as well) [11].
2. tidyverse—v1.3.0 [12].
3. dplyr—v1.0.6 [13].
4. ggplot2—v3.3.3 [14].
5. DESeq2—v1.26.0 [15].
6. phyloseq—v1.34.0 [16].

3 Methods

3.1 Filters Preparation

This section will describe both the retrieval and preparation for the subsequent DNA extraction of disposable bag and cartridge filters. Subheading 3.1.1 can be applied to both types of filter and is followed by the description of the protocol in case it is applied to bag (Subheadings 3.1.2 and 3.1.3) or cartridge (Subheadings 3.1.4 and 3.1.5) filters.

The first steps of this methodology must be carried out in close collaboration with the logistic personnel responsible for the activities conducted inside the desalination plant, as they will have the proper experience with the characteristics of that specific facility and will often be aware ahead of time about any event and activity that will take place in the desalination plant. This step is crucial in order to obtain fresh samples and avoid possible contaminations due to filters' handling and storage by non-trained people. The methodology here described is well suited for the characteristics of the desalination plant of the Italian Antarctic research station "Mario Zucchelli" and most of the steps here described were performed and refined during the preparation of the dataset reported in Cecchetto et al. [5]. The exchange of information with the logistic personnel should be conducted to acknowledge the technical specifications of the investigated desalination plant (e.g., filter's mesh size, additional prefiltration steps), as those characteristics could influence the results with unexpected outcomes if these aspects are not taken into account. The filters used for the prefiltration steps are usually deployed inside specific casings that only an instructed technician can operate, meaning that most of the activities that will be performed in the desalination plant should necessarily be

Fig. 1 Desalination plant of the Italian Antarctic research station "Mario Zucchelli". (**a**) Filter housings of the 25 μm (green arrow, left) and 5 μm (yellow arrow, right) filters. 25 μm bag filter during the deployment (**b**) and right before the retrieval (**c**). 5 μm cartridge filters during the deployment (**d**) and right before the retrieval (**e**)

conducted together with the logistic personnel. Accession to these facilities is thus restricted and has to be done in close collaboration with the plant logistics. Finally, the sampling activity could be planned in advance [3, 4] or could depend on the characteristics of the treated seawater and thus change frequency over time [5].

3.1.1 Sampling and Metadata Collection

1. Collect the filters by extracting them from the housing with nitrile lab gloves, place them in sterile plastic bags and bring the filters to the lab (*see* **Note 1**). Take note of the date and time of the filter's retrieval and, if needed, take note of the filter's spatial orientation inside its housing by labeling the plastic bag according to the chosen labeling design of the samples (Fig. 1d, e) (*see* **Note 2**).
2. Collect additional data regarding environmental and sampling variables, possibly with an equal or greater frequency in respect to the filter's replacement rate (*see* **Note 3**).

3.1.2 Subsampling of the 25 μm Bag Filters

1. Extract the filter from the plastic bag and place it on a clean table covered with an opened sterile plastic bag.

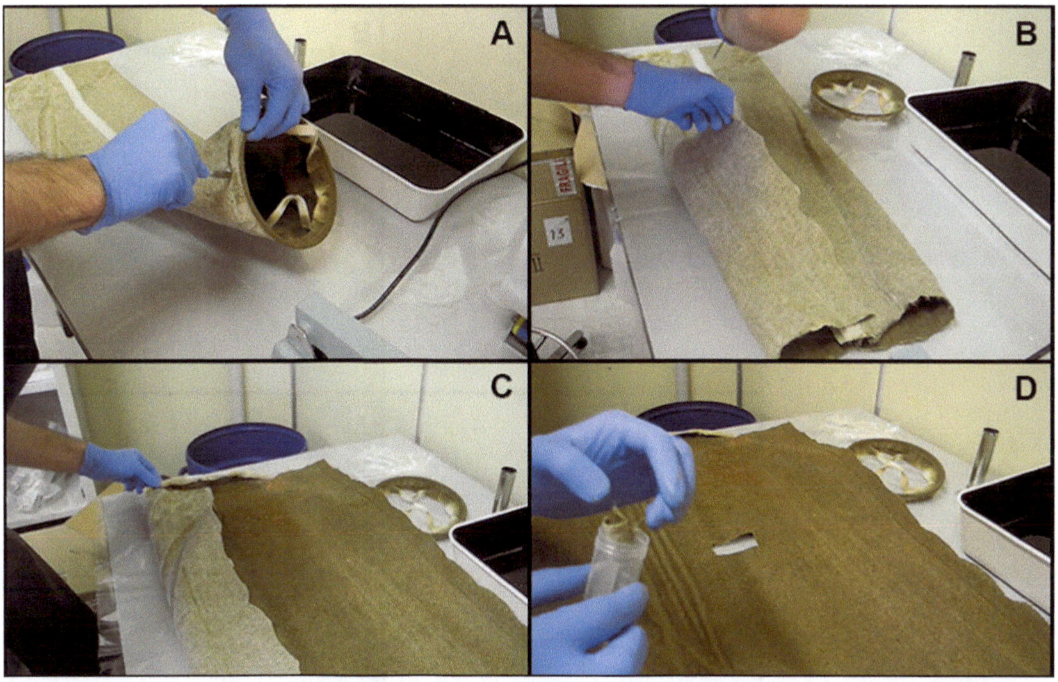

Fig. 2 Subsampling of the 25 μm bag filters. (**a**) removal of the metal ring at the opening of the filter. (**b**) and (**c**) opening of the filter prior to the subsampling. (**d**) subsampling of the filter

2. Using a scalpel with sterilized, disposable blade, cut the bag filter right below the metal ring corresponding to the opening of the filter and dispose the ring (Fig. 2a).
3. Make a "T" cut on one side of the filter, with the longest cut conducted longitudinally and the shortest cut on the opposite side of the opening of the filter (Fig. 2b, c).
4. Open the filter on the table with the inner side now facing upward.
5. Using a scalpel with a new, clean blade, cut three different rectangles of the filter from the center of the filter, one closer to the opening of the filter, one on the opposite side and another one of the center, in order to have three different replicates of that filter. The dimensions of the rectangles are not particularly important, because only a portion (1 cm^2) of that material will be used for the DNA extraction (Fig. 2d).
6. Place each rectangle of the filter in a different 50 mL falcon tube and label it unambiguously (*see* **Note 2**). The falcon tube can be stored at −20 °C until the DNA extraction will take place.

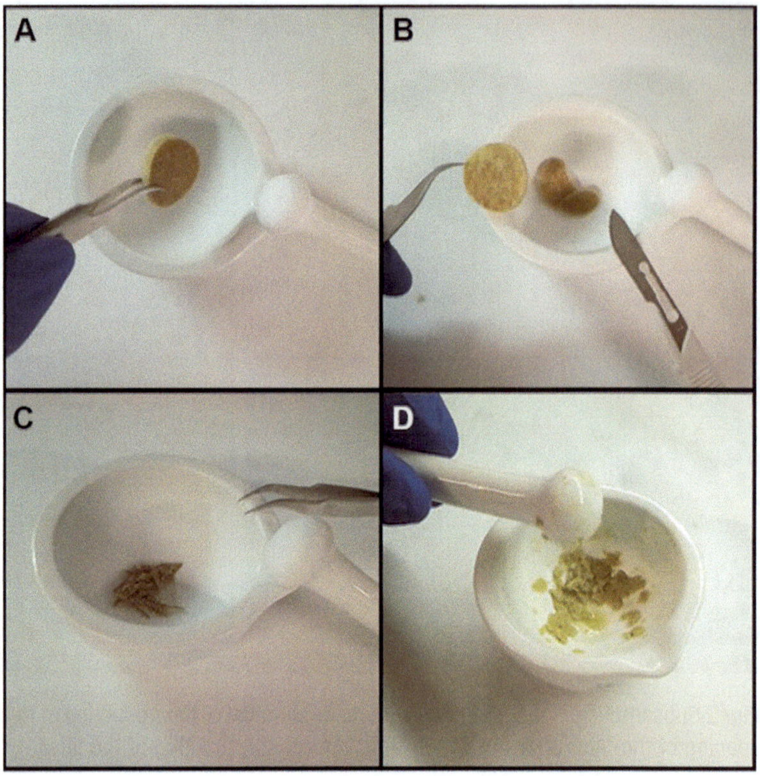

Fig. 3 5 μm processing of the subsamples prior to the DNA extraction. (**a**) and (**b**) separation of the most superficial layers of the filter from the entire subsample. (**c**) the layers cut in small strips, followed by the grinding of the strips using the pestle (**d**)

3.1.3 Preparation of the 25 μm Subsamples for the DNA Extraction

1. Let the falcon tube with the filter rectangle to thaw outside of the freezer for approximately 10 min.
2. Use a pair of sterilized, stainless-steel scissors to cut a 1 cm^2 subsample from the filter rectangle and place it in the mortar.
3. Cut the subsample in thin strips of less than a millimeter of width, eventually continue to cut the stripes to create pieces with approximately the same size (Fig. 3c).
4. Add the appropriate amount of sterilized quartz sand to the subsample in the mortar and start to grind the stripes with the pestle until they lose the original toughness (Fig. 3d) (*see* **Note 4**).
5. The crushed stripes can now be transferred into a centrifuge tube for the DNA extraction using a pair of stainless-steel tweezers with smooth ends, which allow for an easier transfer in the centrifuge tube of the filter's stripes (*see* **Note 5**).

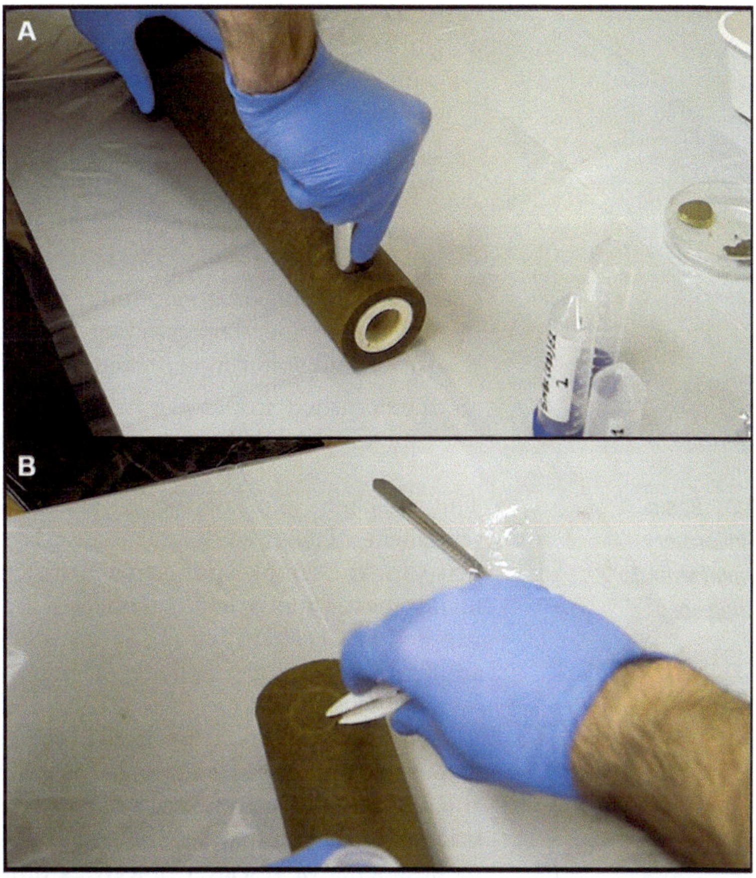

Fig. 4 Subsampling of the 5 μm cartridge filters. (**a**) Cut on the filter's surface using a cork borer. (**b**) Extraction of the subsample

3.1.4 Subsampling of the 5 μm Cartridge Filters

1. Extract the filter from the plastic bag and place it on a clean table covered with an opened sterile plastic bag.
2. Using three autoclave sterilized cork borers, carve a circular cut half a centimeter deep on the two opposite ends of the filter and in the center, in order to have three different replicates of that filter (Fig. 4a) (*see* **Note 6**).
3. Using a pair of heat sterilized tweezers, grab the circular subsample, extract it from the filter and place it in a 50 mL falcon tube (Fig. 4b), remember to keep track of the filter's spatial orientation in the housing (*see* **Note 2**).
4. The falcon tube can now be stored at −20 °C until the DNA extraction will take place.

3.1.5 Preparation of the 5 μm Subsamples for the DNA Extraction

1. Let the falcon tube with the filter rectangle to thaw outside of the freezer for approximately 10 min.
2. Using a pair of heat sterilized tweezers, gently insert one of the sharp ends of the tweezers right below the first upper layers of

the subsample, grab the layers and peel them off from the subsample (Fig. 3a, b). If necessary, repeat the process on the same subsample, until reaching the appropriate amount of material for the DNA extraction (*see* **Note 7**).

3. Place the layers in the mortar and cut the layers with a pair of heat-sterilized scissors in small stripes of less than a millimeter in length. Continue to cut the stripes until they all have approximately the same size (Fig. 3c).
4. Add the appropriate amount of sterilized quartz sand to the subsample and start grinding the stripes with the pestle until they result softer in consistency (Fig. 3d) (*see* **Note 4**).
5. The crushed stripes can now be transferred into a centrifuge tube for the DNA extraction (*see* **Note 5**).

3.2 Basic Exploratory Bioinformatic Analyses

This pipeline was specifically tested for the Illumina MiSeq sequencing dataset of the eukaryotic 18S and bacterial 16S rRNA genes for the samples published in Cecchetto et al. [5]. The command specifications and parameters are set to those that were adopted in Cecchetto et al. [5] for the 18S rRNA sequences and are here showed for simplicity, meaning that they should not necessarily be used.

This pipeline is here reported in the following two subheadings and each command should be executed in the order here reported and provided by the numeration in each subheading. The specifications of each command might be changed to accommodate the requirements of research objectives different from those investigated in Cecchetto et al. [5], but the overall structure should be maintained to perform the analyses here reported. The scripts and pipelines here described are available at https://github.com/MatteoCe/filter_pipeline, and each command separated by a newline character, excluding those that are specifically meant to be run as entire scripts, here reported starting with the "she-bang" (e.g., #!/bin/bash) and mentioned in the comment to each section, should be executed one at a time. The first part consists in commands and small scripts that can be executed in a Unix environment with the installed programs listed in Subheading 2.2.1 and will process the original Illumina sequences fastq files to create the abundance table and taxonomy file. The second part will be executed on these files and will perform the first exploratory analyses of this kind of samples (e.g., alpha diversity indices, principal component analysis).

3.2.1 Sequence Analyses of Metabarcoding Samples (Unix and R Commands)

The pipeline starts after the demultiplexing step, which is not here reported as many sequence services providers usually perform it before sending the results to the client. Nonetheless, if a demultiplexing process is required by the reader, we suggest to adopt the first steps of the tutorial pipeline reported by the obitools programs suite authors [17] (and specifically the command ngsfilter, https://pythonhosted.org/OBITools/wolves.html). The reader should then rename each file according to the name chosen with the labelling strategy followed by and underscore and R1 or R2, for the forward and reverse reads respectively.

1. The following command will create the directory structure for the entire analyses using the command "mkdir". The project directory (filter_proj) will contain four different directories, one for the analyses listed in this section (analyses_preASV), another for the analyses that are listed in the next session (analyses_postASV), another that will store the scripts executed in this section (scripts) and the last that will store the demultiplexed and renamed fastq files (raw). The directory "analyses_preASV" will also contain a variety of subdirectories, one for each particular analyses. Next, use the command "cd" to move to the "raw" directory.

```
mkdir -p filter_proj/{scripts,raw,analyses_postASV,
analyses_preASV/{fastqc,temp,log,cutadapt,merging,fasta/dis-
carded,fastq_prefilter,fastq_prefilter/dataset_prefilter/
quality_log,filterlen/discarded,dereplication}}
```

```
cd filter_proj/raw
```

2. A simple "for" loop will create a "filenames" file with all the sample's labels. This loop will iterate the command "echo" on each file in the "raw" directory corresponding to the forward read, removing the characters in the filenames indicating the orientation of the read (i.e., R1) and the file extension (i.e., .fastq), and print the results (i.e., only the chosen label of each sample) to a new file. This file will be used by the next "for" loop to iterate the analyses on each sample.

```
for i in *_R1.fastq; do echo "${i%_R1.fastq}" >> ../filenames;
done
```

```
cd ../analyses_preASV
```

3. The next lines of code can be copied in a new file, for example "loop1.sh", moved in the directory "scripts" and executed with the command "bash ../scripts/loop1.sh". The loop will first execute the fastqc program and store the resulting files in a new directory "fastqc/sample_label". The command will be executed for both the forward and reverse read. Next, the vsearch program will perform the merging between the two reads and store the resulting fastq files in the "merging" directory. The cutadapt program will be used to remove the primers

from the merged reads (the primers here showed are those reported in Hugerth et al. 2014 [18] and used in Cecchetto et al. 2021 [5], with the reverse complement of the reverse primer). Finally, each read in the fastq files will be renamed with the respective sample name followed by a period character and a ticker (1, 2, 3, etc.), using the vsearch command "--fastq_filter". The standard output and standard error of the entire "for" loop are written to the same log file. These commands will be iterated on each sample identified by the "filenames" file.

```
#!/bin/bash

set -ue pipefail

(

for i in $(cat ../filenames)

do

mkdir fastqc/${i}

echo "FastQC on ${i}"
fastqc ../raw/${i}_R1.fastq ../raw/${i}_R2.fastq -outdir
fastqc/${i}/ -d temp/ -q

mkdir merging/${i}
mkdir merging/${i}/not_merged

echo "Vsearch mergepairs ${i}"
vsearch --fastq_mergepairs ../raw/${i}_R1.fastq --reverse ../
raw/${i}_R2.fastq --fastq_maxns 0 --fastq_maxdiffs 3 --fas-
tqout_notmerged_fwd merging/${i}/not_merged/${i}_R1_not-
merged.fastq --fastqout_notmerged_rev merging/${i}/
not_merged/${i}_R2_notmerged.fastq --fastqout merging/${i}/
${i}_merged.fastq --log merging/${i}/${i}_merge.log

mkdir cutadapt/${i}
mkdir cutadapt/${i}/untrimmed

echo "Cutadapt trimming adapter ${i}"
cutadapt -a ^CGGTAAYTCCAGCTCYV...AYTTRAAGDAATTGACGG$ -o cuta-
dapt/${i}/${i}_merged_adapt.fastq merging/${i}/${i}_merged.
fastq --untrimmed-output cutadapt/${i}/untrimmed/${i}_mer-
ge_noadapt.fastq > cutadapt/${i}/${i}_report.txt
```

```
echo "Vsearch relabel with ${i}.#"
vsearch --fastq_filter cutadapt/${i}/${i}_merged_adapt.fastq
--relabel ${i}. --fastqout fastq_prefilter/${i}.fastq

done

) 2>&1 | tee log/output_loop.log
```

4. Another script, reported in https://github.com/MatteoCe/filter_pipeline as "loop2.sh" and run with the command "bash ../scripts/loop2.sh", can be used to inspect the results obtained from the quality control conducted using fastqc (i.e., "per base sequence quality") and the statistics on the merging and primer removal operations. This script will simply extract the information that were written to the log file "output_loop.log" based on a combination of string matches (*see* **Note 8**). The program xdg-open is used to open the png images of the "per base sequence quality" analyses conducted by fastqc for both the forward and reverse reads. The number of total read pairs detected by vsearch and the proportion of merged and not merged sequences is shown, followed by the three most recurring issues encountered by vsearch during the merging process. Finally, the proportion of reads for which cutadapt could detect the presence of the primer sequences is reported.

```
#!/bin/bash

set -ue pipefail

for i in $(cat ../filenames)

do

unzip -qq fastqc/${i}/${i}_R1_fastqc.zip -d fastqc/${i}/
unzip -qq fastqc/${i}/${i}_R2_fastqc.zip -d fastqc/${i}/

mv fastqc/${i}/${i}_R1_fastqc/Images/per_base_quality.png
fastqc/${i}/${i}_R1_fastqc/Images/per_base_quality_${i}_R1.
png
mv fastqc/${i}/${i}_R2_fastqc/Images/per_base_quality.png
fastqc/${i}/${i}_R2_fastqc/Images/per_base_quality_${i}_R2.
png

xdg-open fastqc/${i}/${i}_R1_fastqc/Images/per_base_quality_
${i}_R1.png
xdg-open fastqc/${i}/${i}_R2_fastqc/Images/per_base_quality_
```

```
${i}_R2.png

echo 'Stats for '${i}''

echo
sed -n '/^Vsearch mergepairs '${i}'$/,/^Cutadapt trimming
adapter '${i}'$/p' log/output_loop.log | sed -n '/^Merging
reads/,/^$/p' | tail -n +2

echo "The three most recurring issues during merging with
number of not merged seqs"

echo
sed -n '/^Vsearch mergepairs '${i}'$/,/^Cutadapt trimming
adapter '${i}'$/p' log/output_loop.log | sed -n '/^Pairs that
failed/,/^$/p' | tail -n +2 | sort -r | head -n 3

echo "Cutadapt stats"

echo
sed -n '/^Total reads processed:/,/^$/p' cutadapt/${i}/${i}
_report.txt

read -p ""

rm -rf fastqc/${i}/${i}_R[1,2]_fastqc

done
```

5. The fastq files are then concatenated using cat, as the information on the sample to which each read belongs is stored in each sequence's header (*see* **step 3** with vsearch command "--fastq_filter"). The command "--fastq_stats" of vsearch is run and the resulting log file is formatted using a combination of unix programs (i.e., sed, cut, grep, tail, and sort) to retain only the information on the "effect of expected error and length filtering" and on the length distribution of the entire dataset. This information will be stored in two new files, "stats_maxee" and "stats_length".

```
cat fastq_prefilter/*.fastq > fastq_prefilter/dataset_prefil-
ter/dataset_prefilter.fastq

vsearch --fastq_stats fastq_prefilter/dataset_prefilter/data-
set_prefilter.fastq --log fastq_prefilter/dataset_prefilter/
quality_log/stats_datasetfastq.txt
```

```
cat fastq_prefilter/dataset_prefilter/quality_log/stats_data-
setfastq.txt | sed -n '/L   1.0/,/^$/p' | sed -e 's/ \+/\t/g'
-e 's/^\t//' -e 's/%//g' | cut -f 1,6-9 | grep -v '-' | tail -n
+2 | sort -V > fastq_prefilter/dataset_prefilter/quality_log/
stats_maxee

cat fastq_prefilter/dataset_prefilter/quality_log/stats_data-
setfastq.txt | sed -n '/^Read length distribution$/,/^$/p' |
sed -e 's/ \+/\t/g' -e 's/>=//' -e 's/^\t//' -e 's/%//g' | tail
-n +3 | grep -v '-' | grep -v '^$' | sort -V > fastq_prefilter/
dataset_prefilter/quality_log/stats_length
```

6. The following lines of code are supposed to be copied in an R script file, reported in https://github.com/MatteoCe/filter_pipeline as "stats.R", and run using the program Rscript ("Rscript ../scripts/stats.R"). This script will use the package collection "tidyverse" and more specifically the packages ggplot2 [14] and dplyr [13] to aggregate the information obtained earlier, plot it together and save it. This plot will show the length distribution of the dataset using bars and the number of reads that would be retained by the "--fastq_filter" command of vsearch if the reads were truncated at each base position and filtered to have a maximum expected error of 1.0, 0.5, 0.25, and 0.1, respectively. This will help to "quantify" the effects of the "maxee" parameter that will be set in the following "--fastq_filter" command, to be run using vsearch (*see* **Note 9**).

```
#!/usr/bin/Rscript

packages <- c("tidyverse")
install.packages(setdiff(packages, rownames(installed.
packages())))
library(tidyverse)

tab <- read.table(file="fastq_prefilter/dataset_prefilter/
quality_log/stats_maxee")
tab_length <- read.table(file="fastq_prefilter/dataset_pre-
filter/quality_log/stats_length")

ma <- max(tab_length$V2)

ma_x <- max(tab_length$V1)

si = ma/100

joined <- dplyr::full_join(tab, tab_length, by="V1")
```

```
p <- ggplot(joined, aes(x=V1)) +
    scale_x_continuous(breaks=seq(0,ma_x,50)) +
     scale_y_continuous(name = "Number of reads", sec.axis =
sec_axis( trans=~./si, name="Percentage of sequences kept")) +
     geom_bar(stat='identity', aes(y=V2.y, colour="Number of
reads")) +
    geom_line(aes(y=V2.x*si, colour="1")) +
    geom_line(aes(y=V3.x*si, colour="0.5")) +
    geom_line(aes(y=V4.x*si, colour="0.25")) +
    geom_line(aes(y=V5*si, colour="0.1"))

ggsave("fastq_prefilter/dataset_prefilter/quality_log/maxee_-
plot.png", plot = p,device = png(), path = NULL,scale = 1, width
= 10, height = 7, dpi = 300, limitsize = TRUE)
dev.off()
```

7. Open the saved plot with xdg-open and inspect the results (*see* **Note 10**).

```
xdg-open fastq_prefilter/dataset_prefilter/quality_log/max-
ee_plot.png
```

8. Filter the sequences using "--fastq_filter" of vsearch based on the expected error (parameter "--fastq_maxee") and output the result to a file in fasta format.

```
vsearch --fastq_filter fastq_prefilter/dataset_prefilter/da-
taset_prefilter.fastq --fastq_maxee 1 --fastaout fasta/data-
set.fasta --fastqout_discarded fasta/discarded/discarded.
fastq
```

9. Run the program mothur to obtain more information on the length distribution of the sequences in the dataset (median, quantiles, etc.) and remove the sequences with homopolymers of 8 or more bases in length.

```
mothur

summary.seqs(fasta=fasta/dataset.fasta)

screen.seqs(fasta=fasta/dataset.fasta, maxhomop=8)

quit()
```

10. The information on the length distribution among the sequences of the dataset is then used to perform another

filtering step that will remove those sequences that are shorter or longer than the specified range.

```
vsearch --fastx_filter fasta/dataset.good.fasta --fastq_max-
len 558 --fastq_minlen 480 --fastaout filterlen/dataset_fil-
terlen.fasta --log filterlen/log_filter.txt --
fastaout_discarded filterlen/discarded/dataset_filtlen_dis-
carded.fasta
```

11. At this point, the dereplication can be performed using the command "--derep_fulllength" of vsearch. The sequences are also relabeled with the prefix "Uniq" and the abundance information is retained in the header ("--sizeout" option), which will be used later by the usearch program to remove the singletons (i.e., sequences occurring only once in the entire dataset).

```
vsearch --derep_fulllength filterlen/dataset_filterlen.fasta
--sizeout --relabel Uniq --log dereplication/log_derep.txt --
output dereplication/unique_seqs.fa
```

12. The program usearch is run using the command "-unoise3" [19], which removes sequencing and PCR point errors, as well as chimeras. The command also removes singletons with the "–minsize" parameter set to 2, creating the Zero-radius Operational Taxonomic Units fasta file (ZOTU.fa).

```
usearch -unoise3 dereplication/unique_seqs.fa -zotus ZOTU.fa
-minsize 2 -log unoise3_report.txt
```

13. Download the mothur-formatted version 4.13 of the 18S pr2 database [20] in a new directory using the wget command and unzip the downloaded files using gunzip.

```
mkdir ../database

wget https://github.com/pr2database/pr2database/releases/
download/v4.13.0/pr2_version_4.13.0_18S_mothur.fasta.gz-O../
database/pr2_4.13.0_mothur.fa.gz

wget https://github.com/pr2database/pr2database/releases/
download/v4.13.0/pr2_version_4.13.0_18S_mothur.tax.gz-O../da-
tabase/pr2_4.13.0_mothur.tax.gz

gunzip -k ../database/pr2_4.13.0_mothur.fa.gz
gunzip -k ../database/pr2_4.13.0_mothur.tax.gz
```

14. Then mothur is used for the taxonomic assignment of the ZOTU fasta file. In this case the "classify.seqs" command will be adopted using the mothur-formatted database downloaded in the previous step (*see* **Note 11**).

```
mothur

classify.seqs(fasta=ZOTU.fa, template=../database/pr2_4.13.0_mothur.fa, taxonomy=../database/pr2_4.13.0_mothur.tax, probs=F)

quit()
```

15. The "--usearch_global" command of vsearch will be used to map all the sequences in the ZOTU fasta file to the entire dataset obtained after the length filtering **step 10** and prior to the dereplication process. This will provide the "count file", with the abundance value of each ZOTU sequence for each sample. The pairwise identity is set to 1, in order to align only the exactly identical sequences.

```
vsearch --usearch_global filterlen/dataset_filterlen.fasta --db ZOTU.fa --id 1.0 --otutabout ZOTU.counts --log usearchglobal_report.txt
```

16. This command will modify the count file to remove the string "#OTU ID" at the beginning of the first line, which would cause issues in R. The taxonomy file is modified by adding a line with the column names (level1, level2, level3, etc.) and replacing the semicolon with a tab character. Finally, the taxonomic assignment file (ZOTU.tax) and the count table (ZOTU.counts) will be moved to the "analyses_postASV" directory for the following analyses (Subheading 3.2.2).

```
sed -i 's/^#OTU ID//' ZOTU.counts

cat <(echo -e ";level1;level2;level3;level4;level5;level6;level7;level8") ZOTU.0_mothur.wang.taxonomy | tr ';' '\t' | sed -e 's/\t$//' > ZOTU.tax

mv ZOTU.counts ZOTU.tax ../analyses_postASV/

cd ../analyses_postASV
```

3.2.2 Exploratory Analyses (R Commands)

1. Open R in interactive mode and load the packages "DESeq2", "phyloseq," and "tidyverse." If they have never been installed previously, use the command install.packages("name_of_the_-package", dep=TRUE) to install them on your computer. Depending on the R version used, Bioconductor [21] may be required to install the packages "phyloseq" and "DESeq2."

```
library(DESeq2)

library(phyloseq)

library(tidyverse)
```

2. Read the count table "ZOTU.counts" as a data frame.

```
count_tab <- read.table("ZOTU.counts", header=T, sep="\t",
row.names=1, check.names=F)
```

3. Read the taxonomic assignment file "ZOTU.tax" as a data frame and turn it into a matrix, as it will be required later by the phyloseq package.

```
tax_tab <- read.table("ZOTU.tax", header=T, sep="\t", row.
names=1, check.names=F)

tax_tab_mat <- as.matrix(tax_tab, mode='character')
```

4. Read the sample information file as a data frame. This should be a "tab separated values" file with all the available information on the filter retrieval time, filtering hours, volume filtered, and so on (Table 1). The table should report the name of the sample files used in the previous section and an empty field in the top left corner. After reading, as the labeling system adopted is composed of numeric characters only, the column containing the labels assigned to the filters (the "filter_series" column in Table 1) is encoded as a factor.

```
sample_info_tab <- read.table("filter_info.tsv", header=T,
sep="\t", row.names=1, check.names=F)
sample_info_tab$filter_series <- as.factor(sample_info_tab
$filter_series)
```

5. The count table is encoded as a "DESeqDataSet" class object. The "design" parameter, which is used for the differential abundance analysis and here not performed, is here disabled using the formula "~ 1". A variance stabilizing transformation is applied to the count table to account for differences in the number of sequences, without prior merging of all the

Table 1
Example of sample info table to be read using the R software. The labeling system adopted here, and shown in the first column on the left, identifies the filter's typology (25 μm filter), the number of the filter in temporal order (01, 02, 03, etc.) and the filter's replicate (i.e., 1, 2, and 3) with "1" indicating the top replicate and "3" indicating the bottom one (*see* Note 2)

	filter_series	deployment_time	retrieval_time	filtering_hours	volume	temp
25011	2501	01/18/2019 08:00:00	01/20/2019 08:00:00	64	X	X
25012	2501	01/18/2019 08:00:00	01/20/2019 08:00:00	64	X	X
25013	2501	01/18/2019 08:00:00	01/20/2019 08:00:00	64	X	X
25021	2502	01/20/2019 08:00:00	01/21/2019 18:00:00	34	X	X
25022	2502	01/20/2019 08:00:00	01/21/2019 18:00:00	34	X	X
25023	2502	01/20/2019 08:00:00	01/21/2019 18:00:00	34	X	X
25031	2503	01/21/2019 18:00:00	01/22/2019 08:00:00	14	X	X
25032	2503	01/21/2019 18:00:00	01/22/2019 08:00:00	14	X	X
25033	2503	01/21/2019 18:00:00	01/22/2019 08:00:00	14	X	X

replicates. This is used as an alternative to the rarefaction method [22].

```
deseq_counts <- DESeqDataSetFromMatrix(count_tab, colData=-
sample_info_tab, design=~ 1)

dds_vst <- varianceStabilizingTransformation(deseq_counts)
```

6. The "variance stabilized" count table is then converted into a matrix object and the negative values, which in the context of a variance stabilizing transformation indicate that in the original count table those values were more likely to be zero, or in any case negligible, were approximated to 0, as suggested by the phyloseq authors (https://www.bioconductor.org/packages/release/bioc/vignettes/phyloseq/inst/doc/phyloseq-FAQ.html#negative-numbers-in-my-transformed-data-table).

```
dds_vst_count_tab <- assay(dds_vst)
```

```
dds_vst_count_tab[dds_vst_count_tab<0] <- 0.0
```

7. "Construct" the taxonomy table, the count table and the sample information table as phyloseq objects and merge them into a single object.

```
TAX <- tax_table(tax_tab_mat)

OTU <- otu_table(dds_vst_count_tab, taxa_are_rows = TRUE)

sampledata <- sample_data(sample_info_tab)

physeq <- phyloseq(OTU, sampledata, TAX)
```

8. Perform a Principal Coordinates Analysis (PCoA) using the phyloseq package, calculating the distance matrix using the default method "bray" (i.e., Bray–Curtis dissimilarity), and plot the results using "plot_ordination" from phyloseq, which creates a "ggplot" object (*see* **Note 12**). The replicates from each sample are then united by a colored triangular polygon using ggplot2, indicating the dissimilarity between the different filters and the replicate of each filter.

```
ord <- ordinate(physeq, "PCoA")

plot_ord <- plot_ordination(physeq, ord, type="sample")

plot_ord2 <- plot_ord + geom_polygon(aes(fill=filter_series))
+ geom_point(size=3)

plot_ord2
```

9. The following analyses will inspect more closely the differences between each filter (filter_series in Table 1), and for this reason, for each ZOTU, all the sequence abundance values of each replicate of the same filter will be summed up using the "collapseReplicates" command of DESeq2. As the obtained object is of class "DESeqDataSet," the "assay" command will be used to convert it into a matrix object.

```
Colldds <- collapseReplicates(deseq_counts, groupby=sam-
ple_info_tab$filter_series, renameCols=TRUE)

Colldds_tab <- assay(Colldds)
```

10. As the count table has been "collapsed," the sample information file must be changed accordingly. Considering that each

replicate of a single sample should contain the same information, then the table read earlier can be modified by eliminating the duplicate lines using the "unique" command (*see* **Note 13**). The row names of the table are changed with those of the filter labeling system adopter (e.g., "filter_series" in Table 1).

```
sample_info_coll <- unique(sample_info_tab)

rownames(sample_info_coll) <- sample_info_coll$filter_series
```

11. The deployment and retrieval time information is converted to a "POSIXct" class vector and a plot showing the hours of filtering time for each filter through time is created. The command "geom_smooth" will plot the smoothing line, with the grey areas depicting the 95% confidence intervals around the smoothing.

```
sample_info_coll$retrieval_time <- as.POSIXct(sample_info_-
coll$retrieval_time, tryFormats='%m/%d/%Y %H:%M:%S')

sample_info_coll$ deployment_time <- as.POSIXct(sample_info_-
coll$deployment_time, tryFormats='%m/%d/%Y %H:%M:%S')

plot_filterhours <- ggplot(sample_info_coll, aes(x = sam-
ple_info_coll$retrieval_time, y = sample_info_coll$filterin-
g_hours)) + geom_point() + geom_smooth()

plot_filterhours
```

12. Another phyloseq object is created by "constructing" the collapsed count table and the simplified sample information file as "phyloseq" objects and merging those with the taxonomy table constructed earlier (**step** 7).

```
OTU_coll <- otu_table(Colldds_tab, taxa_are_rows = TRUE)

sampledata_coll <- sample_data(sample_info_coll)

physeq2 <- phyloseq(OTU_coll, sampledata_coll, TAX)
```

13. In order to create the "taxa barplot," the abundance counts of each sample are transformed as relative abundance values (i.e., percentage), whereas the abundance counts of each ZOTU are aggregated by the fourth taxonomic level (in this case corresponding approximately to the Order). These commands

will thus create the relative abundance counts as a percentage for each taxon at the fourth taxonomic level.

```
physeq2_rel <- transform_sample_counts(physeq2, function(x) x
/ sum(x))

glom_lev4 <- tax_glom(physeq2_rel, taxrank = 'level4')
```

14. The phyloseq object is then transformed, using the "psmelt" function of phyloseq, into a data frame that will store all the sample information, the relative abundance and the taxonomic identification from the first to the fourth taxonomic level. The fourth taxonomic level column is then encoded as a character vector rather than a factor, as the latter would give issues with the following commands. Instead, the "Sample" column, which is automatically created by the "psmelt" function, will be formatted as a factor.

```
data_glom_lev4 <- psmelt(glom_lev4)

data_glom_lev4$level4 <- as.character(data_glom_lev4$level4)

data_glom_lev4$Sample <- as.factor(data_glom_lev4$Sample)
```

15. Finally, as many of the "fourth level taxa" are represented by ZOTUs with a relatively low sequence counts, all the "fourth level taxa" with a relative abundance smaller than 1% of the total dataset sequence count will be aggregated and indicated as "<1%", to simplify the "taxa barplot" legend. Before that, three commands will search the taxonomic names that will be replaced with "<1%", and thus not indicated by their name in the "taxa barplot," and print them in the terminal.

```
small <- unique(data_glom_lev4$level4[data_glom_lev4$Abun-
dance < 0.01])

large <- unique(data_glom_lev4$level4[data_glom_lev4$Abun-
dance > 0.01])

setdiff(small,large)

data_glom_lev4$level4[data_glom_lev4$Abundance < 0.01] <- "<
1%"
```

16. Create a taxa barplot.

```
plot_taxbar <- ggplot(data_glom_lev4, aes(x=filter_series,
```

```
y=Abundance, fill=level4)) + geom_bar(aes(), stat="identity",
position="stack") + xlab("Samples") + ylab("Relative abun-
dance") +
scale_fill_manual(values = c(rep(c("#000000", "#E69F00",
"#56B4E9", "#009E73","#F0E442", "#0072B2", "#D55E00",
"#CC79A7"), length(unique(data_glom_lev4$level4))))) +
theme(legend.position="bottom") + guides(fill=guide_legend
(nrow=5))

plot_taxbar
```

17. Finally, estimate the alpha diversity indices "Shannon" and "Simpson" and plot the results.

```
plot_rich <- plot_richness(physeq, x="filter_series", col-
or="filter_series", measures=c("Shannon", "Simpson")) +
geom_point(size=3) + geom_boxplot(aes(x=filter_series, y=va-
lue), alpha=0.1, width=0.4) +
theme(axis.text.x = element_text(angle = 45, hjust = 1, vjust =
1, size= 10))

plot_rich
```

4 Notes

1. Different types of filters can be used with different housings, which in consequence may be operated differently. Regarding the desalination plant of "Mario Zucchelli" station, prior to the replacement of the filter, the connection with the upstream components of the plant is closed, and the water remaining in the housing is let to flow downstream, meaning that a certain amount of time may be required before the filters can be sampled. This is especially true for filters in the form of cartridges, which cannot be removed prior to the complete removal of water from the housing. Other filters, such as those in form of bags, are difficult to handle if they still contain a certain amount of water, but can be removed from the housing if necessary. Nonetheless, as mentioned in Subheading 3.1, a thorough exchange of information between logistic personnel and researchers is recommended, in order to have a clear understanding of the plant functioning and timings prior to the devising of the following processing approach. Moreover, the filters are usually too big to store before the subsampling, meaning that this operation should be performed inside the desalination plant or at least in a location close to it. This would also reduce the chances of contamination and allow a

proper disposal of the remaining parts of the filter, which would be in any case conducted by the logistic personnel of the desalination plant.

2. Due to the shape of some filters, such as those in form of cartridges, it is impossible to discern the spatial orientation they had inside their housing, if they are placed vertically inside it (Fig. 1d, e). A proper identification of the subsamples extracted from the filters through a clear and useful labeling should include also this information, using a concise definition of the filter's subsample replicate (e.g., "T" for "top" and "B" for "bottom"). This system would eventually allow for a proper investigation of possible "zonation patterns" in the water column inside the filter housing.
3. A multitude of environmental variables can be collected alongside the filters. This will allow for the development of more complex studies investigating the environmental drivers of planktic dynamics. It is highly recommended to collect also a variety of "sampling" variables if possible (e.g., volume of filtered water, temperature of the water during the filtration) as they can be used for the quantification of the collected samples or the detection of dynamics related to the filtration process and the desalination plant (e.g., biofilm formation). One of the most important sampling variables is the volume of filtered seawater, which is recorded by the desalination plant's flowmeter.
4. The stripes might release a certain amount of water, which can be collected with the pestle at the bottom of the mortar and then, always with the aid of the pestle, can be easily poured into the tube for the DNA extraction.
5. The toughness of the filter's matrix suggests the adoption of a stronger mechanic disruption during the DNA extraction process. In this case, a prolonged bead beating disruption step, and/or a higher temperature during the cellular lysis may help considerably.
6. The cork borer should carve a cut deep enough allowing the tweezers to be inserted gently on the side without damaging excessively the surface layers of the filter, which will later be used for the DNA extraction.
7. Going from the outer to the inner layers of the filter, the material that can be found in the matrix sharply decreases, and in consequence the DNA that can be extracted. This means that the majority of the DNA that can be extracted will be obtained from the first layers, which may break in uneven pieces, smaller than the entire width of the subsample. If that happens, be sure that the subsample has thawed enough and

repeat until reaching the appropriate amount of material for the DNA extraction. Be sure to sample the most superficial layers.

8. The output_loop.log file is created by aggregating the log messages that are generated by each program during the execution of the script. Differences in this kind of output might be obtained with different versions of each program, for this reason it is important to use the same versions indicated in Subheading 2.2.
9. Generally, a maximum expected error (--maxee parameter) of 1 can be considered sufficient [23]. Eventually, if too many sequences are removed by that command, a truncation step can be performed earlier using the parameter "--fastq_trunclen" to remove the region of the sequences (starting from the 5′ end) with the highest number of expected errors. Keep in mind that this command will also remove all the sequences that do not reach that length. You can use "--fastq_trunclen_keep" to avoid that condition.
10. The obtained graph will show the number of sequences that will be retained by the "--fastq_filter" command if the reads were truncated at length X and filtered to have a maximum expected error of 1.0, 0.5, 0.25 or 0.1. The progress of each line is influenced by both the quality of the sequences at each base position and the length distribution of the sequences present in the dataset. In case of merged paired reads, the progress of each line could be characterized by a slow descent followed by an inversion of the trend at the start of the alignment region. A drop in the number of sequences retained will appear at the base pair position corresponding to the length distribution, as a result of the "--fastq_trunclen" command (*see* **Note 9**). This graph will not indicate the exact number of sequences that will be removed by the next command, but will give an indication of the overall quality of the dataset and length distribution.
11. In this case the command is set to the default parameters, which means that the classification is performed using the "wang method" [24] and the cutoff value set to 80. This was done in order to provide an easier classification, by increasing the taxonomic assignment success and, for this reason, the bootstrap values were suppressed (probs=F). If needed, the reader should modify the parameters according to the type of project and target organisms investigated, by adopting a more stringent classification that would increase the accuracy in the taxonomic assignment.
12. Different ordination methods can be applied to this kind of dataset. It is therefore suggested to perform multiple analyses in parallel by adopting different ordination and distances

methods and inspect the results. In this example the command performs a PCoA, but the package can adopt other methods.

13. In case of other variables, whose values differ between replicates of each filter are included, a different approach must be adopted. This kind of information can be, for example, transformed using the average value or summed up, based on the type of information.

References

1. Moreira D, López-García P (2019) Time series are critical to understand microbial plankton diversity and ecology. Mol Ecol 28(5): 920–922. https://doi.org/10.1111/mec.15015
2. Taberlet P, Bonin A, Zinger L et al (2018) Environmental DNA: for biodiversity research and monitoring. Oxford University Press, Oxford. https://doi.org/10.1093/oso/9780198767220.001.0001
3. Sewell MA, Jury JA (2009) Desalination plants as plankton sampling devices in temporal studies: proof-of-concept and suggestions for the future. Limnol Oceanogr Methods 7: 363–370. https://doi.org/10.4319/lom.2009.7.363
4. Sewell MA, Jury JA (2011) Seasonal patterns in diversity and abundance of the high Antarctic meroplankton: plankton sampling using a Ross Sea desalination plant. Limnol Oceanogr 56: 1667–1681. https://doi.org/10.4319/lo.2011.56.5.1667
5. Cecchetto M, Di Cesare A, Eckert E et al (2021) Antarctic coastal nanoplankton dynamics revealed by metabarcoding of desalination plant filters: detection of short-term events and implications for routine monitoring. Sci Total Environ 757:143809. https://doi.org/10.1016/j.scitotenv.2020.143809
6. FastQC: A quality control tool for high throughput sequence data. Available at https://www.bioinformatics.babraham.ac.uk/projects/fastqc/Accessed 20 May 2021
7. Rognes T, Flouri T, Nichols B et al (2016) VSEARCH: a versatile open source tool for metagenomics. PeerJ 4:e2584. https://doi.org/10.7717/peerj.2584
8. Edgar RC (2010) Search and clustering orders of magnitude faster than BLAST. Bioinformatics 26:2460–2461. https://doi.org/10.1093/bioinformatics/btq461
9. Martin M (2011) Cutadapt removes adapter sequences from high-throughput sequencing reads. EMBnet J 17:10–12. https://doi.org/10.14806/ej.17.1.200
10. Schloss PD (2009) Introducing mothur: open-source, platform-independent, community-supported software for describing and comparing microbial communities. Appl Environ Microbiol 75:7537–7541. https://doi.org/10.1128/AEM.01541-09
11. R Core Team (2021) R a language and environment for statistical computing R foundation for statistical computing, Vienna, Austria. https://www.R-project.org
12. Wickham H, Averick M, Bryan J et al (2019) Welcome to the Tidyverse. J Open Sour Soft 4(43):1686. https://doi.org/10.21105/joss.01686
13. Wickham H (2021) Dplyr: a grammar of data manipulation. R package version 1.6.0 R found. Stat Comput, Vienna. https://CRAN.R-project.org/package=dplyr
14. Wickham H (2016) ggplot2: elegant graphics for data analysis. Springer-Verlag, New York. ISBN 978-3-319-24277-4, https://ggplot2.tidyverse.org
15. Love M, Huber W, Anders S et al (2014) Moderated estimation of fold change and dispersion for RNA-seq data with DESeq2. Genome Biol 15(12):1–21. https://doi.org/10.1186/s13059-014-0550-8
16. McMurdie PJ, Holmes S (2013) Phyloseq: an R package for reproducible interactive analysis and graphics of microbiome census data. PLoS One 8(4):e61217. https://doi.org/10.1371/journal.pone.0061217
17. Boyer F, Mercier C, Bonin A et al (2016) Obitools: a unix-inspired software package for DNA metabarcoding. Mol Ecol Resour 16(1): 176–182. https://doi.org/10.1111/1755-0998.12428
18. Hugerth LW, Muller EEL, Hu YOO et al (2014) Systematic design of 18S rRNA gene primers for determining eukaryotic diversity in microbial consortia. PLoS One 9(4):e95567. https://doi.org/10.1371/journal.pone.0095567
19. Edgar RC (2016) UNOISE2: improved error-correction for Illumina 16S and ITS amplicon

sequencing. BioRxiv:081257. https://doi.org/10.1101/081257

20. Guillou L, Bachar D, Audic S et al (2012) The Protist ribosomal reference database (PR2): a catalog of unicellular eukaryote small sub-unit rRNA sequences with curated taxonomy. Nucleic Acids Res 41(D1):D597–D604. https://doi.org/10.1093/nar/gks1160
21. Morgan M (2021) BiocManager: access the Bioconductor project package repository. R package version 1.30.15. Available at https://CRAN.R-project.org/package=BiocManager
22. McMurdie PJ, Holmes S (2014) Waste not, want not: why rarefying microbiome data is inadmissible. PLoS Comput Biol 10(4): e1003531. https://doi.org/10.1371/journal.pcbi.1003531
23. Edgar R (2021) Expected error threshold for long reads. Available at. https://drive5.com/usearch/manual/faq_emax.html Accessed 20 May 2021
24. Wang Q, Garrity GM, Tiedje JM et al (2007) Naive Bayesian classifier for rapid assignment of rRNA sequences into the new bacterial taxonomy. Appl Environ Microbiol 73(16): 5261–5267. https://doi.org/10.1128/AEM.00062-07

Chapter 9

Barcoding of Antarctic Marine Invertebrates: From Field Sampling to Lab Procedures

Stefano Schiaparelli, Maria Chiara Alvaro, Matteo Cecchetto, and Alice Guzzi

Abstract

DNA barcoding is a powerful and widespread method used to identify large numbers of species collected in the framework of sampling activities in the field. With the exception of research projects that may count on large teams characterized by tasks' delegation and where many activities may run in parallel, in the majority of cases the barcoding effort is handled by a limited number of persons. The guidelines here reported focus on this second case, with a special attention paid to field procedures, whose efficiency and smoothness are often overlooked.

Key words Antarctica, Coastal sampling, Monitoring, DNA barcoding

1 Introduction

Since its introduction in 2003 [1], DNA barcoding has been increasingly employed as a routine species identification tool in the framework of large-scale surveys to characterize the fauna of a given site or large museum collections [2].

This technique relies on the use of a short standardized sequence of DNA, that is, the mitochondrial cytochrome c oxidase 1 gene (COI), which was found to be effective in discriminating animal species thanks to a between-species sequence variation which is generally higher than the within-species one [1].

When a sequence of an unknown specimen is obtained, this is compared with the pool of existing ones through the Barcode of Life Data System (BOLD) (http://www.barcodinglife.org) [3]. This is one of the major online repositories of COI sequences and provides services such as the registry of Barcode Index Numbers (BINs), which greatly speed up the process of putative species individuation by refining species hypotheses [4] and maintaining a high level of quality standards through the inclusion of additional

Cinzia Verde and Daniela Giordano (eds.), *Marine Genomics: Methods and Protocols*,
Methods in Molecular Biology, vol. 2498, https://doi.org/10.1007/978-1-0716-2313-8_9,

information (e.g., sampling event information) and the collection of images of the barcoded specimens.

The whole barcoding process can even be performed in the field, as long as all required lab equipment can be moved and made available to researchers [e.g., 5]. In the majority of cases however, the molecular work takes place in a second phase, after sample collection in the field and shipping to the lab.

The sampling and lab procedures below described can quicken specimen handling in the field, guaranteeing at the same time the achievement of the best results in a barcoding effort.

Despite the availability of general sampling guidelines for the management of biological samples at sea in the framework of large sampling efforts, which may count on working teams with tasks' delegation [e.g., 6], in the case of small scale, near-shore activities with limited personnel involved there are no available guidelines. Under these circumstances, field operations of a barcoding campaign have to be even more carefully planned and is mandatory to count on clear protocols and have a smooth operational workflow.

The following guidelines provide advices for this second case and are the results of several seasons of Antarctic fieldwork where sampling occurred from small vessels and one or two people only were in charge of all procedures.

The description of DNA barcoding amplification and sequencing steps are omitted from this protocol as these may vary case by case, spanning from fully automated processes which use third sequencing service providers, to those made in own labs where all operations are customized and manually performed. In this context, we describe all operations until tissues' clipping for their later placement in TrakMate sample tube racks (with 96-wells, Fig. 4g) to be sent to BOLD. Details of front-end logistics of DNA barcoding can be found in [7].

2 Materials

2.1 Sample Collection at Sea

1. Nitrile gloves.
2. Digital camera.
3. Plastic label to record station data in the image of the catch.

2.2 Sorting in the Field

1. Nitrile gloves.
2. Forceps.
3. Plastic jars and drums of different sizes.
4. Work bench or a box where to spread the catch for gross or refined sorting.
5. Preprinted labels to be placed in jars/containers to refer single specimens to the sampling stations.

6. Spare labels to be used in the field to mark extra samples/jars.
7. Digital camera.

2.3 Sorting in the Lab

1. Nitrile gloves.
2. Forceps.
3. Trays.
4. Fridge at 4 °C.
5. Magnesium chloride ($MgCl_2$).
6. Optical refractometer.

2.4 Vouchers Labeling

1. Preprinted labels on archive-quality paper with field number codes.
2. Scissors to cut the labels.
3. PC with waterproof keyboard cover.
4. Field notebook.

2.5 Vouchers Photographic Documentation

1. A dSRL (digital Single Lens Reflex) camera equipped with a wireless light system (Fig. 3c).
2. Different photographic sets for organisms having different sizes.
3. Small aquaria and petri dishes for specimen photography.
4. Waterproof scale bars weighed down with small fishing sinkers.
5. Inverted forceps to keep in position the scale bars.
6. Small size fishing weights to keep the scale bars on the bottom of petri dishes.
7. Fridge for maintaining living specimens in a cool environment before photography and subsampling.
8. Supply of clean and cold seawater to fill the small aquaria and petri dishes.
9. PC to rename images according to stations and field codes.

2.6 Subsampling in the Lab

1. Nitrile gloves.
2. Forceps.
3. Scalpel with disposable blades.
4. Dissecting scissors.
5. Bunsen burner.
6. Jars of different volume (from 100 to 1000 mL).
7. Supply of centrifuge tubes (15 and 50 mL).
8. Supply of cryovials.
9. Racks for centrifuge tubes and cryovials storage.

10. 96% ethanol, or RNA later, or solution of 20% DMSO (dimethyl sulfoxide) saturated with sodium chloride (NaCl).
11. Storage facilities for frozen samples.

3 Methods

3.1 Sample Collection at Sea

This part of the activities is the most variable one, relying on the logistic facilities available in the field. It is possible to imagine two typical situations, detailed below, ranging from sampling performed from a boat (Subheading 3.1.1) (Fig. 1a, e) to sampling performed by SCUBA divers (Subheading 3.1.2).

Regardless the specific case, special attention has to be placed in finding the right containers (vials, jars, drums) where to place the sampled specimens that have to be kept alive in seawater in order to obtain a good photographic documentation of the vouchers. The containers need to have different sizes (Fig. 1e) in order to accommodate single specimens or species belonging to the same group, which will be preserved in this way to avoid further damage. All these containers (which may reach high numbers in the case of a successful sampling) must bear station data in order not to get mixed across stations (*see* **Notes 1** and **2**).

Owing to time or facilities restrictions, even a gross sorting may not be feasible on board. In these cases, what is brought back in the lab is the whole sample ("bulk"), hence a variety of small containers of different size is not necessary. However, this latter procedure is never the best solution as barcoding usually targets representative and undamaged specimens in order to crosscheck the coherence of molecular and morphological data or apply a "reverse taxonomy" approach [8]. In this latter situation the bulk sample may even not be kept in seawater, further compromising potential integrity of specimens which may die, intermix with the sediment, and so on, thus turning into bad quality and less representative specimens for the barcoding effort (*see* **Note 3**).

3.1.1 Sampling from a Boat

Sampling from a small boat (i.e., not an oceanographic vessel) can be challenging due to the availability of only small operative spaces, the presence of safety hazards, such as winches in activity, slippery surfaces, and hanging weights. Thus, it has to be carefully planned with the personnel responsible for the safety on board where the sorting operation can be done and where to place and store drums, jars, etc., in order not to interfere with boat operations.

1. Prepare all the small containers to be dedicated to a given station inside a box (*see* **Note 4**) (Fig. 1e).
2. Remove container lids and fill the containers with seawater before reaching the sampling station.

Fig. 1 Overview of field activities starting from sampling at sea to specimens' isolation in the lab. (**a**) The dredge sample has just been released in a stainless steel sorting tray. (**b** and **c**) images of catches from two different stations; note the plastic label on the top of both to uniquely identify them (in this case the roman number refers to the expedition followed by the sampling station name). Now the target will be the isolation of all the species form the catch. (**d**) The sorting has started and some specimens are removed by hand; note the thick gloves worn by the researchers doing the sorting and the two directions taken to cover all the catch. (**e**) A variety of jars and drums have already been filled with seawater while the dredge is working on the bottom; they are now ready to receive larger specimens from the catch. (**f**) Once in the lab the collected species are checked and further divided in trays (**g–i**). (**g**) Holothuroids and hydroids. (**h**) Pycnogonids. (**i**) Crinoids. (**j**) Four specimens of *Chorismus antarcticus* (Pfeffer, 1887) have been isolated from the station catch, one more should be added to reach the number of specimens that will have to be barcoded

3. When the catch is on board, place a label with the station number over the catch.
4. Take a picture of the whole catch with the label on it (Fig. 1b, c).
5. Take additional images of visible species emerging at the surface.

3.1.2 SCUBA Sampling

This kind of sampling may produce single specimens collected by hand by the divers or even a "bulk sampling," for example, when a hard substrate is scraped. All underwater operations are limited by the severe environmental conditions (seawater temperature is generally −1.8 °C) and by the heavy equipment used by divers in polar areas. Here jars and containers, if several depths/stations are investigated during a single dive, have to be obligatorily marked in advance.

If are not the researchers themselves to dive, it is necessary to coordinate the sampling activities with the divers appointed to this task, in order to let them to understand the sampling procedures and field codes in use. This kind of sampling has to be built on top of the existing security restrictions and procedures in use for diving in polar areas (*see* **Note 5**).

1. Prepare all the containers to be dedicated to a given station in advance, in the lab (*see* **Note 6**).
2. Mark with a permanent marker the external side of the containers with large, clearly visible numbers/codes.
3. Arrange the containers within the sampling bag in a logic way, for example, with their numbers ordered according to the expected underwater operations.
4. Arrange sampling bag (if each one is dedicated to a sampling station) in a way that enables an easy recovery from the water (*see* **Note 7**).
5. Once containers are taken out of the water, if mixed, divide them according to the station.
6. Verify that lids are closed not to lose the water inside; if needed, refill with seawater.
7. Organize the containers inside boxes for their transport in the lab.

3.2 Sorting in the Field

Sorting in the field is usually limited to benthos samples brought on board by a dredge or a grab and require a "good eye" to spot as soon as possible those most fragile and delicate organisms that have to be immediately placed in seawater. According to the study targets, the number of requested specimens per species may vary but, above all, it is not known the number of species that will be collected in a given station. It is therefore difficult to foresee in advance how many jars/containers have to be prepared and it is always a good idea to prepare an adequate number of spare containers (*see* **Note 2**). The place where the sorting will be performed has to be decided in advance with the boat crew, in order to avoid any potential risk to people and materials.

1. After having photographed the catch with the station label, spread it carefully on the sorting bench or box.

2. Proceed with the sorting by starting with a corner and then following a direction. Avoid random specimen picking (*see* **Note 8**) (Fig. 1d).
3. Divide specimens according to size, softness etc., or by taxon. The goal here is not to mix groups having different characteristics (*see* **Note 9**).
4. Add station labels in each container regardless its size (*see* **Note 10**).
5. Once sorting has been accomplished, the remaining organisms and sediment can be thrown at sea, possibly at the same sampling station.
6. According to the abundance of specimens in the sample or in the case of time restrictions, sorting may not be feasible and the whole catch has to be placed into a large drum/box and brought back into the lab for sorting operations.

3.3 Sorting in the Lab

This is the part of the whole process where the choice of the final candidates as vouchers for barcoding is done or refined (Fig. 1f). Here the boxes/trays with already sorted taxa are resurveyed and labelled with field code numbers. If a "bulk sample" is present, it is sorted at this stage.

1. Organize the specimens in trays to cross check the quality of specimens and refine sorting wherever necessary (Fig. 1g, i).
2. Count specimens in order to verify if the requested numbers of specimens per taxon to be barcoded has been reached (*see* **Note 11**) (Fig. 1j).
3. Select and start with those specimens that require an immediate treatment and have to be photographed first (*see* **Note 12**).
4. Try to keep a constant temperature at all stages in order not to further increase the thermic shock in the collected specimens and put in a fridge at 4 °C most sensitive taxa (*see* **Note 13**).
5. At this stage it may also be considered to relax some specimens by placing them into $MgCl_2$ [9]. This step will save time at the photographic phase (*see* **Note 14**).

3.4 Vouchers Labeling

This is the crucial part of the whole process as it is here that the link between each physical specimen, the field code and the station data is established. The field codes used to identify specimens have to be established and agreed in advance. They also have to be clearly explained in the field notes and in any data report that will be published after the expedition (*see* **Note 15**). If the sampling activities are performed with the collaboration of a museum, then all these steps would be quite straightforward as they fall within the museum's routines. In these cases, it is possible that preprinted museums labels and data recording sheets containing all the required fields to be filled in are made available by the museum itself (e.g., see Fig. 15.3 in ref. 6). It is a good advice to establish communication with a reference museum to get guidance about these procedures.

In common practice it is much more common that working teams operating in the field will have to think and prepare their own materials and develop their own system of field codes. Here becomes mandatory to rely on an easy and robust system to associate preprinted labels to specimens and record the correspondent associated field data (*see* **Notes 16** and **17**). Official museum voucher numbers will substitute these field codes only in a later stage, that is, when the specimens will be sent to a museum for permanent storing after the barcoding effort and the field data will be entered into the museum's database.

1. Print the labels with the field codes on archive-quality paper (*see* **Note 18**) (Fig. 2a).
2. Cut the labels with the field numbers with scissors and place these in physical contact with the single specimens (*see* **Note 19**).
3. Write all necessary filed notes in the field book or in an Excel spreadsheet (*see* **Note 20**).

3.5 Vouchers Photographic Documentation

Take images of the live specimens already in the field has several advantages and hold a critical role in the sample's validation process. Specimens' colors will fade or will disappear once the organism will be fixed in ethanol, hence potential diagnostic characters will not be available anymore. Some specimens may show dramatically different color patterns but molecular data may later indicate that they belong to the same species [e.g., 10]. If no color documentation is available this variability may fail to be described. Color images of good quality are also mandatorily required by BOLD (https://v3.boldsystems.org/index.php/resources/handbook?chapter=3_submissions.html&section=image_submissions#photo_guide). Despite images can indeed be taken even after the fixation of the samples, the general appearance of these specimens will not be of great utility for comparisons as specimens will be contracted, fragmented or shrunk according to the preservation method adopted. It is therefore a very useful investment of time to get the best at this stage, that is, taking images of the alive specimens. This kind of photography has to be taken in seawater; hence a variety of containers ranging from small aquaria (Fig. 3a) to petri dishes (Fig. 3c) have to be prepared. Then, in order to avoid light reflection into the camera lens it is necessary to use flashes oriented at 45° vs the water surface (Fig. 3c). In this way, the light will reach the subject in the water but there will be no reflection in

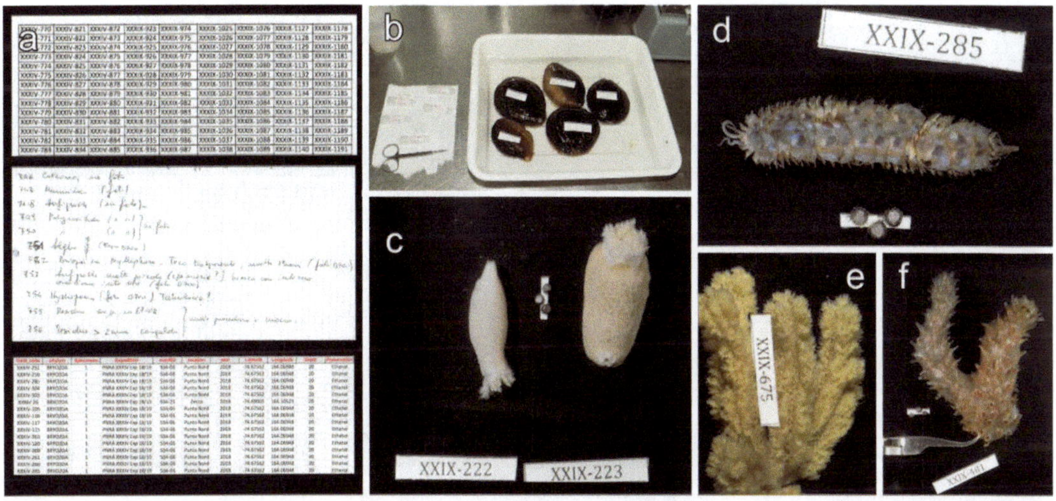

Fig. 2 Specimens labeling. (**a**, top) An image of a preprinted archive-quality museum paper where labels have been sized to fit into the smaller containers in use (i.e., cryovials of 2 mL); the roman number refers to the expedition, while the Arabic number will be associated to a physical specimen. (**a**, middle) An example of field notes: here only the consecutive numbers of the specimens (the right part of the previous code) were reported. (**a**, bottom) the final Excel spreadsheet with all information reported. (**b**) In this image, five holothuroids are prepared for the tissue clipping; each one has received a handwritten label and the cryovials for the tissue bits are ready on the left. In this example preprinted labels have not been used, but this approach is not feasible when there are large numbers of specimens to be processed. In this case, preprinted labels ready to be inserted in the cryovials and associated to the holothuroids (which will be frozen) would have saved much time. (**c**) Two different holothuroid species from the same station are photographed together to save time. The specimens are in seawater hence tentacles are partially expanded. Each specimen is marked by its label and in the middle of them there is the scale bar (1 cm) made with a piece of archive-quality paper kept on the bottom of the tray by small fishing weights. (**d**) A polynoid polychaete photographed with its label and the scale bar; note the natural appearance of the specimen and the possibility to appreciate its colors. (**e**) Here the label has been placed directly over the specimen; a second image for reference will be taken after the label removal. (**f**) The pterobranch *Cephalodiscus*, which would float on the water surface, is kept in place by using inverted forceps. Note in images **c–f** the perfect black background is obtained by using black adhesive plastic layer placed on the bottom of the container; the absence of light reflections is achieved by using flashes oriented at 45° (see explanations of Fig. 3)

the lens. It is good practice to include a measurement scale in the image to provide a size reference (e.g., Fig. 2c, d, f; Fig. 3f, i, k, m), or a color scale to provide color reference. Background in a contrasting color is beneficial where possible (i.e., black). Some of these sets up are described in [6].

1. Place the petri dish with the specimen in the photographic set up or place the specimen in the aquarium (*see* **Note 21**).
2. Put a scale bar close to the specimen and the label with the field code number (*see* **Note 22**).

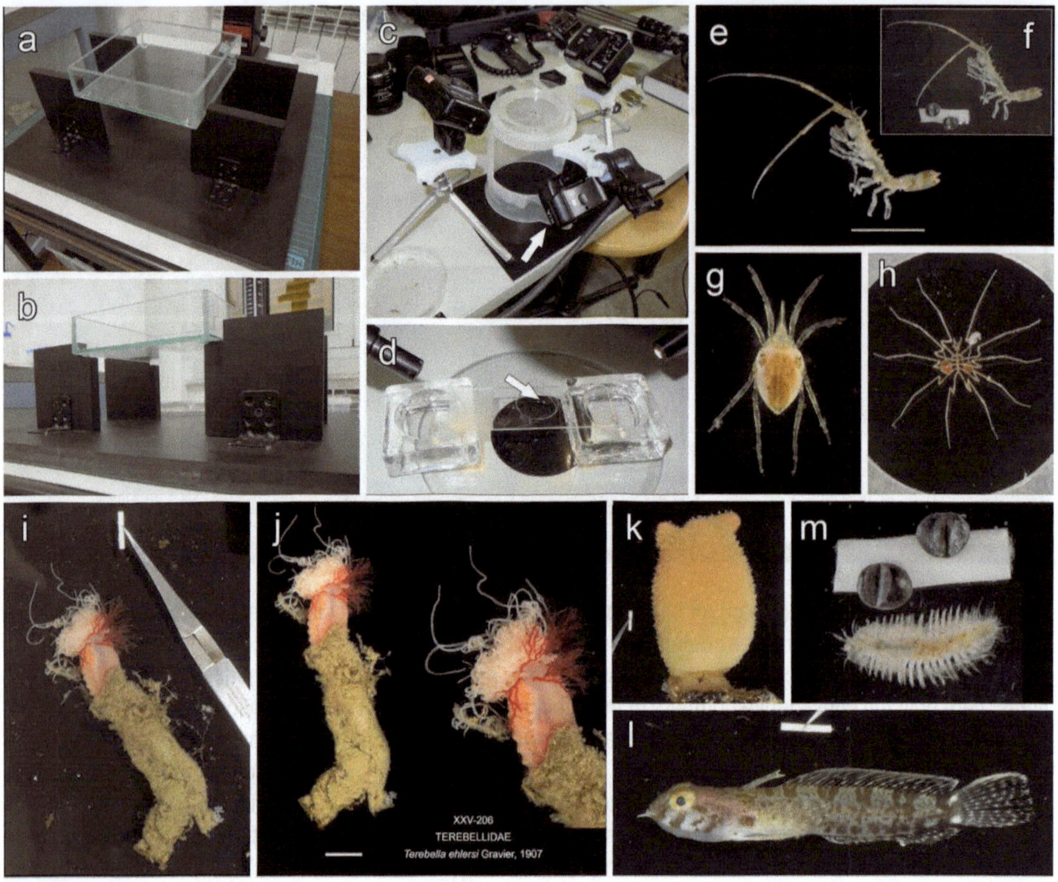

Fig. 3 Different photographic set ups that enable obtaining perfect images with a little effort helping in this way to deal with the high number of images of the specimens selected fort the barcoding that have to be taken. These images will then appear in BOLD "Specimen page" to document each sample morphological features. (**a**, **b**) A small glass rectangular aquarium (size 20 × 20 × 5 cm) is used for photographing larger specimens. The aquarium is suspended by small pillars to allow the positioning of extra-flashes (not figured here) also behind the subject; this extra flash enable capturing finer details of bristles and cirri which would disappear by using only frontal lighting. Pillars and the basement are painted in opaque black to avoid reflections (**c**) An "emergency" photographic set up was obtained by cutting a plastic laboratory pitcher to suspend petri dishes. Three wireless flashes are positioned around the petri dish and with a 45° inclination. The flash marked by the arrow will bring some light from behind the subject. Even here a small black piece of adhesive plastic was attached to the table to provide the black background. (**d**) Similar set up can also be prepared "on the fly" to document tiny species. In this case a microscopic glass slide containing an alive marine mite (arrow) was suspended between two laboratory "salt cellars," the light being provided by optical fibers oriented at 45°. (**e**) An arcturid photographed with the set up shown in **c**. This image was corrected for sharpness and contrast in Photoshop; debris and dirt were digitally removed from the background and a scale bar (=1 cm) added in the desired position. (**f**) This is the original image from which **e** was produced; note the scale bar made with the archive paper that was used as reference to add the final scale bar in **e**. (**g**) The marine mite photographed with the set up shown in **d**. (**h**) A pycnogonid photographed with the set up shown in **c**. (**i**) A tererebellid photographed in a small plastic box with a black background. This is the original image where the scale bar (=1 cm) is kept in place by inverted forceps and there is a lot of debris around the subject. (**j**) The same image of **i** optimized in Photoshop for publication. (**k**) A large ascidian, *Cnemidocarpa verrucosa* (Lesson, 1830) photographed inside a large box with a black background. (**l**) A tiny fish (scale bar =1cm) photographed with the scale bar kept in place by inverted forceps. (**m**) A tiny polynoid polychaete was photographed close to the scale bar; here two small fishing weights were used to keep the scale bar on the bottom of the container (scale bar =1cm)

3. Adjust the position of the scale bar and that of the label in a way that both can be included together with the specimen in the same image.
4. Shoot the general image.
5. Take additional images to document specimen's morphological features by zooming where necessary.
6. Remove the specimen with the label and clean the aquarium by changing the water (*see* **Note 23**).

3.6 Subsampling in the Lab

Subsampling in the field may be required when specimens are large (e.g., in the case of sponges, cnidarians, fishes) and the fixation of the whole organism may thus represent an unfeasible solution. In these cases, therefore, a small bit of tissue is clipped (Fig. 4a, d) and preserved in ethanol (or other media ideal for preserving tissues destined to molecular analyses), while the remaining part of the body is generally frozen. Whenever subsampling has to be performed, this step should follow clear protocols and instructions, decided in advance, where it has to be specified whether whole or partial specimens are required and, when necessary, which parts have to be clipped (*see* **Note 24**).

1. Wear nitrile gloves.
2. Crosscheck labels to verify any possible mismatch before tissue clipping (*see* **Note 25**).
3. Prepare new blades or clean the tools (e.g., scissors) from previous use before dissecting another specimen in order to avoid contamination (*see* **Note 26**).
4. Proceed with the subsampling.
5. Place the tissue bit for the subsampling in a cryovial containing absolute Ethanol, 20% DMSO or DNA later (*see* **Note 27** and **28**) (Fig. 4d).
6. Place the label relative to the subsampling in the cryovial.
7. Place the whole specimen in a centrifuge tube or a plastic container of adequate volume and containing ethanol (or frozen it in a plastic bag if too large for the containers).
8. Place the label relative to the specimen inside the container.
9. Store the cryovials and the other containers at the best temperature for the fixative in use.
10. Discard used blades in proper displacement containers.
11. Clean all the tools for the next use.

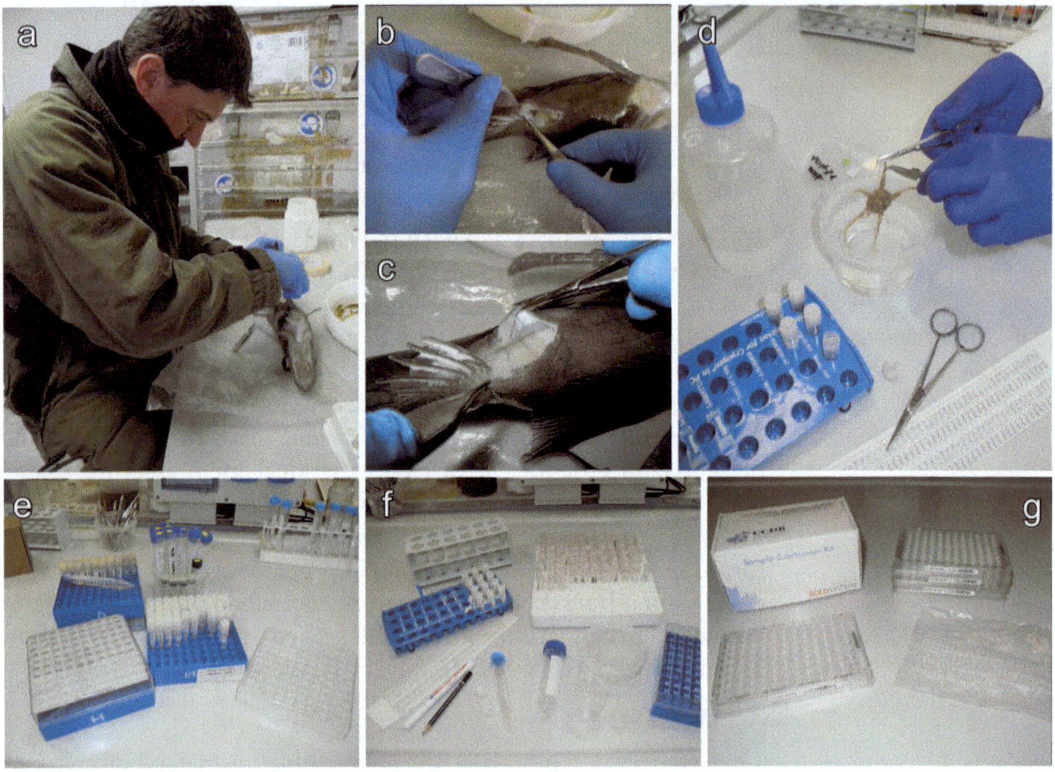

Fig. 4 Tissue clipping and vials to preserve the tissue samples. (**a**) Tissue clipping from a large fish. In some cases logistic constraints impose that these operations are done outside the laboratory facilities. In all cases it is mandatory to wear nitrile gloves to avoid contamination. (**b** and **c**) The fish skin is cut to reach muscle tissue. (**d**) Tissue clipping of the ophiuroid *Ophioplinthus gelida* (Koehler, 1901) with the associated *Iophon* Gray, 1867 sponge covering the disc. In this case, to avoid possible contamination and problems due to the presence of the sponge, a small fragment of an uncovered arm was be used. Note the already opened cryovial in a workstation rack and the labels preprinted on archive-quality paper with a size compatible with that of the cryovials. (**e**) There are several models of racks that enable a compact storage of cryovials and also to trace their position thanks to the numbers reported on the transparent lid. A printed "map" reporting the filed codes can also be added to simplify sample tracking. (**f**) A set of containers (cryovials, centrifuge tubes, and plastic jars) has to be always available in quantity. Tissue subsamples will be stored in the cryovials and the whole range of sizes will be used to accommodate the specimens. Ethanol-proof markers will be used to indicate field codes on the external side of the containers, while archive paper museum labels will be placed inside the containers. These two ways of marking the samples will ensure that all the needed information is not lost. (**g**) The standard BOLD submission kit with microplates racks marked with unique barcodes and cap strips (supplied with the sampling kit). These racks will be used in the fully automated sample processing at BOLD

4 Notes

1. It is rather easy that even the most carefully planned sampling procedure will have to be adapted "on the fly," for example, due to a massive and successful sampling which may run out the available containers for alive specimens storing. In these cases most problems arise from the difficulty to properly mark in a

permanent way those spare, free containers that may be available. This can be resolved in having extra labels to be filled in the field (*see* **Note 2**). The worst case is the complete absence of a sufficient number of containers that may occur when too many stations are sampled or in the case of particularly successful catches. In these latter cases the pragmatic solution is to do just a very general sorting to pick up most fragile organisms or no sorting at all and bring the bulk sample in the lab for further processing.

2. When jars or barrels have to be labelled in the field and "on the fly" a good way is not to mark their exterior with a marker (which may be a difficult operation due to wet surfaces and, if successful, then requires robust cleaning after sampling), but to put inside the container a piece of waterproof, archive-quality or museum specification paper, where the needed information can be written with a simple pencil.
3. In polar areas, thanks to the low or below zero temperatures, even if specimens are sorted out of the sediment hours after sampling and not kept in seawater, they still provide good quality materials for DNA barcoding. This situation is generally not met in the tropics, where tissues degenerate quickly after sampling due to the high air temperatures (Schiaparelli, unpublished observation). In such cases also the photography of the selected specimens is much more difficult since, for example, these may be full of sediment (hence have to be cleaned), air bubbles (hence will float if placed again in water) or will be fragmented and dispersed within the sediment (e.g., in the case of polychaetes).
4. In order to quickly move jars and containers these can be stored in boxes with handles. This will help in moving material as needed and store them when moving to the next sampling station. Boxes must be sturdy and with handles large enough to be grabbed even with heavy polar gloves. If possible, dedicate a single box to each station to avoid containers' mixing during sampling.
5. SCUBA diving in polar areas is a dangerous activity restricted to few, well-trained divers. Safety rules and diving procedures may vary for each country and of course will not be treated here. Regardless the rules and restrictions in use, it is necessary to verify well in advance if the choice of sampling materials used to collect the specimens (*see* **Note 6**) and if the envisaged sampling schemes are appropriate and feasible. All procedures should be ideally tested in a pilot study, to be performed prior to real sampling, in order to adjust procedures and materials wherever needed.

6. Quite often the choice of containers/jars/drums made by the researchers do not meet the requirements of an underwater use, especially if these have to be handled by divers wearing thick gloves operating at subzero temperatures. Thus, it is always a good practice to talk in advance with those who will be in charge of the underwater sampling to verify if the choice of materials is correct. It could be possible that changes will have to be done, for example, by bonding together the jars and their lids with small ropes in order to prevent the lids floating, or by arranging a series of centrifuge tubes into a sort of "cartridge belt" in order to keep them together when these are taken out of the sampling bag. Some jars are also hard to be opened underwater where divers operate in cold water and wear very thick gloves, hence middle size (i.e., 100 to 500 cL) jars with large mouths and heavily ribbed lids (to help grabbing while opening and closing the jar) have to be preferred.
7. Usually sampling bags are taken out of the water sometime after divers have ended the dive, hence these will remain on the bottom or attached to a rope for a while. According to the time of the season, diving can take place from a hole in the sea ice or from a rubber boat when there are open water conditions. Either way, bags will have to be pulled quite strongly and may become heavy once outside the water (e.g., with many jars filled with samples and water). In order not to lose them, the sampling bags have to be sturdy and properly attached to ropes with carabiners in a way that will enable to take these out of the water one after the other and process their content accordingly. Again, the rigid environment where these activities take place is of help in maintaining the tissues in a good quality for molecular analyses, even in the case that sample processing will be performed well after sampling.
8. According to the number of people involved in sorting operations it is possible to begin by opposite the corners. Here the goal is to ensure that the entire sample is sorted. By proceeding without an order some areas will inevitably remain unchecked. The unique exception to this procedure is when particularly fragile or brittle organisms are spotted. These have to be isolated immediately to prevent any further damage (Fig. 1d).
9. To accomplish this step, it is necessary to count on the availability of containers of different sizes and volumes. Antarctic benthos is characterized by sponge species of large size, which will have inevitably to be accommodated into large drums (Fig. 1e). Small invertebrates cannot go inside large drums, otherwise will be dispersed or lost. For these tiny specimens a variety of tubes and small jars has to be made available and at hand. Many organisms may produce large quantity of mucus

(e.g., the sponge *Mycale* (*Oxymycale*) *acerata* Kirkpatrick, 1907 or the fish *Histiodraco velifer* (Regan, 1914)); in such cases single containers have to be dedicated to these taxa.

10. Once closed with their lids, the smaller containers, vials and tubes can also be put inside the drums containing the larger specimens from the same station. In this case it is possible to avoid preparing an extra series of labels for each small container.
11. For example 5 specimens per species; this number has to be decided in advance and may vary according to the study general aims.
12. Many species under stress condition or when placed together with potential predators react by producing copious quantities of mucus or will fragment. Tissues will always be available for analyses, but in some cases body parts could be mixed (e.g., the elytrae of different polynoid polychaetes), hence these mixed fragments will have to be discarded and direct tissue clipping on the specimens has to be done to avoid mistakes.
13. Many Antarctic species are stenothermal and will not survive above zero temperatures for too long. Laboratory conditions are usually designed for researchers' comfort, not to keep alive specimens. It is therefore easy that room temperature will largely exceed species' thermal tolerance if a cool room is not available. Under these unfavorable conditions, for example, nemerteans will suddenly autotomize after some time, while other invertebrates will die. It is therefore a good practice to have access to a fridge at 4 °C in order to avoid any extra stress at this stage.
14. There are several ways to anesthetize invertebrates, but the most common and widespread method is to use of magnesium chloride crystals in the quantity of ~80 g per liter of seawater [9]. The solution has to be made isotonic with the seawater of the sampling area. This step can be checked by using an optical refractometer.
15. Usually it is the same group that has collected the specimens that will process the samples for the molecular study, but it is not always the case. Most problems arise when there is a lack of communication about the meaning and interpretation of the field codes and these are managed after the expedition by a different working group, hence the importance to think of a "rosetta stone" to enable the understanding of acronyms and field codes in use by other people working on these materials.
16. Preprinted labels (Fig. 2a top) save a lot of time by diminishing handling time, samples contamination and, especially, possible writing errors due to tiredness and lack of concentration. Moreover, problems arising from personal calligraphy or

different ways to indicate numbers in different countries will also be eliminated.

17. A simple pattern of field codes could be "number of the expedition-number of the sample" (e.g., XXVIII-00231, Fig. 2a). The first part of the code refers to the number of the expedition and could also include the sampling year (e.g., XXVIII-2021) or the expedition leg (e.g., XXVIII-2021-A). This part will remain stable in all labels. The second part is a consecutive number (i.e., starting with 00001 and going up until the last specimen collected during the expedition). This number has to be linked to station data either in a field book and, later, in an Excel spreadsheet.
18. There are several ways to prepare labels with the field codes to be associated with the physical samples. In any case these have to be prepared by using archive-quality museum paper. The simplest method, in the field, would be to add the codes manually by using a pencil (Fig. 2b). However, this is a time-consuming activity, which does not work when there are hundreds of specimens to be processed. In these cases, pre-printed labels represent the ideal solution. They have to be printed on the same archive-quality paper by using a laser printer and can be produced in sizes that fit those of the containers in use.
19. If a subsampling will be necessary (*see* Subheading 3.6) it is a good practice to print labels in pairs, where one will be placed together with the whole specimen and the other one with the subsample.
20. This operation can be done on the fly, for example, by writing notes in a field book (Fig. 2a in the middle) or, with more calm, by filling an Excel spreadsheet (Fig. 2a, bottom). Usually the first operation precedes the second one. It is mandatory to cross check and revise the matches between labels, vouchers and station data after some entries. It is in fact highly possible that due to the intense activities some mistakes are done in associating labels to specimens or recording the data in the notebook or in the Excel spreadsheets. The photographic documentation (see next section) will also be helpful in resolving possible mismatches in labeling or mistakes in data entry.
21. Petri dishes are the most effective solution to isolate specimens and place all the samples to be photographed in the fridge waiting for their turn.
22. Simple scale bars of a given size (e.g., 1 cm) can be easily made by cutting the archive-quality museum paper of the labels. This paper is strong enough to keep its shape and it is unaffected by immersion. These small-scale bars can be maintained close to the specimen by using inverted forceps (Fig. 3i, k, l) or by using

small fishing weights (Figs. 2c and 3m). The latter solution is perfect when very small containers (e.g., petri dishes of 5 cm of diameter) are used.

23. Water in the petri dishes and aquaria has to be changed frequently. It will progressively become dirty and these small particles will affect image quality or will require adjustments in Photoshop to be eliminated from the background. Moreover, some species will release mucus or other secondary metabolites that may trigger alarm responses in the following specimen to be photographed.
24. It is mandatory to decide in advance, taxon by taxon, which parts of the animal may be removed for separate fixation without the risk of selecting parts that may be crucial for future taxonomic identifications based on morphology. It is also important to avoid contamination, for example, by selecting tracts of the digestive system that may be filled with prey tissues. A good practice is to clip tissues (muscles, if present, are the best) not in contact with the external environment (e.g., potentially carrying debris or other organisms) or the digestive tract.
25. This step represents a final quality control where the correspondence with field codes (especially if a new system of codes is introduced just for the subsampling) has to be crosschecked in order to eliminate any ambiguity or mismatch. It is also a good practice to report on the label that will remain with the physical specimen that a subsample has been taken. Samples from which a subsample has been taken have to be clearly indicated in the Excel file.
26. To properly clean a tool it is necessary to carefully remove tissue fragments present on the tool with laboratory paper. Then place the tool in ethanol. When the tool has to be used, take it from the ethanol and pass it on the flame of the Bunsen burner. These passages will eliminate contamination.
27. There are several options for the conservation of tissue subsamples for molecular analyses. Ethanol is the cheaper and effective solution, while other media are more expensive. Precooled ethanol (e.g., kept at −20 ° C before use) is recommended to make the fixation process more effective.
28. Since the amount of tissue required for analyses can be small, cryovials of 2 mL represent an ideal solution and can be organized in racks for 100 vials. The ratio of fixative vs tissue volume should be around 9:1 (Fig. 4e, f).

References

1. Hebert PDN, Cywinska A, Ball SL et al (2003) Biological identifications through DNA barcodes. Proc Biol Sci 270:313–321
2. Miller SE, Hausmann A, Hallwachs W et al (2016) Advancing taxonomy and bioinventories with DNA barcodes. Philos Trans R Soc Lond B Biol Sci 371:20150339
3. Ratnasingham S, Hebert PDN (2007) BOLD: the barcode of life data system (http://www.barcodinglife.org). Mol Ecol Notes 7(3): 355-364
4. Ratnasingham S, Hebert PDN (2013) A DNA-based registry for all animal species: the barcode index number (BIN) system. PLoS One 8(7):e66213
5. Bucklin A, Steinke D, Blanco-Bercial L (2011) DNA barcoding of marine metazoa. Annu Rev Mar Sci 3:471–508
6. Schiaparelli S, Schnabel KE, Richer de Forges B et al (2016) Sorting, recording, preservation and storage of biological samples. In: Clark MR, Consalvey M, Rowden AA (eds) Biological Sampling in the deep-sea. Wiley, pp 338–367
7. Borisenko AV, Sones JE, Hebert PD (2009) The front-end logistics of DNA barcoding: challenges and prospects. Mol Ecol Resour 9 (suppl):27–34
8. Markmann M, Tautz D (2005) Reverse taxonomy: an approach towards determining the diversity of meiobenthic organisms based on ribosomal RNA signature sequences. Philos Trans R Soc Lond B Biol Sci 360(1462): 1917–1924
9. Lörz AN, Maas E, Linse K, Fenwick GD (2007) *Epimeria schiaparelli* sp. nov., an amphipod crustacean (family Epimeriidae) from the Ross Sea, Antarctica, with molecular characterisation of the species complex. Zootaxa 1402:23–37
10. Williams GC, Van Syoc R (2007) Methods of preservation and anesthetization of marine invertebrates. In: University of California Press (ed) The Light and Smith Manual, pp 37–42

Chapter 10

DNA Barcoding Procedures for Taxonomical and Phylogenetic Studies in Marine Animals: Porifera as a Case Study

Laura Núñez-Pons, Valerio Mazzella, Francesca Rispo, Jana Efremova, and Barbara Calcinai

Abstract

DNA barcoding is a versatile approach that has revolutionized taxonomy and other akin topics in biology and ecology, due to its simplicity and relatively costless procedures. The method consists in the production of one or a few amplicons from informative genetic regions via Sanger sequencing. These markers are selected because they tend to evolve at a similar pace as speciation, allowing to discriminate organismal species. The applicability of this technique is here portrayed for the taxonomical identification of marine sponges (phylum: Porifera) as an exemplification.

Key words Molecular taxonomy, DNA barcode, Sanger sequencing, PCR, Sponge classification

1 Introduction

Taxonomical identification has classically been performed by the comparative study of shared descriptive morphological traits of query organisms in relation to those of holotypes [1–3]. This approach though, bears constraints in the selection of diagnosing elements, and in establishing the criteria for species delimitation in certain groups. This is particularly true in the microorganismal realm and in pluricellular forms with scarce or confounding diagnostic features. In these cases, undeniably, the development of molecular taxonomy has revolutionized systematics [4, 5]. The method by which short fragments of DNA allow for species identification is called DNA barcoding, and works in the same way as the familiar black stripes UPC (Universal Product Code) barcodes catalog items in a supermarket against a reference products list (database). Actually, the gene marker regions used in molecular taxonomy are called "barcodes." Optimal selection of informative

Cinzia Verde and Daniela Giordano (eds.), *Marine Genomics: Methods and Protocols*,
Methods in Molecular Biology, vol. 2498, https://doi.org/10.1007/978-1-0716-2313-8_10,

DNA barcodes is crucial for species discrimination. As a general rule, these genetic markers should have low intraspecific (within species) versus high interspecific (between species) variation [5]. The difference between this inter- and intraspecific genetic distances is referred to as "Barcoding Gap" [6]. Furthermore, for wide taxonomic application the flanking sites of these DNA markers should be conserved for a broad array of groups, allowing for the development of universal PCR primers. Pioneer applications of molecular taxonomy focused on microbial communities using the 5S rRNA gene [7]. Some years later, the cytochrome c oxidase I (COI or COX1) marker was proposed by Folmer et al. [4] as a discriminatory phylogenetic barcode between metazoan invertebrates at the species level. DNA barcoding has evolved into a highly pragmatic technique, suitable for a number of applications including, but not only, verification of new species, recognition of cryptic and sibling species, detection of alien species, identification of endangered and threatened species, linking egg and larval stages to adult species, and in general for taxonomical corroboration [8]. Different marker barcodes are employed for different organismal taxa. For instance, the mitochondrial gene COI is a marker for animals and some protists [4]. Nuclear internal transcribed spacers (ITS) rRNA are used in fungi and microalgal classification [9]. Nuclear ribosomal 16S and 18S rRNA genes are utilized for the identification of prokaryotes [10], and certain microbial and pluricellular eukaryote groups respectively [11]. When barcoding is applied to characterize samples with more than one organism, the procedure used is termed as "DNA metabarcoding." In this case sample preparation follows different procedures regarding molecular tagging for demultiplexing, and downstream analyses are much more complex [12].

The majority of taxonomists are adopting molecular means of classification as a required tool to complement other approaches, and provide more complete descriptions and better phylogenetic reconstructions for cataloguing integrative taxonomy. In this sense, DNA barcoding may also allow nonexpert taxonomists and scientists from other fields to identify their study organisms uncostly by simply sequencing a small portion of material. Several initiatives in the actuality compile genetic data, accompanied, when possible, with phenotypic and sampling information to create universal public access databases, such as GenBank (https://www.ncbi.nlm.nih.gov/genbank/), as well as others more specific by barcode and taxa groups, like BOLD for eukaryotes (https://biodiversitygenomics.net/projects/bold/), UNITE for fungi based on ITS (https://unite.ut.ee), or SILVA for prokaryotes based on aligned ribosomal on 16S or 18S (https://www.arb-silva.de). The more we keep populating these platforms, the more these databases will become accurate and robust for our future research works. In this chapter, we expose the laboratory procedures in the application of DNA barcoding for taxonomical classification in phylum Porifera (marine sponges) as case study. We followed the guidance of the Sponge Barcoding Project database (https://www.spongebarcoding.org), combining histological and morphological studies, with the barcodes of three marker regions obtained through four sets of primers, underlying the importance of integrating molecular and morphological data. These DNA marker regions are a partial fragment of the COI comprising the universal Folmer region [4] plus the downstream Erpenbeck extension recommended for Porifera [13, 14] ranging in total ~ 1000 bp, a fragment spanning from end of the 18S to the beginning of the 28S (this is partial 18S gene, full-length of ITS1, 5.8S and ITS2, and partial 28S gene; [15]), and the region C of the 28S of ~440 bp [16].

2 Materials

2.1 Sampling and Preservation

1. Scuba equipment, underwater camera, GPS.
2. Zip bags 0.5–1 L and/or 50 mL tubes.
3. Permanent markers.
4. Buckets of 3–5 L (1–2) (*see* **Note 1**).
5. Razor and scissors.
6. Ethanol 100° and 70°.
7. Bunsen burner and lighter.
8. Gloves, paper towels, notebook.
9. Tubes, 1.5 mL.
10. Liquid nitrogen (N_2) and freezer −20 °C or −80 °C (*see* **Note 1**).

2.2 Morphological Observations

1. Light optical microscope.
2. Dissecting microscope.
3. Electronic microscope (SEM).
4. Microtome.
5. Hot plate.
6. Petri dishes.
7. Transfer pipette, tweezers, scalpel, teasing needle, glass engraving pen.
8. Slides, cover glasses.
9. Stubs holder, carbon-graphite disc.
10. 65% Nitric acid (under an extractor hood).
11. 10% formalin or 2.5% glutaraldehyde.
12. 5% hydrofluoric acid.

13. Ethanol absolute.
14. Hematoxylin–eosin combined histological stain.
15. Tissue clearing agent (xylene substitute).
16. Paraffin or another embedding medium (epoxy resin).
17. Porifera taxonomy guides, databases, and bibliography (e.g., *see* refs. 17–20).

2.3 DNA Extraction

1. Centrifuge for 1.5 mL tubes.
2. Thermoblock.
3. Extraction kit—Spin-column based (ZYMO Research), according to the instruction manual of supplier. Materials included the following.
   (a) Proteinase K solution: 20 mg/mL sterile molecular water.
   (b) Grind Buffer.
   (c) Genomic Lysis Buffer.
   (d) Columns Zymo-Spin.
   (e) Prewash Buffer.
   (f) Wash Buffer.
4. Tubes, 1,5 mL.
5. Micropipettes for 200 μL and 1 mL.
6. 200 μL and 1 mL sterile, pipette tips.
7. Mini pestles for fitting in extraction tubes.
8. Scissors.
9. Timer.
10. Phosphate buffered saline (PBS; 10%): weigh 80 g NaCl, 2 g KCl, 14.4 g Na_2HPO_4, 2.4 g KH_2PO_4. Add distilled water to final volume of 1 L. Adjust solution to desired pH with NaOH and HCl (typically pH $\approx$ 7.4).
11. Sterile H_2O.
12. Ethanol molecular grade.

2.4 PCR (Polymerase Chain Reaction) Amplifications and Sequencing

2.4.1 PCR Reactions

1. Micropipette.
2. 10 and 200 μL sterile, pipette tips.
3. Vortex mixer.
4. Ice (to store primers while working).
5. 0.2 mL PCR reaction tubes.
6. PCR Reagents: Taq DNA Polymerase (5u/μL); Buffer (5×); 10 μM Primers (specific for the target region, Table 1); 25 nM $MgCl_2$; sterile H_2O.
7. Thermocycler machine to run the reactions.

Table 1
Thermocycling conditions for gene targets COI, PorCOI, 18S-ITS1–5.8S-ITS2-28S, 28S-region

Gene target	Primers	Primer sequence	Thermocycling conditions
COI (~ 640 bp)	LCO1490 HCO2198	5′-GGTCAACAAATCATAAAGAYATYGG-3′ 5′-TAAACTTCAGGGTGACCAAARAAYCA-3′	Initial denaturation: 95 °C 5 min 35 cycles (94 °C 30 s, 50 °C–54 °C 45 s, 72 °C 90 s) Final extension: 72 °C 10 min
PorCOI (~ 560 bp)	PorCOI_fwd PorCOI_rev	5′-AATATGNGGGCNCCNGGNATNAC-3′ 5′-ACTGCCCCCATNGATAAAACAT-3′	Initial denaturation: 95 °C 5 min 36 cycles (94 °C 30 s, 53–57 °C 45 s, 70 °C for 90 s) Final extension: 72 °C 10 min
18S–ITS1–5.8S–ITS2–28S (~850 bp)	RA2 ITS2.2	5′- ACTGCCCC CATNGATAAAACAT-3′ 5′-CCTGGTTAGTTTCTTTTCCTCCGC-3′	Initial denaturation: 95 °C 5 min 38 cycles (94 °C 30 s, 52.8 °C 45 s, 72 °C 60 s) Final extension: 72 °C 10 min
28S–region c (~ 430 bp)	28S-C2-fwd 28S-D2-rev	5′-GAAAAGAACTTTGRARAGAGAGT-3′ 5′-TCCGTGTTTCAAGACGGG-3′	Initial denaturation: 95 °C 5 min 35 cycles (94 °C 30 s, 50 °C–54 °C 45 s, 72 °C 90 s) Final extension: 72 °C 10 min

2.4.2 Agarose Gel Preparation

1. Balance.
2. Microwave.
3. Erlenmeyer flask.
4. Agarose.
5. EDTA solution: weigh 93.05 g EDTA disodium salt (MW = 372.24 g/mol) and add distilled water to final volume of 500 mL. Adjust solution to pH 8 with NaOH.
6. Stock TAE—Tris–acetate–EDTA: weigh 242 g Tris base and add 57.1 mL glacial acetic acid, 500 mM EDTA, pH 8.0. Add distilled water to final volume of 1 L.
7. Ethidium bromide or RedSafe™ Nucleic Acid Staining Solution (for safer practices).
8. Casting trays and combs.
9. Electrophoresis unit.

2.4.3 PCR Purification

1. AMPure XP magnetic beads solution.
2. Magnetic rack.
3. Tubes 1.5 mL.
4. Micropipettes.
5. Pipette tips (20–200 μL; 1–10 μL).
6. Ethanol 70% (fresh prepared).
7. Sterilized water (molecular biology grade).
8. Permanent marker.

2.5 Sanger Sequencing

1. Sterile water.
2. Lambda DNA/HindIII Marker 2.
3. NanoDrop® spectrophotometer.
4. DNA template (PCR products).
5. Primers (forward and reverse) 4.5 μM.
6. Double strand sense (forward and reverse) Sanger Sequencer.
7. Software for visualization and editing of output sequences.

2.6 Gel Elution (OPTIONAL: See Note 2)

1. Agarose gel containing target DNA.
2. Pipettes and tips.
3. Vortex.
4. Water bath at 50 °C.
5. Centrifuge.
6. Tubes (1.5 and 2 mL).
7. Gel dissolution buffer: 5.5 M guanidine thiocyanate (GuSCN), 20 mM Tris–HCl, pH 6.6).

8. Washing buffer: 20 mL NaCl, 2 mM Tris–HCl, pH 7.5 (final concentration after adding ethanol) with 80% EtOH.
9. Elution buffer: 10 mM Tris–HCl, pH 8.5.

2.7 Plasmid Cloning (OPTIONAL: See Note 2)

1. 2× Rapid Ligation buffer.
2. pGEM®-T or pGEM®-T Easy Vector (50 ng).
3. PCR product to be transformed.
4. T4 DNA Ligase (3 Weiss units/μL).
5. Deionized water to a final volume of 10 μL.
6. N,N-dimethylformamide (DMF) or dimethylsulfoxide (DMSO).
7. High Efficiency Competent cells (typically the JM109 strain).
8. 100 mM IPTG: 12 g in 5 mL sterile distilled water. Dissolve and filter-sterilize. Store in aliquots at - 20 degrees C.
9. 3% X-gal in DMF: 0.15 g in 5 mL DMF (dimethyl formamide) or dimethylsulfoxide (DMSO). Store at −20 °C.
10. 100 mg/mL ampicillin: 5 g in 5 mL sterile distilled water. Dissolve and filter-sterilize. Store in aliquots at −20.
11. L-Broth: 5 g tryptone; 2.5 g yeast extract; 2.5 g NaCl; distilled water to 500 mL. Dissolve solids in the water, then divide the L-broth into 100 mL aliquots. Autoclave for 30–45 min. LB and L-agar must be autoclaved on the same day they are made up. Make sure you put autoclave tape on the bottles before autoclaving.
12. L-Agar: 5 g tryptone; 2.5 g yeast extract; 2.5 g NaCl; 1.5% Agar (7.5 g in 500 mL); distilled water to 500 mL.
13. Petri agar plates prepared with LB/ampicillin/IPTG/X-Gal: Add 1 mL (per 1 L of L-Agar) each of the Amp, IPTG, and X-Gal stocks (e.g., 500 microliter each in 500 mL L-agar). Invert to mix thoroughly. Carefully pour ~25 mL into each sterile plastic petri dishes. Allow to set, and store wrapped in plastic at 4 °C.
14. Shaking incubator and stationary incubator at 37 °C.
15. Sterile spreading device.
16. Ice bucket filled with ice.
17. Water bath at 42 °C.
18. Sterile 1.5 mL tubes.

3 Methods

3.1 Sample Collection and Preservation

1. When in the field, retrieve all environmental data relevant to the site location (GPS coordinates, temperature, depth, salinity, pH, etc.).
2. Collect samples ~1–3 cm^3 (for molecular and morphology approaches) in the most sustainable friendly procedures. If working on animals of clonal growth, if possible, grab only a portion to allow regeneration. If target organisms are not clonal, try to leave a significant population on site.
3. Take a picture of the entire sample in situ and also of the sampled portion to match information.
4. Place samples in labeled zip bags 0.5–1 L and/or 50 mL tubes individually separated and with enough amount of seawater as to allow survival until processed. If possible, maintain the samples cool in coolers, in water tanks, or at least in the shade to avoid undesirable reactions and DNA degeneration (*see* **Note 1**).
5. Back in the lab process samples quickly to avoid reactions and DNA degeneration. Cut small portions 0.5 cm^3 or less and place them in labeled 1.5 mL tubes for molecular analysis. Preserve preferentially in order of preference: snap-frozen in liquid nitrogen and then −80 °C, −80 °C, −20 °C, ethanol absolute (in 10 times the sample volume) at 4 °C, ethanol absolute at room temperature (Fig. 1).
6. Keep a larger piece of material of at least 2 cm^3 in ethanol 70° for morphological studies.

3.2 Morphological Observations

Indicated for sponge specimens possessing mineral skeleton (siliceous). Ideal method for not losing rare spicules and microscleres, and only for siliceous sponges.

3.2.1 Spicule Preparations for Optical Microscope

1. Take small portion of the sample ~2/3 mm^3 (including the external and internal parts).
2. Place it on a glass slide.
3. Under an extractor hood, add some drops of nitric acid.
4. Using a tong, let it boiling under a flame until organic parts dissolve.
5. Repeat if necessary **steps 3** and **4** (e.g., if by eye the organic parts are still detectable).
6. Let glass slide cool down.
7. Put some ethanol drops.
8. Gently heat the glass over a flame.
9. Let it dry.

Fig. 1 Underwater sample collection. Collection of biological samples (e.g., marine sponges) underwater by scuba and sample preservation at −80 °C until used for DNA extraction for barcoding procedures

10. Put some drops of mounting medium and cover with a cover glass.

3.2.2 Spicule Preparations for Optical Microscope and SEM

Ideal for getting clean preparations:

1. Take a small portion of sponge ~5 mm^3 (including the external and internal parts).
2. Put it inside a glass, testing tube.
3. Under an extractor hood add 2 mL of nitric acid.
4. Heat the solution over a flame until bubbling.
5. Check boiling preventing the solution from escaping from the tube by moving the tube away from the flame.
6. When the organic material is digested, spicules presence is detectable by gently shaking the tube and checking that the solution appears clear.
7. Let it cool off and then add water to fill the tube and wait 15 min for spicule to settle on the testing tube bottom (× 2 times).
8. Remove the supernatant with a pipette and add ethanol 70° to fill the tube.
9. Wait 15 min for spicules to settle (× 2 times).
10. Add ethanol 95° or 100° and wait 15 min.

11. Remove the supernatant.
12. Gently shake the tube to resuspend the spicules and add a drop of the resuspended solution on a slide.
13. Let the liquid dry at room temperature or put it on a hot plate to accelerate.
14. Put some drops of mounting medium and cover with a cover glass.

For SEM (scan electronic microscopy) observation

1. Follow steps from 1 to 11.
2. Let spicules settle on the tube bottom (15 min).
3. Using a pipette suck up the spicules concentrated on the bottom and drop them on a small piece of glass cut with a glass engraving pen to fit it with stub holders (~1 cm^2). Alternatively use round microscope coverslip.
4. Let it dry at room temperature or put it on a hot plate to accelerate the process.
5. Put a carbon-graphite disc on the stub holder.
6. With a tweezers put the small glass on the stub holder.
7. Proceed for SEM.

3.2.3 Preparations of Skeleton Organization for Optical Microscopy

By hand-cut sections for optical view of mineral skeleton organization.

1. Let the sample dry at room temperature.
2. Using a scalpel, cut a section tangential to the surface as thin as possible.
3. Cut a perpendicular section throughout the sponge to get the external and the internal parts.
4. Put the slices on a glass and cover them with tissue clearing agent. For very pigmented, and dark samples leave, for about half a day, the slices into a small well, covered by the clearing agent.
5. Take the slices and seal the preparation with the cover glass and the mounting medium.
6. Alternatively, after **step 3**, with the help of a tweezers gently put the slices into sodium hypochlorite for some seconds to rapidly digest and eliminate the pigmented, organic part.
7. To stop the digestion, put the slices into ethanol.
8. Wash them in water and again in ethanol.

9. Repeat if necessary (e.g., the section is not sufficiently clear), but pay attention in avoiding skeletal break down due to digestion of the organic parts.
10. Follow **steps 4** and **5**.

3.2.4 Preparations of Skeleton Organization for SEM

By hand-cut sections of mineral skeleton organization for SEM observations.

1. Let the sample dry at room temperature.
2. Cut thicker sections of the sample of approximately few millimeters (tangential and perpendicular throughout the sponge) using a scalpel.
3. Follow previous **steps 6–8**.
4. Let the sections dry at room temperature.
5. Prepare SEM stubs putting a sticky carbon-graphite disc on the stub holder.
6. Put the slice on the sticky disc.
7. Alternatively, you can also use a common vinyl glue pouring some drops directly on the stubs.
8. Let it harden a little to the extent that the sample adheres, without been drowned with the glue.
9. Proceed for SEM.

3.2.5 Preparations for Organic Skeletons (Spongin Fibers)

Indicated for specimens lacking siliceous skeleton (spicules) and possessing instead organic skeletons formed by fibers of spongin.

1. Take a portion of sponge including the external and internal parts (~ 0.5 mm^3).
2. With a tweezers put the fragment into sodium hypochlorite for few seconds to digest and eliminate the organic part present among fibers (mesohyl). Stop the digestion, putting the fragment into ethanol and then wash it into water.
3. Put the fragment into a petri dish.
4. Under the stereomicroscope, with tweezers and needle, release the fibers from the organic part.
5. Repeat if necessary the **step 2** (e.g., often it is not easy to remove the mesohyl).
6. Pay attention to avoiding fibers break down.
7. When the portion of skeleton is ready (no, or scanty organic part among fibers obscuring the skeleton organization) wash it in water and dehydrate in ethanol.
8. Put the fragment on a glass and let it dry.
9. Add a clearing agent.
10. Close the preparation with the cover glass and the mounting medium.

3.2.6 Histology for Water-Canal System

1. Fix a small portion of sample (few millimeters in thickness) in 10% formalin or in glutaraldehyde 2.5% solution in filtered sea water, and buffered to a final pH of 7.6–7.8.
2. For paraffin embedding samples thickness may be wider.
3. Rinse it in filtered sea water (×2 for 30 min).
4. Dehydrate the portion in increasing concentrations of alcohol to remove the water and fixative.
5. For thin portions (about 5 mm) follow these typical dehydration sequences:

   For paraffin inclusion.

   50% ethanol: about 4–6 h.

   70% ethanol: about 4–6 h.

   80% ethanol: about 4–6 h.

   95% ethanol: about 4–6 h.

   100% ethanol: about 4–6 h.

   For Epoxy resin inclusion time is extended:

   50% ethanol: 6 h.

   70% ethanol: 24 h.

   80% ethanol: 24 h.

   90% ethanol: 24 h.

   100% ethanol: 24 h.

   If the sponge has spicules, after **step 3**, follow this procedure.
6. Dissolve siliceous spicules by immersion in 5% hydrofluoric acid in filtered sea water for ~1 h and a half.
7. Rinse in filtered sea water.
8. Then dehydrate following **step 4**.

   Follow the classical procedure for histological embedding using paraffin or epoxy resins.

   Before embedding, the sample must be put into increasing mixture of ethanol and resin (epoxy resin inclusions) or ethanol–tissue clearing agent, and tissue clearing agent–paraffin (paraffin inclusions):

   For paraffin inclusion.
9. 1:1 mixture of 100% ethanol–tissue clearing agent: 4–6 h.
10. 100% tissue clearing agent: 4–6 h.
11. 1:1 mixture of tissue clearing agent–paraffin: 4–6 h.
12. 100% paraffin: 4–6 h.
13. 100% paraffin: 4–6 h.

The same procedure for tissue examination can be also used for observing skeletal organization in silicified sponges. In this case:

14. Use paraffin.
15. Avoid **steps 6** and 7.
16. Cut, with the microtome, thicker sections ~50 μm.

    For resin inclusions, a typical procedure after dehydrate steps and before embedding is:

17. 1:1 mixture of 100% ethanol-pure resin for 24 h.
18. Pure resin for 24 h.

    Follow the specific procedure according to the type of embedding medium for embedding the sample.

19. Cut the embedded sample using a microtome getting ~5–7 μm thick histological sections.
20. Stain the sections with hematoxylin–eosin. For paraffin section, paraffin removal is necessary.

Classical sponge species identification is based mainly on the recognition of the skeletal organization and evaluation of spicule size and shape; external morphological features (i.e., color, growth form, texture and consistence, superficial ornamentations etc.) are also important accessory characters. As a consequence, information in situ, provided also by pictures, are extremely helpful. For species devoid of the skeleton (both organic and mineral), analysis of the choanocyte chambers, or collagen pattern etc. is compulsory (*see* Fig. 2 for examples on morphological diagnostic features). In supporting the recognition of these specific characters and as a guide in understanding taxonomic terms the Thesaurus of Sponge Morphology [17] can be used.

The Systema Porifera [19] is the main tool that, via dichotomic keys, assists in the assignment of the order, family, and genus. When the identification at the genus level is achieved, checklists related to the specific geographic area of the examined sample should be consulted (e.g., World Porifera Data Base—WPDB [20]) in order to check and compare all the species known for the area with the examined sample. The examination of the holotype is often necessary when doubts persist in the identification and species assignment.

3.3 DNA Extraction

If samples are stored in ethanol or other preserving chemicals, add washing steps in PBS (10%) and H_2O prior DNA extraction:

1. Place the specimen sample to be extracted (~ 0.5 cm^3 or ca. 25 mg) into a tube (1.5 mL).
2. Add 1 mL PBS (10%) and incubate at room temperature for 1 h.

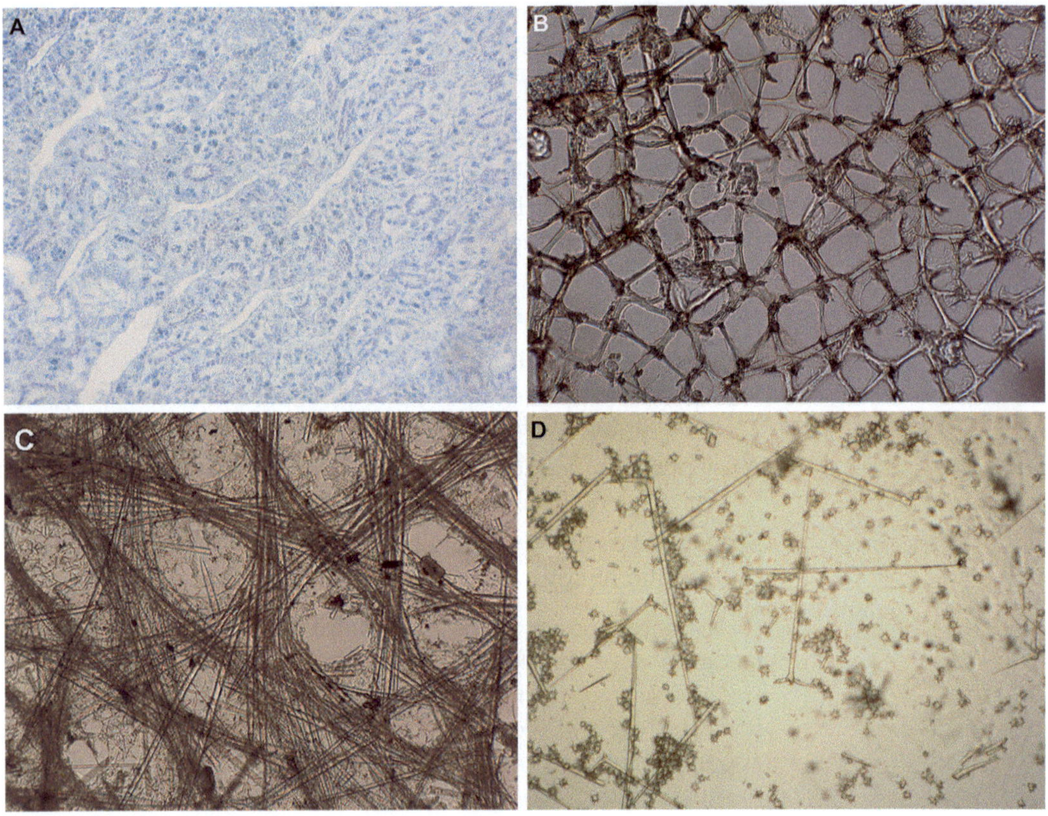

Fig. 2 Spicules preparation and histology for sponge morphological taxonomy. Optical microscope observations of: (**a**) histological section of sponge mesohyl; (**b**) spongin reticulum with embedded spicules; (**c**) skeletal arrangement with spicule organization; (**d**) slide of dissociated spicules showing micro- and macroscleres

3. Discard the supernatant and cover the sample with sterile H_2O. Incubate at room temperature for 40 min. Repeat this step.
4. Discard H_2O and proceed with DNA extraction.

DNA extraction based on Spin-column kits.

1. Heat the Grind Buffer in a Thermoblock to 65 °C.
2. In a petri dish, cut a small piece (ca. 25 mg) from your biological specimen using sterilized scissors and tweezers. Sterilize scissors and tweezers by washing with water and soap and then burning with ethanol and Bunsen burner between samples (*see* **Note 3**).
3. Transfer your biological material into a tube (1.5 mL).
4. Add 200 μL preheated Grind Buffer and 10 μL Proteinase K (20 mg/mL).

5. Pestle and eventually mince the sponge sample with scissors inside the tube.
6. If using the same scissors and/or pestle: Take care to sterilize well the scissors by washing and burning it with alcohol between samples.
7. Put the tube into a Thermoblock (56 °C/65 °C) for 1 up to 3 h or at 37 °C overnight (<15 h). The duration depends on the specific compactness of the tissues.
8. When tissue is digested: 800 μL of Genomic Lysis Buffer is added and the mixture is vortexed thoroughly for at least 6 s. The mixture is incubated for 10 min at room temperature.
9. In the meantime, prepare the Zymo columns and tubes for DNA elution.
10. Transfer the sponge lysis mixture into the column ("Zymo-Spin") in two steps: 500 μL each and centrifuge for 1 min at 13,400 × *g*, room temperature.
11. Take the column and discard the flow-through.
12. Wash the column with 200 μL DNA Prewash Buffer: centrifuge for 1 min at 13,400 × *g*, room temperature.
13. Wash the Column with 500 μL DNA Wash Buffer: centrifuge for 1 min at 13,400 × *g*, room temperature.
14. Take the column and discard the buffers.
15. Take a new tube and place the column inside.
16. Add 50 μL preheated H_2O as elution buffer (60 °C) and incubate for 3 min at room temperature.
17. Centrifuge for 30 s at maximum speed (15,500 × *g*).

3.4 PCR (Polymerase Chain Reaction) Amplifications

A basic PCR setup requires a mixture of reagents that allow the enzymatic polymerase reaction to occur as it occurs naturally in cells.

1. Calculate the amounts of required reagents for your PCR reactions. Take care to include a negative blank reaction with no DNA template and a positive sample of DNA template that you know normally amplifies for the conditions to be performed in your calculations. See as an example the master mix formula in point 4 of this section, and extrapolate for your calculations the quantities for each reagent according to the number of reactions adding 1–2 extra reactions every 10 samples (this is to address pipetting errors).
2. Thaw all reagents at room temperature, while keeping them in ice.
3. Vortex all reagents after thawing, except the Taq.

Table 2
Common PCR problems and protocol optimization

Problem	No band	Unspecific band (multiple, smeary band)
Possible explanations	(a) Primer could not bind to the target fragment (b) Insufficient DNA (c) Degraded DNA	(a) Primer unspecific. Binds to several areas of the DNA, not only on the target gene (b) Contamination with other unwanted DNA.
Solution	(a) Lower annealing temperature. This will decrease binding specificity of the primer, but increase general binding chance. Also reduce mg concentration, or add additives that favor amplification (e.g., BSA) (b) Increase amount of DNA; or sometimes dilute DNA template 1: 10 or more with sterile H_2O. Dilution can at times increase the PCR success because inhibitors present in the extraction are also diluted (c) Reextract your sample to get new fresh DNA	(a) Increase annealing temperature to boost primer binding specificity. As a result, lower amount of PCR product will be produced (smaller band), but with higher specificity (b) Check if your blank indicates contamination, for example, if blank shows a band, or if your positive control has untargeted bands. If so use new reagents and primer aliquots (c) Perform plasmid cloning or gel elution to troubleshoot (*see* Subheadings 3.6 and 3.7)

4. Prepare the master mix in a tube (1.5 mL). For each 25 μL reaction by mixing: 2,5 μL of 2 mM buffer, 1 μL of 2 mM dNTPs, 0,8 μL of 20 μg/μL BSA, 0,8 μL of 10 μM of each primer set (*see* Table 1) forward and reverse, 0,3 μL of 5u/μL Taq DNA polymerase, sterile water until reaching 23.5 μL final volume (final volume variable) (*see* **Note 4**). Keep the tube on ice while working. Mix the master mix thoroughly.
5. Add 23.5 μL of your master mix into your PCR tubes (you may use the same pipette tip).
6. Add 1.5 μL of your target DNA, using different pipette tip for each sample, and making sure you include it into the master mix liquid in the PCR tubes (*see* **Note 5**).
7. Place PCR tubes into thermocycler and program it according to your specific temperatures and times (Table 1).
8. Optimize your PCR protocol if necessary (Table 2).

3.5 Agarose Gel Electrophoresis

Check the success the PCR and purity of your products through visualization of DNA amplicon bands on agarose gel run by electrophoresis [21].

1. Dissolve agarose powder in the same buffer (TAE buffer) used for the running session of the gel in an Erlenmeyer flask (e.g.,

put 1 g of powder in 100 mL of buffer to obtain 1% agarose gel, 2 g in 100 mL to obtain 2% agarose gel).

2. Place the suspension in a dedicated microwave oven for about 2 min. Check that the solution is clear, and the agarose dissolved. As analternative, place the flask on a laboratory hot plate for about 15 min, until the solution is clear, and all the agarose is dissolved.
3. Let cool the gel solution for 2 min at room temperature.
4. Add ethidium bromide or RedSafe™ (3 μL per 100 mL of gel solution, under chemical hood).
5. Pour the gel solution in a casting tray; add the comb and let polymerize the gel 30 min at room temperature (opacity of the gel indicates optimal consistency), preferentially under the chemical hood.
6. Fill the electrophoresis unit with the running buffer (TAE buffer).
7. The DNA is negatively (−) charged. Be sure to place correctly the gel in the electrophoresis unit. Remember that black pole is the negative (−) and the red pole is the positive (+) and that the DNA is attracted by the positive pole.
8. Open the casting trays and place the gel in the electrophoresis unit. Remove the comb.
9. Load the PCR products in the gel wells.
10. Run the gel applying the electric current based on the research needs (e.g., slower running to perform a gel elution, faster running just for a visualization) (*see* **Notes 6** and 7).

3.6 Troubleshooting by Gel (See Notes 8 and 9)

Gel elution is an approach that can be applied to separate our target PCR product in the agarose gel from untargeted DNA bands that are coamplified, when bands are strong and distant enough. If your target gene band is strong (abundant amplification of target gene) and distant from coamplified unwanted DNA fragments, you can cut the target fragment from an agarose gel and extract the DNA from the gel matrix (e.g., Fig. 3).

1. Run your sample on an agarose gel.
2. Excise the DNA fragment from the gel using a sterile scalpel, using for few seconds a UV lamp to visualize and cut a small portion of gel containing the band.
3. Weight the gel slice and add to a clean tube.
4. Add 200 μL of washing buffer for every 100 mg of agarose gel fragment <2%. As an alternative, for every 100 mg of agarose gel >2% the volume of the washing buffer is doubled (i.e., 400 μL).

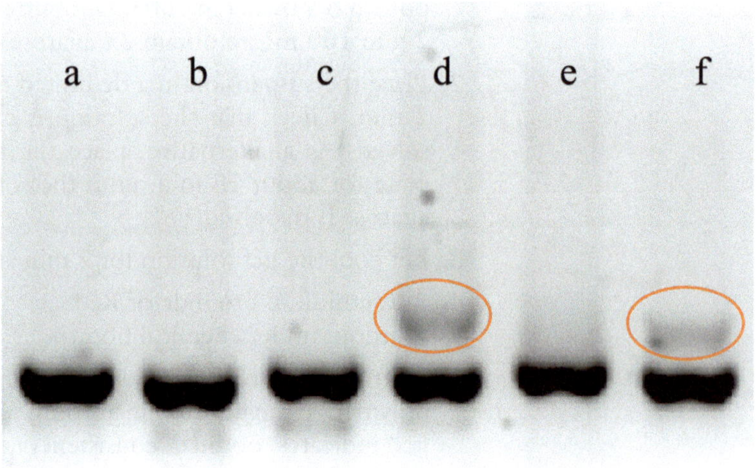

Fig. 3 Troubleshooting coamplification by gel elution. Acceptable bands for sequencing: **a**, **b**, **c**, **e**. Candidates for troubleshooting with plasmid cloning: **d**, **f**. Reason: double band, if as in the examples we have a strong and distant band of the target gene we can also apply gel elution. The resulting target band could be carefully cut and DNA can be extracted from the gel fragment

5. Incubate at 50 °C for 10 min, vortexing every 2 min until all the gel is dissolved.
6. Load the sample in a DNA binding silica spin-column (previously placed in a collection tube).
7. Centrifuge at 11,000 × *g* speed and discard the flow-through. (Repeat the step until all the sample has been passed through the column.)
8. Add 700 μL of washing buffer and centrifuge for 30 s at 11,000 × *g* and discard the flow-through.
9. Repeat previous step.
10. Centrifuge for 1 min at 11,000 × *g* to dry the silica column.
11. Place the column in a new sterile 1.5 mL tube. Add 15–30 μL of elution buffer (as an alternative also molecular grade H_2O).
12. Incubate for 1 min at room temperature.
13. Centrifuge for 1 min at 11,000 × *g* and preserve your DNA at −20 °C.

3.7 Troubleshooting by Plasmid Cloning (See Notes 8 and 9)

Plasmid cloning can be used to address coamplification issues of untargeted (unwanted) DNA fragments that are cloned along with our target PCR product. If your PCR result in multiband outcomes and/or Sanger Sequencing results give multiple sequences it is possible to isolate single DNA sequences by cloning them into bacterial cells. The principle is based on each transformed bacterial

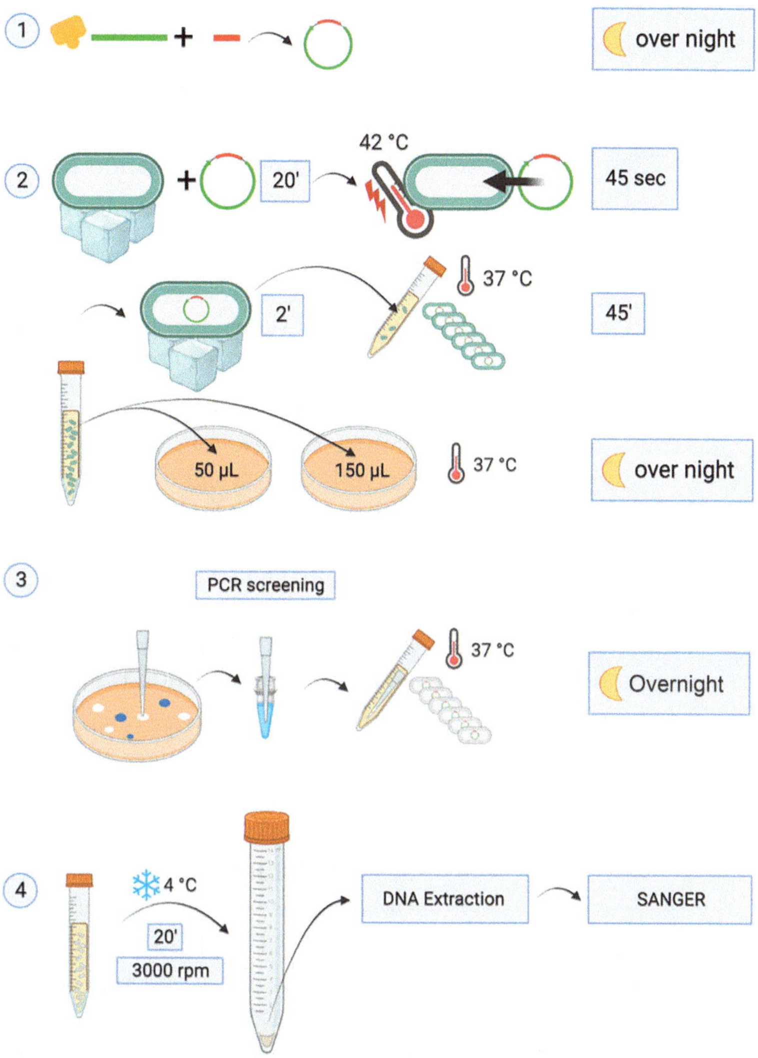

Fig. 4 Procedure of plasmid cloning with pGEM®-T Easy Vector. (1) Ligation. The linear Vector (green) and DNA-Insert (red) are connected via enzymatic reaction using a ligase (yellow). (2) Transformation. The competent bacterial cells take up the vector during a heat-shock reaction. After cultivation, the transformed bacteria are plated on a LB-Amp-IPTG-X-Gal plates. (3) PCR Screening. White colonies are picked and used for PCR screening and cultivation to augment the specific DNA-products of interest. (4) DNA Extraction and Sequencing. DNA is extracted from bacterial pellets, using an according DNA isolation Kit, and Samples are prepared for Sanger Sequencing

colony containing only one single insert (Fig. 4). This step is regularly performed with gel-elution DNA products.

3.7.1 Ligation

1. Defrost and briefly centrifuge the components of the pGEM®-T Easy Vector Kit to collect contents at the bottom of the tube (*see* **Note 10**).

2. Vortex the 2× Rapid Ligation Buffer vigorously before each use.
3. Prepare Mastermix for your ligation reaction with following amounts per sample.
   - 5 μL of 2× Rapid Ligation Buffer.
   - 1 μL of pGEM®-T Easy Vector.
   - 1 μL of T4 DNA Ligase.
   - x μL of PCR product (calculation based on PGEM protocol; amount depends on DNA quantity and length of target gene).
   - x μL of deionized water (to a final volume of 10 μL).
4. Incubate the reactions for 1 h at room temperature and proceed with an overnight incubation at 4 °C for maximizing the number of successful transformants (incubations can be shortened at higher temperatures, but may entail lower probabilities of success).

3.7.2 Transformation

1. Centrifuge the ligation reactions and thaw competent cells. Add 5 μL of each ligation reaction to a sterile 1.5 mL tube.
2. Transfer 40 μL of Competent cells to the ligation reaction tube from step 5. Flick the tubes and incubate on ice for 20–30 min.
3. Heat-shock the cells for 45–50 s in a water bath at 42 °C, and return tubes to ice for 2 min.
4. Add 250 μL LB to the ligation reaction transformations (bacteria) and grow in 37 °C shaking incubator for 45 min. Meanwhile, dry the petri plates in the incubator.
5. Add 20 μL of X-Gal and 5 μL of IPTG to the agar plate and plate until absorbed.
6. Grow the resulting cells (150 μL and 50 μL of each transformation culture) on agar petri plates prepared with LB/ampicillin/IPTG/X-Gal.
7. Incubate plates overnight at 37 °C.

3.7.3 Screening and Bacterial Cultivation

1. Select ~5 white colonies (each containing one insert, as opposed to the blue colonies, that should not have inserts).
2. Prepare a PCR Mastermix for PCR screening with M13 universal primers (*see* Subheading 3.4).
3. Distribute the Mastermix in PCR tubes.
4. Dip the pipette tip with the picked colony first into the PCR Mastermix and then into a Falcon containing 3 mL LB-Medium +3 μL Ampicillin.
5. Amplify the Vector containing your insert by using following thermocycling protocol: initial denaturation: 95 °C 5 min;

30 cycles (94 °C 1 min, 52 °C; 1 min, 72 °C 1 min); final extension: 72 °C 5 min.

6. Culture the picked bacterial colonies in LB-Amp containing falcon at 37 °C in the shaking incubator.

3.7.4 Bacterial DNA Extraction and Sequencing

1. Take your falcons from the incubator and centrifuge at 4 °C, 4000 × *g*, 20 min.
2. Discard the subnatant and carefully dry the pellet by inverting the falcon upside down for 5 min.
3. Extract the plasmidic DNA from the bacterial pellet using a Mini-Prep Kit.
4. Prepare DNA for sequencing (*see* Subheading 3.9).

3.8 Purification of PCR Products

Purify your PCR products from primers, buffer salts, and interfering reagents, and other potential untargeted DNA prior to sequencing for optimal results (*see* **Note 11**).

1. Prepare the PCR products (usually 22 or 45 μL based on your reactions).
2. Place in a new sterile tube a volume 1.2× of magnetic bead solution in respect to the volume of our reactions (e.g., so in case our reaction volume is 22 μL: 22 × 1.2 = 26.4 μL of beads solution). In general, a volume of beads' solution 1.8x in respect to the sample is recommended from the official protocol, but 1.2 works very well.
3. Mix the PCR sample with the beads' solution, pipetting ten times gently, in order to have a perfect mix.
4. Leave the tubes 5 min at room temperature.
5. Place the tubes in the magnetic rack and leave 5 min at room temperature. You can see beads clearly forming a spot nearby the magnet in each well (Fig. 5).
6. Aspire and discard the supernatant, avoiding to touch the clustered beads.
7. Add 200 μL of ethanol 70% and wait 30 s at room temperature.
8. Aspire and discard the solution avoiding the beads.
9. Repeat the washing step adding 200 μL of ethanol 70% and waiting 30 s at room temperature.
10. Aspire and discard the supernatant, being particularly careful to eliminate all the solution from the tubes in order to avoid an ethanol contamination. If needed, aspire and discard the solution in two steps.
11. Leave 10 min at room temperature, tilting a little bit the magnetic rack on a side in order to help the complete evaporation of the ethanol solution. Check that the beads are dried and

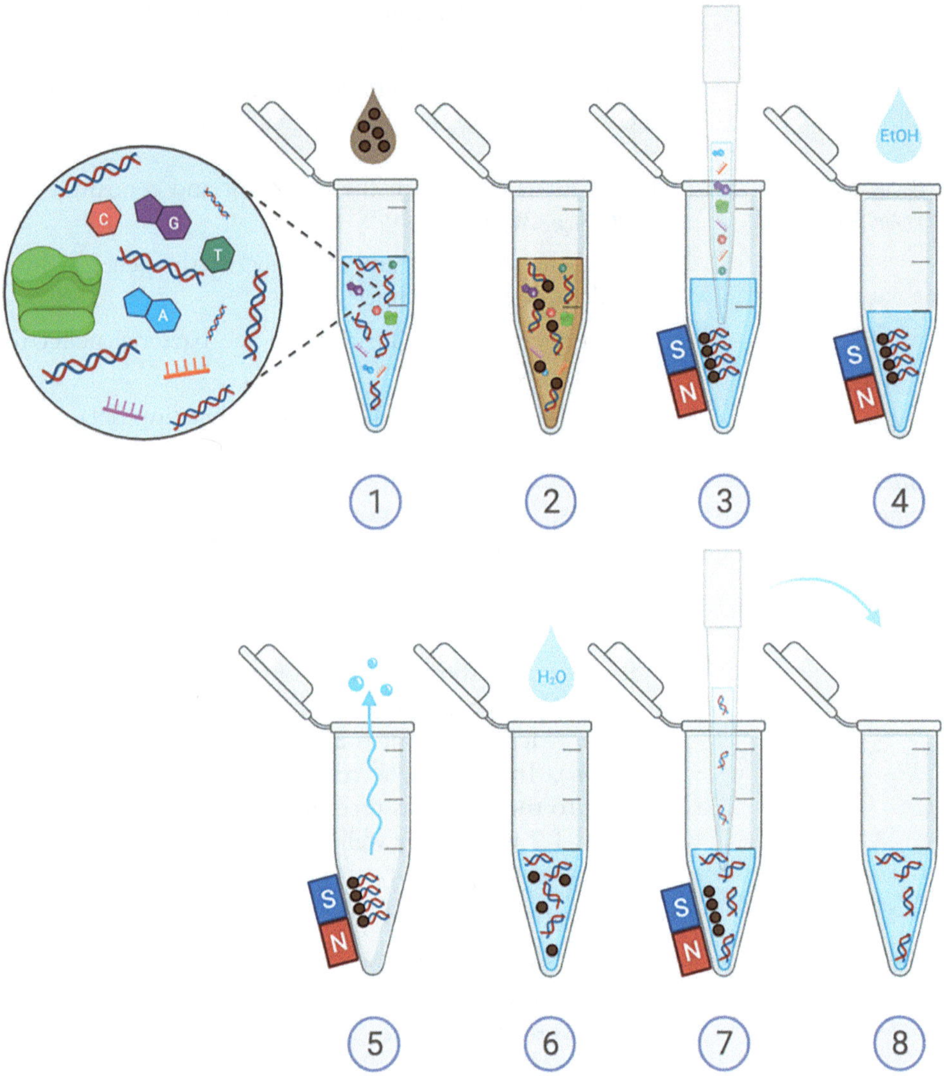

Fig. 5 Purification of PCR amplicons using the Agencourt AMPure XP PCR* purification system. (1) Add paramagnetic beads to your PCR sample. (2) The paramagnetic beads bind selectively to DNA fragments of 100 bp and larger. (3) The paramagnetic beads and bound DNA are separated from the excess reactives by placing the tube into a magnetic rack. (4) Excess primers, nucleotides, salts, and enzymes are removed and the magnetic pellet is washed carefully with 70% ethanol. (5) Ethanol is evaporated inside of the magnetic rack. (6) DNA is eluted from the paramagnetic beads with H_2O. (7) The paramagnetic beads are separated from the aquatic DNA solution by again placing the tube into the magnetic rack. (8) The purified DNA solution is transferred carefully into a new tube

that there is no ethanol on the bottom and on the walls of the before proceeding (*see* **Note 11**).

12. Remove the tubes from the magnetic rack and place it in a normal rack.

13. Elute the bead's spot in 12 μL of molecular grade water, pipetting gently at least ten times, until the beads are all resuspended.
14. Leave 3 min at room temperature.
15. Place the tubes in the magnetic rack and leave 3 min at room temperature.
16. Collect 10 μL of each purified sample and place it in a new labeled tube.
17. Visualize the result of the purification by gel electrophoresis (*see* Subheading 3.5).

3.9 Sanger Sequencing

1. Visualize your purified samples by gel electrophoresis (remember to run only 1 μL of purified product). Visual quantification is performed by comparison to 1 μL of Lambda2 marker which contains a total DNA amount of 500 ng distributed in fragments of different sizes (Fig. 6).
2. As an alternative, check an approximate concentration by a Nanodrop spectrophotometer. Place 1 μL of each sample on the sampling spot and read the ng/μL values. Nanodrop and other visual quantification measurements have to be considered as indicative, as values are not precise.

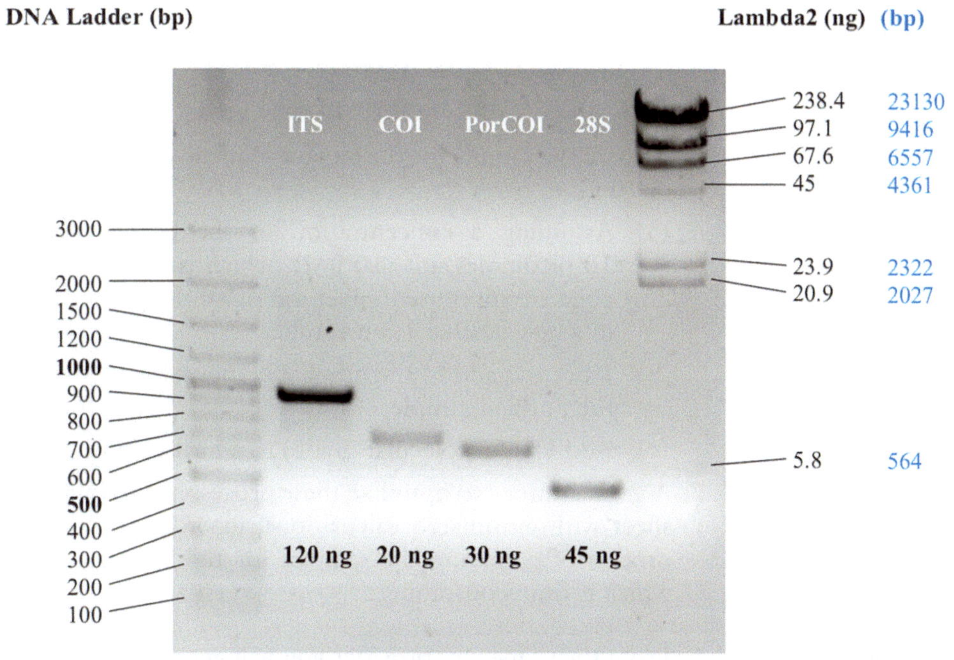

Fig. 6 Example of an agarose gel with our target PCR products and a Lambda2 ladder for PCR quantification. The size of each of our target PCR products (ITSs, COI, PorCOI, and 28S-C) can be approximated by the position of the bands compared to the labeled ladder on the left; whereas the weight of each band is estimated by visual comparison of the width of each band with the lambda2 ladder

3. The optimal concentration to perform a Sanger sequencing is 15 femtomoles per microliter (fmol/μL) (quantomolar) of the purified samples and a final quantity of 4.5 picomoles/μL (=4.5 μM) of each primer over the total volume of the purification mix, which is usually 5–10 μL.
4. Calculate the femtomoles for every sample using the following formula.

$$\frac{(\mathrm{A})}{(\mathrm{B}) \times (\mathrm{C})} \times 1000.000 = \text{total femto} - \text{moles}$$

where (A) is the estimated weight of our sample calculated for a 1 μL volume, (B) is a constant number (660) calculated based on 660 pg/pmol, which is the average molecular weight of a nucleotide pair, and (C) is the targeted length of the amplified fragment (*see* **Note 12**).

5. Example. Calculate how much of our purified sample is needed, assuming that we need to prepare it in a final volume of 10 μL, that our sample's concentration is 50 ng/μL and our fragment length is 650 bp.

$$\frac{50\ \mathrm{ng}/\mu\mathrm{L}}{660 \times 650\ \mathrm{bp}} \times 1000.000 = 116.55\ \mathrm{fmol}/\mu\mathrm{L}$$

Considering that in 10 μL of reaction's volume we will have 150 total femtomoles of purified sample (15 fmol/μL).

$$\frac{150\ \mathrm{fM}}{116.55\ \mathrm{fmol}/\mu\mathrm{L}} = 1.287\ \mu\mathrm{L}$$

6. Prepare the sample for the Sanger sequencing in a final volume of 10 μL.
   (a) Assuming a concentration of the primer solution of 10 picomoles/μL (10 μM), which is the most common concentration used: pipet 4.5 μL of the primer's solution in a new labeled 1.5 mL tube.
   (b) Pipet and add to the mix 1.28 (approximate to 1.3) μL of the purified sample.
   (c) Add H_2O (molecular grade) to a final volume of 10 μL.
7. A good strategy to optimize these processes is to build an excel sheet with automated calculations, in order to optimize the process of preparing the samples for the Sanger sequencing, which is time-consuming.

3.10 Sequence Processing and Taxonomy Annotation

1. From the sequencer provider you will get sequence chromatogram files in ".ab1" format. Load these files into a compatible Sanger sequence software for view and processing (*see* **Note 13**).

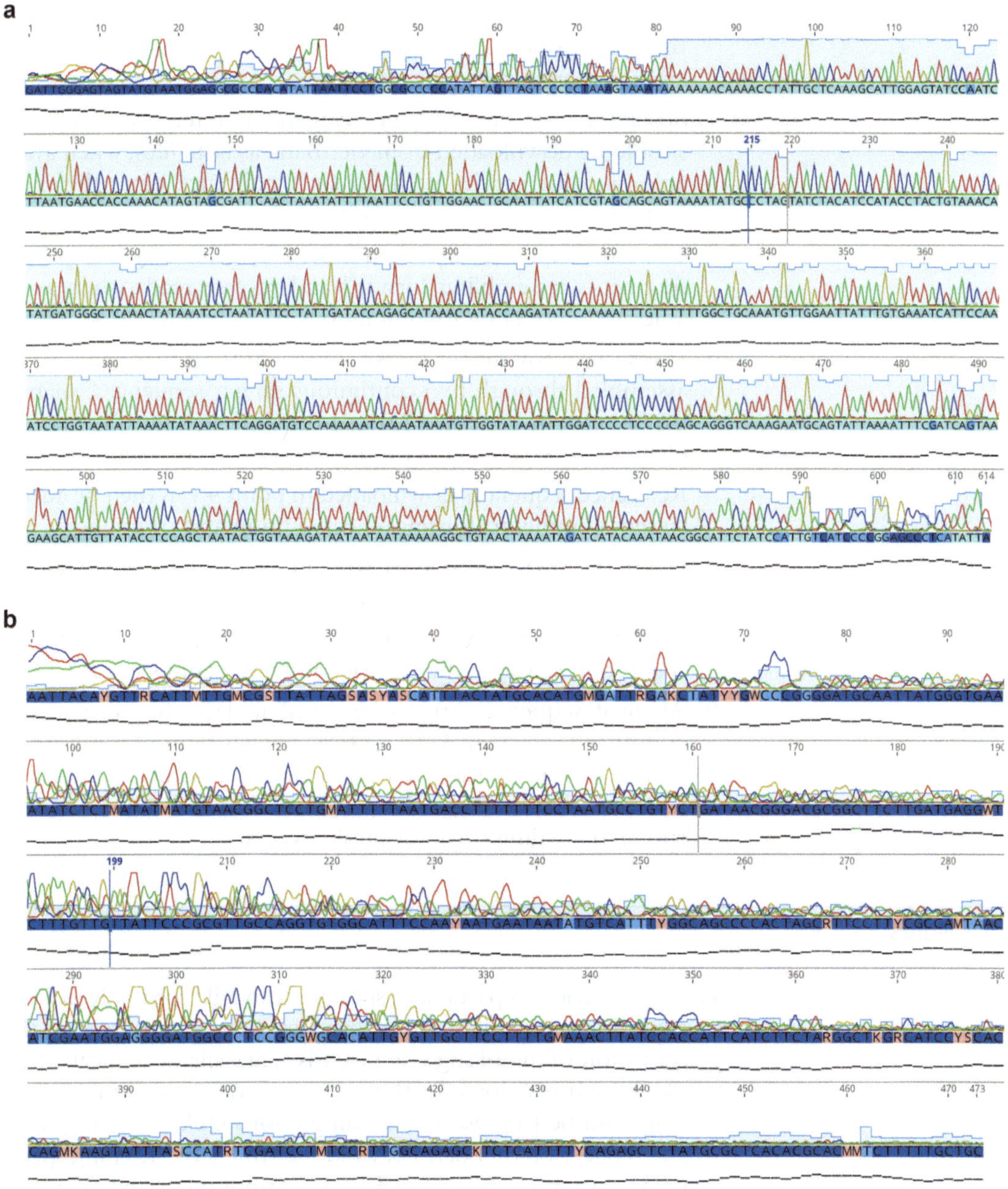

Fig. 7 Example of a good quality Sanger chromatogram (**a**) and a bad quality chromatogram (**b**) from sponge samples for the marker PorCOI. Visualization on Geneious Prime software

2. Check for high quality sequencing outputs by observing clean, round, unambiguous peaks in the chromatogram spectra (Fig. 7).
3. Raw sequences can then be Blasted using the online plugin from NCBI nucleotide database (http://www.ncbi.nlm.nih.

gov/BLAST/), by selecting the sequence and copy-paste on the submission square, to confirm target taxon origin of the amplified sequences (e.g., Porifera).

4. Several conspecific and/or congeneric reference sequences should be downloaded from GenBank as fasta files, when available, and for species with no such reference data, the most matching records are to be downloaded.
5. The raw sequences are then edited by trimming the primer fragments, assembling forward and reverse reads when available, and correcting for ambiguous bases (nucleotides).
6. If you have any longer marker over >800 bp that needs to be assembled through two contiguous sanger sequences, such as for instance in our case COI and PorCOI, you need to concatenate the two sequences by "de novo assemble" actions.
7. Reference sequences downloaded in point 4 can then be added to our alignments, for final check-editing and trimming.
8. From the final alignments we can proceed to phylogenetic studies, genetic distance and barcoding gap calculations.
9. Edited sequences with primers removed will be checked for translation sense, in the case of codifying genes such as COI.
10. All final sequences are to be saved in fasta file format and submitted to GenBank (https://www.ncbi.nlm.nih.gov/genbank/) and other general or specific databases, such as Sponge Barcoding Project database (https://www.spongebarcoding.org), for public access.

4 Notes

1. Some species, in particular, sponges, can fail repeatedly in PCR amplifications, and this has been at times attributed to the activation of inhibiting compounds (e.g., allelochemicals) that block such reactions. In these cases, a way to avoid this challenge can be to preserve the sample for DNA barcoding right after collection on board, or as sooner the better keeping specimens alive in buckets of 3–5 L (1–2) until processed. This can be done with portable liquid nitrogen (N_2) tanks taken on the boat, or with certain buffering solutions taken on site as well, which preserve block inhibiting agents (e.g., SDS (sodium dodecyl sulfate) based, C-TAB (cetyl trimethylammonium bromide) buffers).
2. When PCR reactions reveal multiband outputs even after trying more stringent conditions (e.g., increasing annealing temperatures), or sequencing spectra exhibit background noise from our PCR products, we are probably coamplifying

undesired DNA combined with our target fragment. In such cases, this challenge can be addressed troubleshooting with plasmid cloning on our target PCR product, or with gel elution (Subheadings 3.6 and 3.7).

3. During the DNA extraction procedures be aware to wash any material in contact with the biological material samples between each sample to avoid contamination (tweezers, scissors, mini pestles, plates).
4. You can check the success of DNA extraction quantitatively and qualitatively by measuring the DNA concentration with Nanodrop and performing a PCR of your target gene followed by visualization with gel electrophoresis, respectively.
5. Concentrations of the original PCR reagents added to the master mix may change with the brand of reagents used. Optionally you may as well add additives to your master mix that facilitate PCR amplification. For instance, for Porifera, BSA is often used (20 mg mL^{-1}); other additives may include DMSO, betaine, formamide, and TMAC, among others.
6. Basic tips in PCR reactions: never vortex enzymes (polymerase)—you risk to denature the protein structure; you may use the same pipette as long as you do not touch at all the tube walls; use always different pipette tips when to contact with different DNA templates between samples or reaction mixes.
7. If our PCR reveals no band or very weak band in the gel, we can try to decrease the annealing temperature by 2–3 °C for less stringent conditions, down to ~45 °C (as a general rule).
8. In the case of obtaining several bands (multiband) in our gel for the same sample, we can try to increase by 2–3 °C the annealing temperature for more stringent conditions (but see next note point).
9. For the troubleshooting with plasmid cloning kit we can, once we have some usage management, run the reactions with half of the recommended quantities, as most of these kits are relatively costly.
10. Prepare excel sheets with automated calculations in order to optimize the process of preparing the samples and safe time.
11. Tips for handling the magnetic beads and ethanol preparation (https://www.beckmancoulter.com/wsrportal/techdocs?docname=B37419).
    - Always keep the beads in solution at room temperature or 4 °C. Never freeze the beads as this irreparably damages them.
    - As a recommendation, use tips with filter or if using without filter, do it with extreme care during the pipetting to avoid

contamination of micropipette with beads being sucked up beyond the tip.

- It is important to remove all the ethanol in this step, as any tiny amount can interfere with the sequencing output quality.
- Take care to always keep the beads in solution. Dry beads will be nonfunctional.
- Always resuspend beads thoroughly by vortexing before usage.
- Never reuse beads, discard after usage.
- For additional information see the manufacturer's web page: https://www.thermofisher.com/it/en/home/references/protocols/nucleic-acid-purification-and-analysis/dna-extraction-protocols/dna-extraction-from-tissue.html
- Fresh 70% ethanol should be prepared for optimal results.

12. Examples of software for sequence visualization and editing are Geneious Prime, Sequencher, BioEdit, Traceviewer, Genedoc, EditView, and Sequence Scanner.

Acknowledgments

Thanks are due to R. Pannone, E. d'Aniello, C. La Vecchia, E. Mauriello, and M. L. Palomba for laboratory advice and support.

References

1. Huxley J (1940) The new systematics. Oxford University Press, London
2. Schuh RT (2000) Biological systematics: principles and applications. Cornell University Press, New York
3. Simpson GG (1961) Principles of animal taxonomy. Columbia University Press, New York
4. Folmer O, Black M, Hoeh W et al (1994) DNA primers for amplification of mitochondrial cytochrome c oxidase subunit I from diverse metazoan invertebrates. Mol Mar Biol Biotechnol 3(5):294–299
5. Hebert PDN, Cywinska A, Ball SL et al (2003) Biological identifications through DNA barcodes. Proc R Soc Lond B Biol Sci 270(1512):313–321
6. Paulay G, Meyer Christopher P (2005) DNA barcoding: error rates based on comprehensive sampling. PLoS Biol 3(12):e422
7. Woese CR, Kandler O, Wheelis ML (1990) Towards a natural system of organisms: proposal for the domains archaea, bacteria, and Eucarya. Proc Natl Acad Sci U S A 87(12):4576–4579
8. Savolainen V, Cowan RS, Vogler AP et al (2005) Towards writing the encyclopaedia of life: an introduction to DNA barcoding. Philos Trans R Soc Lond Ser B Biol Sci 360(1462):1805–1811
9. Schoch CL, Seifert KA, Huhndorf SRV et al (2012) Nuclear ribosomal internal transcribed spacer (ITS) region as a universal DNA barcode marker for fungi. Proc Natl Acad Sci U S A 109(16):6241–6246
10. Hugenholtz P, Goebel BM, Pace NR (1998) Impact of culture-independent studies on the emerging phylogenetic view of bacterial diversity. J Bacteriol 180(18):4765–4774
11. Creer S, Fonseca VG, Porazinska DL et al (2010) Ultrasequencing of the meiofaunal biosphere: practice, pitfalls and promises. Mol Ecol 19(SUPPL. 1):4–20

12. Creer S, Deiner K, Frey S et al (2016) The ecologist's field guide to sequence-based identification of biodiversity. Methods Ecol Evol 7(9):1008–1018
13. Xavier JR, Rachello-Dolmen PG, Parra-Velandia F et al (2010) Molecular evidence of cryptic speciation in the "cosmopolitan" excavating sponge *Cliona celata* (Porifera, Clionaidae). Mol Phylogenet Evol 56:13–20
14. Erpenbeck D, Hooper JNA, Wörheide G (2006) CO1 phylogenies in diploblasts and the "barcoding of life" - are we sequencing a suboptimal partition? Mol Ecol Notes 6: 550–553
15. Wörheide G, Nichols S, Goldberg J (2004) Intragenomic variation of the rDNA internal transcribed spacers in sponges (Phylum Porifera): Implications for phylogenetic studies. Mol Phylogenet Evol 33(3):816–830
16. Chombard C, Boury-Esnault N, Tillier S (1998) Reassessment of homology of morphological characters in tetractinellid sponges based on molecular data. Syst Biol 47:351–366
17. Boury-Esnault N, Rützler K (1997) Thesaurus of sponge morphology. In: Smithsonian contributions to zoology. Smithsonian Institution Press, Washington, DC, pp 1–55
18. Hooper JNA (2003) Sponguide. Guide to sponge collection and identification. http://www.qmuseum.qld.gov.au/organisation/sections/SessileMarineInvertebrates/index.asp
19. Hooper JNA, Van Soest RWM (eds) (2002) Systema Porifera. A guide to the classification of sponges, vol 1&2. Kluwer Academic/Plenum Publishers, New York, p 1708
20. de Voogd NJ, Alvarez B, Boury-Esnault N, Carballo JL, Cárdenas P, Díaz MC, Dohrmann M, Downey R, Hajdu E, Hooper JNA, Kelly M, Klautau M, Manconi R, Morrow CC, Pisera AB, Ríos P, Rützler K, Schönberg C, Vacelet J, van Soest RWM (2022). World Porifera Database. https://www.marinespecies.org/porifera on 2022-03-03. https://doi.org/10.14284/359. Accessed May 2021
21. Towbin H, Staehelin T, Gordon J (1979) Electrophoretic transfer of proteins from polyacrylamide gels to NC sheets: procedure and applications. Proc Natl Acad Sci U S A 76: 4350–4354

Chapter 11

Environmental DNA from Marine Waters and Substrates: Protocols for Sampling and eDNA Extraction

Dominique A. Cowart, Katherine R. Murphy, and C.-H. Christina Cheng

Abstract

Environmental DNA (eDNA) analysis has emerged in recent years as a powerful tool for the detection, monitoring, and characterization of aquatic metazoan communities, including vulnerable species. The rapid rate of adopting the eDNA approach across diverse habitats and taxonomic groups attests to its value for a wide array of investigative goals, from understanding natural or changing biodiversity to informing on conservation efforts at local and global scales. Regardless of research objectives, eDNA workflows commonly include the following essential steps: environmental sample acquisition, processing and preservation of samples, and eDNA extraction, followed by eDNA sequencing library preparation, high-capacity sequencing and sequence data analysis, or other methods of genetic detection. In this chapter, we supply instructional details for the early steps in the workflow to facilitate researchers considering adopting eDNA analysis to address questions in marine environments. Specifically, we detail sampling, preservation, extraction, and quantification protocols for eDNA originating from marine water, shallow substrates, and deeper sediments. eDNA is prone to degradation and loss, and to contamination through improper handling; these factors crucially influence the outcome and validity of an eDNA study. Thus, we also provide guidance on avoiding these pitfalls. Following extraction, purified eDNA is often sequenced on massively parallel sequencing platforms for comprehensive faunal diversity assessment using a metabarcoding or metagenomic approach, or for the detection and quantification of specific taxa by qPCR methods. These components of the workflow are project-specific and thus not included in this chapter. Instead, we briefly touch on the preparation of eDNA libraries and discuss comparisons between sequencing approaches to aid considerations in project design.

Key words Environmental collection, eDNA, DNA extraction, Metazoan, Metabarcoding, Metagenomic, Biodiversity monitoring

1 Introduction

Environmental DNA (eDNA) refers to DNA fragments found in water, sediment, snow, and air [1, 2]. Metazoan eDNA originates from the processes of excretion, shedding and decay of tissues, cells, and other secretions from animals or their remains. These genetic fragments can be captured and matched to the organism through DNA sequencing [3, 4]. Given its ubiquity in natural

Cinzia Verde and Daniela Giordano (eds.), *Marine Genomics: Methods and Protocols*,
Methods in Molecular Biology, vol. 2498, https://doi.org/10.1007/978-1-0716-2313-8_11,

environments, the relative ease of sampling, and the sensitivity of the approach for uncovering species, eDNA has enabled scientists to gain valuable insights into marine community functioning and aided conservation programs [5]. Within a mere decade, eDNA has become widely adopted for marine biodiversity monitoring, often in conjunction with "traditional" survey methods [6–8], or as an independent tool [9–12]. The breadth of marine eDNA studies have included predator–prey relationships, trophic linkages, seasonal shifts in biological assemblages [13–15], tracking of endangered and invasive species [16–19], investigations of patterns in genetic diversity [20, 21], and more. The expanding range of applications across marine habitats and taxonomic groups demonstrates that eDNA will continue to advance the field of ecological genetics for years to come.

The general process of an eDNA study involves environmental sample acquisition, processing and preservation, and eDNA extraction; this is then typically followed by eDNA library preparation, Next-Generation sequencing (NGS) and sequence data analysis, or by alternative genetic methods such as qPCR to detect specific taxa of interest. The success in addressing the study question hinges on appropriate execution of each component. First, eDNA is vulnerable to loss and contamination through improper handling at early stages of the workflow (Fig. 1), which can influence the accuracy and reliability of the results [5, 22]. To avoid these pitfalls, eDNA projects require careful and appropriate experimental designs, sample preservation methods, inclusion of controls, and the use of sterile techniques both in the field and in the laboratory [23, 24]. Secondly, the choice of methods for laboratory processing and extraction of eDNA from environmental samples influences the amount of eDNA that can be recovered, which in turn affects the ability to detect species [25]. A variety of extraction protocols have been used and compared [24, 26, 27]. Some investigators have advocated for the optimization and standardization of sampling and extraction protocols to reduce biases and increase the efficiency and applicability of eDNA surveillance in marine habitats [25, 28]. Until the establishment of a standard approach, the choice of method continues to be dependent upon the study system, collection method and target taxa, among other factors. Third, sequencing library preparation methods and NGS platforms appropriate for the research question need to be selected to produce suitable data types, such as metabarcoding to obtain resolution at lower taxonomic levels, or metagenomic sequencing to assess higher order taxonomic diversity.

eDNA metabarcoding, a "targeted" approach, uses one or several conserved DNA markers depending on the taxonomic groups of interest. Common markers include mitochondrial Cytochrome Oxidase I (COI) and 12S ribosomal RNA (12S) genes, as well as fragments of nuclear 18S ribosomal RNA (18S) and large

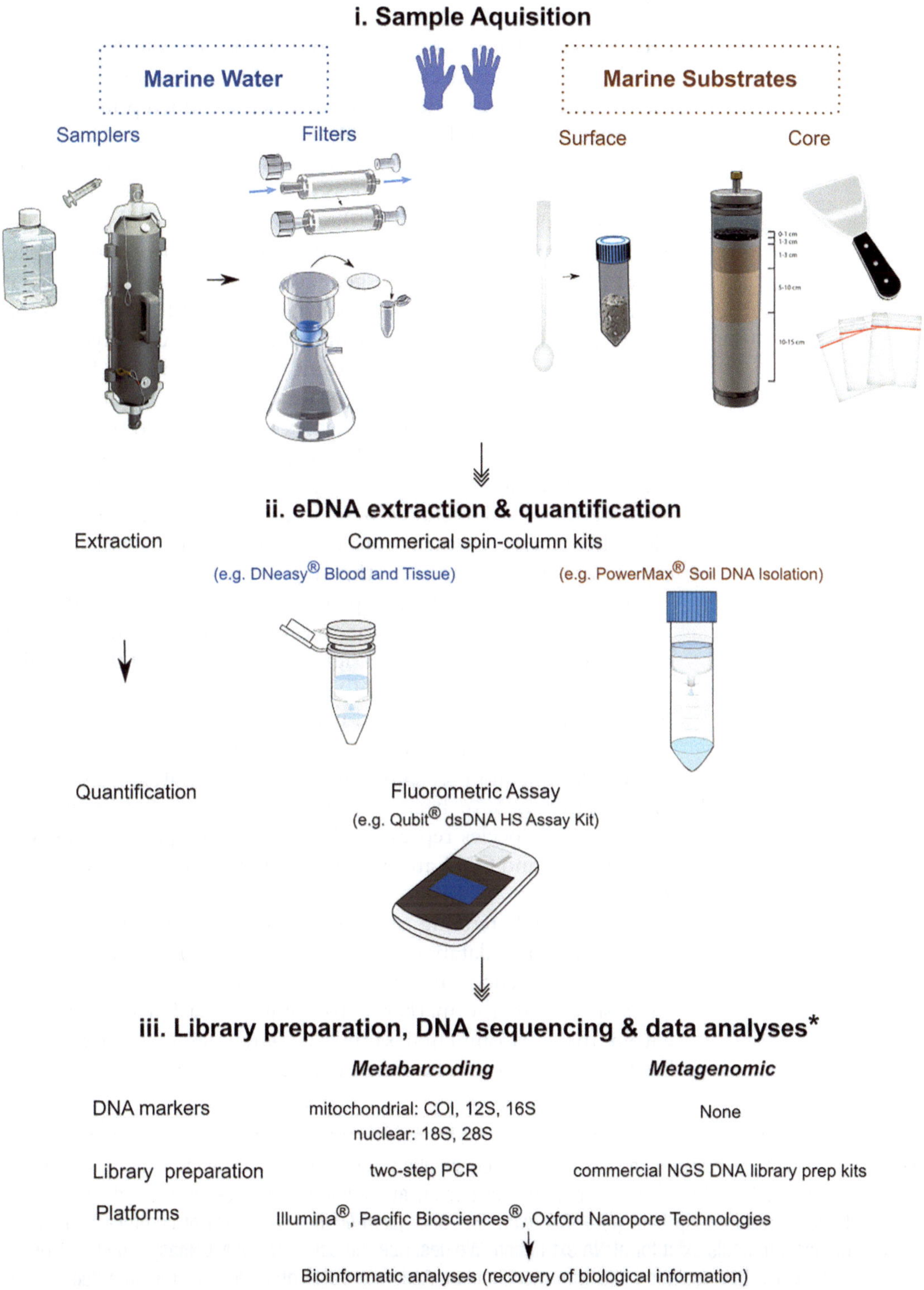

Fig. 1 General workflow for eDNA studies. *(i) Sample acquisition*. Marine water can be collected using small volume syringes (<100 mL), larger bottles (several hundred mL), or Niskin oceanographic water samplers (one to several liters). Water samples are then filtered through Sterivex™ filter units containing an enclosed filter membrane, or onto open membrane using a filtration apparatus. Marine substrates can be collected by using a scooping tool to obtain small quantities of the surface (~5 cm) superficial layer, whereas both surface

subunit ribosomal RNA (28S) genes, among others (Fig. 1, [26, 29–32]). The most common and budget-friendly library preparation method involves a two-step PCR amplification protocol that requires only generic reagents and self-designed primers rather than specifically branded kits [33, 34].

Alternatively, eDNA metagenomic sequencing – an "environmental shotgun" sequencing (ESS) approach, circumvents the use of conserved markers by the direct sequencing of total eDNA, expanding taxon detection ability beyond the limit of a few marker genes [35, 36]. This method generally relies on commercial NGS DNA library preparation kits, such as the NEBNext® Ultra II library prep kit (New England BioLabs Inc.), KAPA HyperPrep® kit (Roche Sequencing), or Nextera® library prep kit (Illumina). For metazoan communities, eDNA metagenomic sequencing can bypass limitations of taxonomic biases of the selected primers, and PCR amplification artifacts [37, 38]. To avoid an unduly long chapter, the details on constructing the different types of sequencing libraries are not included here. Typically, core sequencing facilities offer library construction advice and services, which is one expedient way of completing this step in the workflow.

Regardless of whether a project requires metabarcoding or ESS, a major component for uncovering taxa is the use of powerful, massively parallel NGS technologies, most frequently the Illumina short read platforms (MiSeq, HiSeq, NovaSeq), and more recently the Third generation long read platforms from Pacific Biosciences (PacBio) and Oxford Nanopore Technologies (ONT). eDNA metabarcoding and metagenomic sequencing strategies using Illumina platforms have been the popular methods for biodiversity surveys of all species represented in the sample [15, 36, 39–45], while PacBio and ONT are emerging potential alternatives [34, 46, 47].

Both metabarcoding and metagenomics will require a well-curated reference database to make accurate identifications. Metabarcoding sequence data are matched to reference sequence databases specific to the marker(s) to identify taxa having signatures within the environmental samples, sometimes including species level identifications. The metagenomic approach generates great

Fig. 1 (continued) and subsurface substrates can be simultaneously collected by employing a coring device driven into the seafloor. The core is then sliced horizontally at predetermined thicknesses, and each slice is retrieved and stored in a zip bag. *(ii) eDNA extraction and quantification.* A variety of commercial kits and noncommercial protocols exist for eDNA extraction. We describe the use of QIAGEN DNeasy Blood and Tissue and PowerMax Soil DNA Isolation kits with modification. Accurate eDNA quantification requires fluorometric assays, and the Invitrogen Qubit dsDNA HS Assay Kit is described. *(iii) Library preparation, DNA sequencing and data analysis.* Methods and strategies depend on the research objective. Common choices of DNA markers, relevant library preparation kits and sequencing platforms are indicated. Asterisk (*) denotes these components are discussed in this chapter, but detailed methods are not included

depth of sequencing coverage of all represented eDNA molecules in the sample, affording a broad view of taxonomical diversity. However, the tremendous magnitude of metagenomic sequence data requires high-capacity computational resources to analyze. In addition, the recovery of macrofaunal signatures can be challenging as the current reference databases contain predominantly prokaryotic sequence data, while eukaryotic sequences are far more limited in comparison [36, 39, 48]. Undoubtedly, both metabarcoding and metagenomic sequencing strategies have advantages and disadvantages and deciding to use one or both concomitantly depends on the investigator's research goal.

In addition to massively parallel sequencing methods, a common smaller scale approach is genetic monitoring or detection using quantitative PCR (qPCR) [6, 49, 50], digital droplet PCR (ddPCR) [51], and even conventional PCR [52–54] assays. Genetic monitoring is often implemented when there is a specific species or group of interest, such as when the project aims to detect a rare or invasive species [55]. As with metabarcoding, genetic monitoring requires the amplification of DNA markers. The amplification requires primers that are highly specific to the taxon of interest, often utilizing fluorescent probes with strict testing and quality control to avoid any nonspecific or cross-amplification [49, 51, 56]. Since the input template is eDNA, the collection and extraction methods described in this chapter apply.

Finally, there are a plethora of strategies and programs for bioinformatic analyses of resulting eDNA sequence datasets. The nature and scope of analyses depend on the study question(s), experimental design, and availability of computational resources. These aspects of eDNA studies are therefore beyond the scope of this chapter. Readers can consult available resources [e.g., [57–62]] for detailed information on various open source bioinformatic tools and pipelines.

This chapter focuses on the methods prior to the library preparation and sequencing steps, particularly those for environmental sample acquisition, preservation, and eDNA extraction and quantification for two commonly studied sample types in the marine environment: marine water and substrate (Fig. 1). The cosampling of marine water and substrate to screen for metazoan eDNA is becoming more prevalent, given that each milieu potentially hosts differing assemblages. Therefore, simultaneous surveys may improve detectability of taxa and deliver more comprehensive assessments of biodiversity [63–66]. We aim to supply adequate instructional details to aid scientists considering eDNA as a new study approach.

The methodologies we describe are broadly used, though there is not yet an established community standard for either marine water or substrate. The eDNA extraction methods presented here are largely based on the most commonly used commercial items and kits. Readers are advised that comparable commercial kits may exist, as well as noncommercial options that may serve as lower cost alternatives.

2 Materials

2.1 Decontamination of Sampling Tools and Workspaces (See Note 1)

1. Lab disinfectant (e.g., 10% bleach).
2. 70% ethanol (EtOH).
3. Ultrapure water.
4. Scientific cleaning wipes (e.g., Kimwipes).
5. Disposable regular-length nitrile gloves.

2.2 Surface Marine Water Sampling (See Note 2)

1. 0.22 μm Sterivex™ filter units (Millipore SVGPL10RC) (*see* **Note 3**).
2. Disposable shoulder-length gloves.
3. Disposable regular-length nitrile gloves.
4. 1 L sterile sampling bottles.
5. Individual 60 mL Luer lock syringes (*see* **Note 4**).
6. Sterile or decontaminated 3/16″ ID tubing (*see* **Note 5**).
7. Female Luer fitting caps, to serve as Sterivex outlet caps.
8. Male Luer fitting caps, to serve as Sterivex inlet caps.
9. Sterile, single use 1.0 mL or 2.0 mL transfer pipettes.
10. QIAGEN ATL buffer (*see* **Notes 6** and **7**).
11. Parafilm strips.
12. Sealing bags (e.g., zip or self-adhesive).
13. Field disinfectant (e.g., 10% or 50% bleach solution).
14. Ultrapure water.
15. Permanent marker.
16. Lab tape.
17. Optional: Storage cooler with ice packs (*see* **Note 8**).

2.3 Subsurface Marine Water Sampling (See Note 2)

In addition to everything needed for surface water sampling, you will also need the following:

1. Water sampler (e.g., GO-FLO bottle, Niskin bottle).
2. Messenger weight (triggers remote closure of sampling bottle).
3. Rope that fits your water sampler and weight (*see* **Note 9**).

2.4 Marine Water eDNA Extraction and Quantification Materials

1. QIAGEN DNeasy Blood & Tissue kit (*see* **Note 10**).
2. Proteinase K solution (20 mg/mL).
3. QIAGEN Buffer AL (*see* **Note 7**).
4. 10% bleach.
5. 70% EtOH.
6. 100% EtOH.
7. 3 mL syringes.
8. 2 mL DNA LoBind tubes.
9. 5 mL DNA LoBind tubes.
10. Parafilm strips.
11. Pipettes and filter tips of 1–10, 2–200, 100–1000 μL.
12. Ultrapure water.
13. Markers.
14. Disposable gloves.
15. Vortex.
16. Microfuge.
17. Incubator for heating at 56 °C.
18. Heat block.
19. Qubit® fluorometer.
20. Qubit® dsDNA HS Assay Kit.
21. 0.5 mL thin-walled tubes for fluorometry.
22. Recommended: Laminar flow hood or PCR workstation with air filtration and UV decontamination capabilities (*see* **Note 11**).

2.5 Surface Marine Substrate (~ 5 cm) Sampling and Storage (See Note 12)

1. Sterile 50 mL centrifuge tubes (e.g., Falcon tubes).
2. Small single-use or stainless-steel spatulas or medicine spoons.
3. Parafilm strips.
4. Sealing bags (e.g., zip or self-adhesive).
5. Field disinfectant (e.g., 10% or 50% bleach solution).
6. Ultrapure water.
7. Markers.
8. Disposable nitrile gloves.
9. Transport cooler with ice packs or other cooling supplies.

2.6 Surface and Subsurface Marine Substrate (Core) Sampling and Storage (See Note 13)

1. Sterilized sealing bags (e.g., zip or self-adhesive).
2. Wide stainless-steel spatulas or disposable spatulas.
3. Stainless-steel or single-use forceps.
4. Field disinfectant (e.g., 10% or 50% bleach solution).

5. Ultrapure water.
6. 70% EtOH.
7. Markers.
8. Disposable nitrile gloves.
9. Heavy duty aluminum foil.
10. Scissors.
11. Paper towels.
12. Sample labels (optional).
13. Sterile container for water (optional).
14. Transport cooler with ice packs or other cooling supplies (optional).

2.7 Marine Substrate eDNA Extraction and Quantification Materials

1. QIAGEN PowerMax® Soil DNA Isolation Kit (*see* **Note 14**).
2. Sterile 50 mL centrifuge tubes (e.g., Falcon tubes).
3. Single-use or stainless-steel spatulas or medicine spoons.
4. Parafilm strips.
5. Lab disinfectant (e.g.,10% bleach).
6. Markers.
7. Pipettes and filter tips of 2–200, 100–1000 μL, 5 mL.
8. Container to collect liquid flow through.
9. Disposable nitrile gloves.
10. Balance.
11. Vortex.
12. Vortex adapter.
13. Centrifuge equipped with rotor to spin 50 mL tubes.
14. Recommended: Laminar flow hood or PCR workstation with air filtration and UV decontamination capabilities (*see* **Note 11**).
15. UV Crosslinker for sterilization of tools (optional).
16. Refrigerator cooling at 4 °C.
17. Incubator for heating to 60 °C.
18. Qubit® fluorometer.
19. Qubit® dsDNA HS Assay Kit.
20. 0.5 mL thin-walled tubes for fluorometry.

3 Methods

3.1 Replicates and Controls

eDNA samples are prone to degradation and contamination, and improper handling practices would compromise the integrity of the samples and results. It is therefore a best practice to always collect replicate eDNA samples, specifically, at least in triplicate for each collection site. Control samples are also crucially essential for evaluating potential contamination as well as confirming that the protocols proceeded properly. Thus, negative controls should be incorporated at each stage, to monitor which step or steps may have contributed to any contamination. At a minimum, this generally includes negative field controls and negative DNA extraction controls, as well as other negative controls for downstream steps such as library preparation and sequencing. Negative field controls are typically comprised of sterile or ultrapure water taken to the field in containers of the same type and volume as those used for field sample collection. Alternatively, empty containers of the same type used for field sample collection may also serve as negative field controls, depending on the specific application. All negative field controls are to be processed alongside the field samples either in the field or on return to the laboratory using the same protocol and equipment (*see* **Note 15**).

3.2 Decontamination of Sampling Bottles and Other Reusable Equipment (See Note 1)

1. Rinse bottles and wipe equipment with water.
2. Submerge bottles, caps, and other appropriate labware completely in 10% bleach for 10 min minimum.
3. Remove submerged labware and rinse thoroughly with ultrapure water.
4. Wipe clean electrical and other nonsubmersible equipment with 10% bleach, followed by ultrapure water.
5. Optional: rinse and/or wipe with 70% ethanol to speed up drying.

3.3 Cleaning and Preparation of Laboratory Workspaces (See Note 1)

1. Wipe the inside of the microfuge or centrifuge with 10% bleach, followed by 70% ethanol to remove bleach residue and prevent corrosion.
2. Clean the lab bench, all work surfaces, and all equipment such as pipettes with 10% bleach followed by 70% ethanol.
3. When available, we recommend 20 min of UV sterilization of the PCR workstation or laminar flow hood prior to and after protocols are completed (*see* **Note 11**).
4. Gloves should be changed frequently. If the researcher suspects that some sample (liquid or substrate) has spilled onto the gloves, pipettes or working surface, change gloves, clean, and sterilize the pipettes and surfaces.

5. Nondisposable spatulas and other scraping/scooping tools that are used to transfer substrate from the 50 mL tubes and zip bags should be sterilized before protocols, between samples, and after use.

3.4 Surface Marine Water (0–1 M Depth) Sample Collection (See Note 2)

1. Prepare the field negative control by filling a decontaminated 1 L bottle with ultrapure water. Take this control to the field and filter alongside environmental samples.
2. Don a pair of regular nitrile gloves upon reaching your water sampling site.
3. Pull a shoulder-length glove over on the arm you will use to hold the sampling bottle.
4. Carefully open your sampling bottle, ensuring neither the bottle nor cap contact any contaminating surfaces.
5. Using your arm covered with shoulder-length glove, reach into your water source with the open bottle. Flush the bottle with the field water once, and then fill with 1 L of field water. Do not submerge your arm past the top of your long glove.
6. Immediately cap the bottle.
7. Repeat **steps 4–6** for each additional replicate at that site.
8. Remove and dispose of your long glove.
9. If you will not be immediately filtering the sample before collecting additional samples, label the sample bottle with permanent marker and transport in a cooler with ice.
10. Change all gloves between sites, repeating the sampling process above.
11. Proceed to water filtration (*see* **Note 8**).

3.5 Subsurface Marine Water (>1 M Depth) Sample Collection (See Note 2)

1. Before each usage and between sampling sites, decontaminate the water sampler (e.g., Niskin, GO-FLO), its messenger, and any work surfaces with 10% bleach, followed by subsequent rinses with ultrapure water and 70% ethanol (*see* Subheadings 3.2 and 3.3).
2. Connect your water sampler to a rope or cable, ensuring that the sampler is tied open.
3. Lower the water sampler into the water until it reaches your target depth.
4. Once reaching the target depth, slide the messenger weight down the rope to activate closure of the water sampler in situ.
5. Retrieve the water sampler and set it on a decontaminated surface.
6. Put on clean nitrile gloves.
7. Carefully empty 1 L of water from the water sampler into a new 1 L sample holding bottle.

8. Repeat **steps 2–7** for each additional replicate at that site.
9. If you will not be immediately filtering the sample before collecting additional samples, label each sample holding bottle with permanent marker and transport in a cooler with ice.
10. Proceed to water filtration (*see* **Note 8**).

3.6 Marine Water Filtration and Storage

1. Find a stable surface for filtering and don nitrile gloves. Clean your surface as described in Subheading 3.3.
2. Wipe your gloves and the outside of your 1 L negative control or sample bottle with 10% bleach.
3. Carefully open the package containing the Sterivex filter unit, leaving the unit inside the opened package until needed.
4. Carefully open the 1 L bottle, making sure not to touch the lip of the bottle. Insert the length of 3/16″ ID tubing into the bottle.
5. Carefully open the package containing a sterile 60 mL syringe.
6. Attach the end of the tubing to your syringe.
7. Fill syringe with sampled water, then carefully detach tubing from syringe.
8. Attach filled syringe to the inlet port of the Sterivex unit and push the syringe plunger to filter the water through the unit.
9. Repeat **steps 6–8** until your entire water sample has been filtered through the same Sterivex unit. If you need to set the filter down between syringe fills, temporarily replace it into its original packaging to prevent it from being contaminated.
10. Using the now empty syringe, push air through the Sterivex unit to express out any residual water.
11. Cover the outlet of the Sterivex unit with an outlet cap.
12. Using a sterile disposable pipette, add 1.5 mL of Buffer ATL into the Sterivex unit via the inlet port. Dispose of the pipette.
13. Cover the inlet port using an inlet cap.
14. Tightly seal the inlet and outlet by parafilming around both caps.
15. Vigorously invert the Sterivex unit multiple times to ensure even buffer dispersal.
16. Label the Sterivex unit with date, location, and filtered water volume using permanent marker.
17. Place the Sterivex unit into a labeled bag and seal. Replicates from the same site can be placed together into one bag.
18. Store the bags of Sterivex filters until needed for extraction. Filters filled with Buffer ATL may be stored at room temperature (*see* **Note 6**).

3.7 Marine Water eDNA Extraction and Quantification

This extraction protocol is adapted from those given in [80–82], using the QIAGEN DNeasy Blood & Tissue kit (*see* **Note 7**).

1. Wipe the outside of the Sterivex unit with 10% bleach, followed by 70% EtOH. Allow to dry.
2. Ensure the outlet cap is firmly in place and parafilmed to prevent leakage.
3. Carefully remove the inlet cap from the Sterivex unit and set aside.
4. Add 150 μL of Proteinase K.
5. Recap the inlet port and seal with Parafilm to prevent leakage.
6. Invert 5 times to mix.
7. Incubate at 56 °C for 24 h with shaking/agitation; this can be accomplished by use of a shaking incubator or placing a rotator or mixer within a standard incubator.
8. Remove the Sterivex unit from incubation and shake vigorously by hand five times, then carefully remove the inlet cap.
9. Withdraw all liquid lysate from the Sterivex unit using a 3 mL syringe attached to the inlet port. Record the volume and transfer to a 5 mL LoBind tube.
10. Mix 100% EtOH and Buffer AL in a 1:1 ratio, ensuring you have enough to process all samples (*see* **Note 16**). This should be prepared fresh each day.
11. Add 2 volumes of EtOH–AL to 1 volume of lysate in each 5 mL tube. Vortex vigorously to mix.
12. Pipette 650 μL of the mixture into a labeled DNeasy spin column.
13. Spin at 6000 × *g* for 1 min, then discard flowthrough.
14. Repeat **steps 12** and **13** until entire sample volume has been filtered through the same spin column, discarding the flow-through each time.
15. Place the spin column into a new 2 mL collection tube (from kit).
16. Ensure ethanol has been added to the stock bottles of Buffers AW1 and AW2, as indicated on the bottles.
17. Add 500 μL Buffer AW1 to spin column, then centrifuge for 1 min at 6000 × *g*.
18. Discard flowthrough and collection tube.
19. Place the spin column into a new 2 mL collection tube. Add 500 μL Buffer AW2, then centrifuge for 3 min at 20,000 × *g*.
20. Discard flowthrough. Place the spin column onto the same collection tube and centrifuge again for 1 min at 17,000 × *g*.

21. Transfer the spin column to a new 2 mL DNA LoBind tube.
22. Add 100 μL of Buffer AE (preheated to 56 °C) directly onto the DNA binding silica matrix in each spin column.
23. Incubate at room temperature for 10 min.
24. Centrifuge for 1 min at 6000 × *g*.
25. Pipette eluate back to the spin column, then incubate at room temperature for another 10 min.
26. Centrifuge for 1 min at 6000 × *g*.
27. Discard the spin column.
28. Transfer the DNA eluate to a new, labeled 1.5 or 2.0 mL DNA LoBind tube.
29. Use Qubit® dsDNA HS Assay Kit and 0.5 mL thin-walled tubes to quantify extracted DNA samples, following manufacturer's instructions.
30. Store DNA at −20 °C or −80 °C.

3.8 Surface Marine Substrate (~5 cm) Sampling and Storage

1. Don a pair of nitrile gloves prior to handling supplies and label all tubes and zip bags.
2. Change gloves and prepare an empty, sterilized 50 mL tube to serve as a negative field control by making sure that the cap is well closed and encircled tightly with a strip of parafilm (*see* **Notes 17** and **18**).
3. Take this control to the field in a cooler to be transported alongside the environmental samples.
4. Upon reaching the sampling site, don a pair of gloves and carefully scoop substrate from the benthic surface into a 50 mL tube using a collection instrument. Avoid setting the tube cap on a surface to prevent it from being contaminated.
5. Carefully close the tube and encircle the cap tightly with a strip of parafilm.
6. Place tube in its corresponding zip bag.
7. Either sterilize the collection instrument or switch to a new single use instrument for the next sample.
8. Repeat **steps 4–7** for each additional replicates at that site.
9. Place any replicate samples together in their corresponding bag; seal the bag and place it in the transport cooler.
10. Be sure to change gloves frequently, particularly in between sampling locations (*see* **Note 19**).
11. Transport the cooler back to the laboratory and immediately store samples at −80 °C until ready for DNA extraction (*see* **Note 20**).

3.9 Surface and Subsurface Marine Substrate (Core) Sampling and Storage

1. Don a pair of nitrile gloves prior to handling supplies.
2. Prepare all necessary bags, containers, and labels (*see* **Note 21**).
3. Cut several sheets of aluminum foil, one for each sample slice to be collected.
4. Wear two differently size gloves on each hand for processing. Place the smaller glove first, then add the larger glove. Discard the larger gloves after handling each sample and change smaller gloves as needed (*see* **Note 22**).
5. To sample a core by cutting horizontal slices, first wipe clean the sides of the core prior to removing its top.
6. Push the bottom core stopper upward to drain the surface water from the top; if seeking to analyze the water, decant water into a sterile container of your choice.
7. For each slice, push the bottom stopper upward so that the substrate is raised above the rim of the coring device (*see* **Note 23**).
8. Once the desired thickness is reached, use a wide spatula to carefully slice the substrate from the across the core.
9. Use the spatula to place the slice onto an aluminum sheet and homogenize the sample (*see* **Note 24**).
10. Push the substrate from the aluminum sheet into a corresponding labeled zip bag and discard the aluminum sheet.
11. Flatten the zip bag to remove air space just prior to sealing it and store the sample in the transport cooler if applicable (*see* **Note 25**).
12. Either sterilize the spatula or switch to a new sterile spatula and repeat **steps 7–11** for each subsequent slice of substrate until the entire core is processed (*see* **Note 26**).
13. An empty, sterilized and sealed zip bag should serve as the negative field control. The empty bag should be transported and stored along with the experimental sample bags (*see* **Note 17**).
14. Once core sampling is complete, store all samples at −80 °C until ready for DNA extraction (*see* **Note 20**).
15. If using a cooler, transport the cooler back to the laboratory and immediately store samples at −80 °C.
16. Clean and sterilize the processing space, equipment, and supplies, including coring devices.

3.10 Marine Substrate eDNA Extraction and Quantification

This extraction protocol follows the PowerMax® Soil DNA Isolation Kit manufacturer's manual, therefore, all solutions are named in accordance with those instructions.

1. Don a pair of nitrile gloves.
2. Check Solution C1 and if it has precipitated, heat the solution at 60 °C until the precipitate has dissolved.
3. Prepare new 50 mL tubes, separate from those provided by the kit, and label them with the appropriate sample name (*see* **Note 27**).
4. Remove all samples from the freezer.
5. For surface substrate samples stored in 50 mL tubes, thaw the entire sample (*see* **Note 28**).
6. Once thawing is complete, determine approximate wet sediment weight by weighing an empty 50 mL tube and subtracting that weight from the total weight of the sample tubes.
7. If a sample weighs more than 10 g, transfer 10 g or less to the new, labeled 50 mL tube using the thin metal spatula or medicine spoon, remembering to sterilize the spatula in between samples. Alternatively, use single-use spatulas (*see* **Note 29**).
8. For core slices, remove portions of the homogenized substrate from the zip bag and transfer 10 g or less to a new, labeled 50 mL tube and wait for thaw (*see* **Note 30**).
9. Once samples have been weighed and are ready to process, change gloves again for the next steps.
10. Add 15 mL of PowerBead Solution into each PowerBead Tube labeled corresponding to each sample. The bead tubes containing the solution are referred to as the PowerMax® Bead Solution Tubes in the manufacturer's instructions. We abbreviate them as "bead tubes" hereon.
11. Add up to 10 g of previously weighed substrate to each bead tube. Reserve bead tubes without sample to accommodate negative field and negative extraction controls (*see* **Note 31**).
12. Mix each bead tube by vortex for 1 min at the highest speed.
13. Add 1.2 mL of Solution C1 to each of the bead tubes.
14. Mix each bead tube by vortex for 30 s at the highest speed.
15. Further secure the caps of each bead tube by wrapping them with Parafilm to prevent leakage (*see* **Note 18**).
16. Add the tubes to a vortex adapter and mix for 10 min at the highest speed.
17. Centrifuge the bead tubes at 2500 × g for 3 min (*see* **Note 32**).
18. Transfer the supernatants to new and clean collection tubes provided with the kit (*see* **Note 33**).
19. Discard the bead tubes.

20. Add 5 mL of Solution C2 to the supernatants and invert the tubes twice to mix.
21. Next, incubate the tubes at 4 °C for 10 min.
22. Centrifuge the tubes at 2500 × *g* for 4 min.
23. Avoiding the resulting pellet, transfer supernatants to new and clean collection tubes provided with the kit and discard the tubes with pellets.
24. Add 4 mL of Solution C3 to the supernatants and invert the tubes twice to mix.
25. Incubate the tubes at 4 °C for 10 min.
26. Centrifuge the tubes at 2500 × *g* for 4 min.
27. Avoiding the resulting pellet, transfer the supernatants to new and clean collection tubes provided with the kit and discard the tubes with pellets.
28. Mix Solution C4 by shaking.
29. Add 30 mL of Solution C4 to the supernatants and invert twice (*see* **Note 34**).
30. Label spin filter tubes from the kit with appropriate sample names and fill each with the corresponding sample solution from the previous step (*see* **Note 35**).
31. Centrifuge spin filter tubes at 2500 × *g* for 2 min.
32. Discard flowthrough to a liquid discharge container.
33. Add a second volume of the sample solution to the corresponding spin filter and centrifuge tubes at 2500 × *g* for 2 min.
34. Discard flowthrough and repeat this process until the entire volume of the sample solution has been centrifuged through the spin filter (*see* **Note 36**).
35. Add 10 mL of Solution C5 to the spin filters and centrifuge at 2500 × *g* for 3 min.
36. Discard the flowthrough.
37. Centrifuge the spin filter tubes at 2500 × *g* for 5 min.
38. Carefully place each spin filter into a new and clean collection tube provided by the kit (*see* **Note 37**).
39. Add 5 mL of Solution C6 to the center of each spin filter membrane.
40. Centrifuge at 2500 × *g* for 3 min.
41. Discard the spin filter. The liquid remaining in the final collection tube is the eluted eDNA to be stored at −20 °C.
42. Use Qubit® dsDNA HS Assay Kit, fluorometer and 0.5 mL tubes to quantify extracted DNA samples prior to library preparation (*see* **Note 38**).

4 Notes

1. When collecting samples in the field as well as processing them either in the field or in the laboratory, gloves must be always worn. In the field, change to new gloves when handling field negative control samples, as well as between field samples, for every sampling site. To prevent or minimize contamination of eDNA samples in field sampling, we recommend single use supplies whenever possible; such supplies should not be unpackaged until immediately before use to minimize contamination risk. Otherwise, tools and equipment that are to be reused must be sterilized using a solution of sodium hypochlorite (household bleach) between uses. Cleaning items with bleach diluted to 10% with sterile water or ultrapure water (Type I water with resistivity $\geq$18 MΩ-cm) is customary in eDNA field and laboratory protocols [11, 25]. A 50% bleach solution has also been used to remove extraneous DNA products during eDNA processing [23]. Reusable field supplies and equipment should be thoroughly wiped with or soaked in the 10% bleach for a minimum of 10 min and well rinsed with ultrapure water. An optional final rinse or wipe with 70% ethanol helps speed up drying. During laboratory processing, gloves should be changed whenever they are soiled or suspected to have been contaminated, and clean clothing and laboratory coats dedicated to eDNA processing should be worn. All reagent and sample containers should be secured from potential contamination prior and after usage.
2. The marine water sampling protocol we describe is for a 1 L sampling volume. The actual volume and method of water collection and processing depends on the purpose of your study, the turbidity of the site, and anticipated availability of eDNA. Therefore, a pilot study may be required, or at least a thorough literature search to determine appropriate parameters relevant to your study system [23, 67]. Water collection vessels can range from small volume syringes (5–10 mL) to larger bottles (several hundred mL), to Niskin type oceanographic water samplers (one to many liters) (Fig. 1). The syringes are sterile one-use samplers, whereas reusable large bottles and expensive Niskin water samplers must be thoroughly decontaminated before and after sampling. We recommend processing a standardized volume across all samples but filtering the same volume for every sample may sometimes prove difficult. Particulates in turbid water can clog filters resulting in reduced filtration capacity, in which case prefiltration and/or increasing the number of collection replicates may be required [68]. If prefiltering, ensure that every sample is prefiltered, including controls. Lastly, consider filter membranes with larger pore size (e.g., 0.45 μm) to reduce clogging [69, 70].

3. Self-contained membrane filter cartridges such as Sterivex™ units (Millipore) (Fig. 1) are increasingly popular for water filtration in eDNA studies as they greatly reduce the potential for contamination, and thus are described in our protocol. Traditional filtration funnels using open membrane filters (Fig. 1) can also be used if proper decontamination and sterile techniques are followed.
4. The filtration protocol we describe is tailored to remote field sites with limited access to electrical power and cold storage. Thus, the filtration step is manual, using syringes rather than a powered pump. For faster processing, a battery powered pump or hand pump can be used instead of syringes.
5. Length of tubing should be sufficient to extend to the bottom of the water collection bottle while attached to the syringe.
6. Following filtration, filters can be preserved by immersion in lysis buffer such as QIAGEN buffer ATL or lab-made Longmire's buffer, immersion in 100% ethanol, or stored dry either at −80 °C or on silica gel. Preservation by immersion in lysis buffer is common practice as it has been demonstrated to stabilize DNA and limit degradation [67, 71], and allows for extended storage at room temperature [72]. As the filtration protocol we describe applies to remote field sites without cold storage or electricity, sample preservation using lysis buffer is the most appropriate. Our protocol uses QIAGEN ATL Buffer suitable for the downstream extraction we describe. Longmire's buffer that can be easily prepared in the lab is a suitable alternative, but additional processing steps are required during extraction. The lysis buffer preservation method can be applied to eDNA filters prepared from any other filtration method.
7. This protocol requires more volumes of Buffer ATL, Buffer AL, and Proteinase K per sample than those indicated by the QIAGEN DNeasy kit manual. Ensure that you have sufficient stocks of these components prior to starting filtration and DNA extractions.
8. Filtration of collected water can take place in the field or upon return to the lab but ideally should occur as soon as possible to limit eDNA degradation – something that can happen rapidly, depending on conditions [23, 73]. When immediate filtration is not feasible for some samples, plan to process all samples in the same manner to minimize extraneous variables. If water must be transported prior to processing, we recommend storing water samples on ice and protected from sunlight, such as in a cooler.

9. Ensure you have enough rope to lower your water sampler to the desired depth(s). For shallower water sampling, the target depth can be marked on the rope. For deeper depths such as oceanic sampling, deploying the rope or cable over a suspended meter wheel will be useful to determine depth reached by the water sampler.
10. While we describe the QIAGEN DNeasy Blood & Tissue kit in our protocol as it is the popular kit of choice and suitable for many marine samples, alternative kits and/or extra purification steps may be better suited for samples that are expected to have high levels of PCR inhibition [70, 74]. This is generally a more prominent issue for freshwater studies as compared to marine, but it is still something to take into consideration.
11. We strongly recommend performing laboratory processing of eDNA in a dedicated PCR workstation or a laminar flow hood with positive air pressure, air filtration and UV decontamination capabilities to minimize potential for contamination. If these facilities are unavailable, we recommend dedicating a lab bench space not previously exposed to PCR operations, and thoroughly bleach-clean it before first use for eDNA and again after each use. If using a PCR workstation or laminar flow hood, all labware to be used inside should also be UV sterilized, including tubes, pipettes, trash containers, racks, tips, and so on, before and after use each day. Alternatively, one can use a UV crosslinker to sterilize individual tools, such as spatulas and pipettes.
12. Small quantities (~10 g) of surface substrate samples are often retrieved by hand using small scooping instruments to remove material directly from the superficial layer of the benthos (~ 5 cm). The material is then placed in 50 mL tubes for preservation and later laboratory processing. If not using single use tools, it is critical to thoroughly clean and sterilize reusable collection instruments between samples. Once collected, all samples should be placed on ice or frozen immediately after collection to reduce the risk of eDNA contamination and rapid degradation.
13. Surface and subsurface substrate can be collected using coring devices driven into the sediment floor, retrieved, and by cutting horizontal slices down the length of the sediment core at predetermined slice thicknesses. Each slice is homogenized and then placed in sterilized sealing bags for preservation (Fig. 1). It is critical to thoroughly clean and sterilize reusable equipment such as coring devices and their associated parts, prior to their first use, between sampling sites, and after the sampling campaign is completed. Once collected, all samples should be processed and placed on ice or frozen immediately to reduce the risk of eDNA contamination and rapid degradation.

14. The PowerMax® Soil DNA Isolation Kit (QIAGEN) is commonly used to extract eDNA from different substrate types in various environments, including seafloor canyons [75]; rivers [76]; marinas [65]; abyssal plains [12]; hydrothermal vent regions [31]; deep- sea continental slopes, mud volcanoes and seamounts [32, 66]. Some studies have employed potentially lower cost alternatives under other contexts including modified CTAB for freshwater sediment [63]; Mu-DNA method for various soils [77]; NucleoSpin® Soil kit for alpine lake sediment [78]; modified high salt protocol for swamps [27]. A succinct review on extraction methods for various media including soils, sediments, and substrates is presented in reference [26].

15. Since negative controls and experimental samples are processed in an identical manner side-by-side, the absence of end products in the former validates absence of contamination in the latter [23]. With regards to the order of water filtration or processing of samples, there is currently no common standard on whether experimental samples should precede negative controls or vice versa. In principle, the order should make no difference if careful practice against contamination is strictly observed across all samples. However, one may consider processing the negative control last as the most stringent monitoring strategy for potential contamination. In addition to negative controls, some studies have used positive controls. Positive controls typically consist of samples spiked with known nontarget DNA that will produce a consistent and predictable outcome useful for troubleshooting and validation of procedures. For example, a nontarget DNA as an internal positive control during library preparation can serve to monitor for PCR inhibition and library construction success [23].

16. Add up the total volume of ATL + Prot. K lysate across all your samples. You will need 2 volumes of the EtOH–AL mixture to 1 volume of ATL + Prot. K lysate (e.g., for lysates totaling 20 mL, you would need 40 mL of EtOH–AL mixture).

17. Negative field controls for sediment can consist of either empty containers or containers filled with sterile or ultrapure water (*see* Subheading 3.1). If choosing 50 mL tube of water as negative field controls, please note that eDNA extraction by the PowerMax® Soil DNA Isolation Kit (QIAGEN) described here requires adding reagents from the kit, which will result in a total volume exceeding the carrying capacity of the tube. Thus, we recommend empty containers as negative field controls when using this kit. Readers can consult alternative methods [63, 79] that would accommodate sterile or ultrapure water negative controls.

18. Wrapping the tube caps with Parafilm is an extra level of protection against spillage, but not required. This step would require precutting thin sections of Parafilm for each sample.
19. Some environments will require reaching into the water column to make a collection. In this case, we recommend using shoulder length gloves.
20. Cooling or cryopreservation of samples as soon as possible after collection is commonly practiced. There may be, however, cases where access to −80 °C storage conditions are not immediately available. A variety of other cooling and alternative strategies have been summarized in [26], including short-term preservation at 4 °C and −20 °C and longer-term preservation at −80 °C.
21. In addition to writing sample names on the zip bags, you may also choose to prepare sample labels using acid-free archival paper that is cut into small tabs, prepared with sample information, and placed inside of zip bags along with each substrate sample. Remember to sterilize the sample labels prior to use and use sterile tools to handle them.
22. Two gloves on each hand are worn to help reduce the potential for cross- contamination while working with wet substrate. Using this strategy allows the researcher to change the outer, larger glove between samples, while keeping the smaller, inner glove (unless soiled). Please also consider that the entire process of slicing cores is better performed when two people are present; one person dons on gloves, holds the core and pushes up the stopper, while another person slices and bags the samples, cleans the spatulas, and changes their gloves throughout the process.
23. Slice thickness varies with the investigator's questions; however, common size fractions include 0–1 cm, 1–3 cm, 3–5 cm, and so on. Desired thickness size can be marked directly on the core prior to the sampling campaign to facilitate ease of processing and reduction of contamination risk.
24. "Homogenizing" or chopping substrate using the spatula can be done to avoid unfreezing the entire sample for laboratory processing, unless the entire sample will be processed in the lab. The ability to homogenize substrate will depend on the texture and type of substrate being collected.
25. Flattening the zip bags to remove air space prior to completely closing them is optional and done to conserve space in freezers.
26. One method of sterilizing spatulas between slices is arranging sequential bins or buckets containing bleach solution, double distilled water (ddH_2O) or type I ultrapure water and 70% EtOH. Once a spatula has been used to sample, wipe off visible

substrate and then place the spatula in the first bin while processing the next sample with an alternate spatula. Next, move the spatula to the second then third bins to "rinse" it until ready for use once more.

27. The number of new 50 mL tubes should match the number of surface (~ 5 cm) substrate samples that weigh more than the 10 g, as the PowerMax® Soil DNA Isolation Kit protocol suggests processing of up to 10 g of wet soils, such as those obtained from marine environments. These tubes allow the researcher to weigh a maximum of 10 g by transferring substrate from the 50 mL sample tube into a new tube without using tubes provided by the kit that are necessary for downstream steps.
28. Thaw time varies depending on the weight and type of the sample. Our prior experience from a project analyzing heterogeneous rocky substrates identified a maximum thaw time of 3 h for up to 10 g.
29. The weight of substrate to be collected and processed depends on the type and availability of the substrate, as well as the investigation. Researchers have recovered stable and high DNA content from 1 g samples obtained from brackish substrates for the purposes of detecting benthic invertebrates [79]. Other researchers used 10 g of sediment for investigating marine benthic communities [12, 65, 75]. Reference [32] suggests the use of 10 g as being satisfactory for benthic biodiversity surveys. With this in mind, we recommend processing a standardized weight across all samples, though some researchers may have difficulties collecting the same weight for every sample depending on the environment.
30. This methodology can be used for those seeking to process a portion of a sample that weighs >10 g without fully thawing the entire sample. If the researcher wishes to process the entire sample having a weight much more than 10 g, the entire sample could be thawed prior to transfer into multiple 50 mL tubes at 10 g and processed as normally.
31. For negative field controls that consist of an empty 50 mL tube or an empty zip bag, add the 15 mL of PowerBead Solution into the empty tube or zip bag and close tightly. Invert the tube or zip bag several times, then transfer the liquid from the tube or zip bag into a new bead tube and continue forward to the next vortex step. For negative extraction control, add the 15 mL of PowerBead Solution directly to a new bead tube and continue forward to the next step vortex step. The extraction set should therefore include bead tubes for a field negative control, an extraction negative control, and each experimental substrate sample.

32. All centrifugations should occur at room temperature.
33. It is normal if the supernatants are dark in color as they may still contain some substrate particles. The transfer of supernatant to a clean tube can be done by carefully pipetting from the bed tube to the clean tube. Try to avoid pipetting packed substrate material but subsequent centrifugation steps will aid in the removal of solid material carry-over.
34. Some tips for avoiding cross-contamination while adding 30 mL Solution C4: we recommend the use of 5 mL filtered pipette tips, which would require the changing of each tip 6x. Otherwise, using extreme caution, the solution can be transferred to each sample tube via a sterile 50 mL graduated cylinder or conical tube, but neither item must come in contact with the sample tubes nor their contents, else they must be resterilized to avoid cross-contamination. As usual, all equipment must be sterilized prior to use. After the addition of Solution C4, each collection tube will be nearly full of liquid except the negative control tubes. Therefore, use extreme caution handling the tubes as the liquid can spill.
35. The filters appear to hold a maximum of 16 mL, but we suggest working with about 14 mL to avoid spillage and cross-contamination.
36. The entire volume can require about 4 centrifugations.
37. Avoid splashing any remaining Solution C5 onto the spin filter.
38. When preparing the eDNA extractions for quantification, accurate quantification can be identified using only 2 μL of the eDNA extract and 198 μL of working solution (WS) for a total volume of 200 μL in each tube.

Acknowledgments

We thank Cinzia Verde for inviting this submission. We are grateful to colleagues at Ifremer for their advice with sampling substrates, as well as to the Laboratories of Analytical Biology at the National Museum of Natural History, Smithsonian Institution, for their support in marine water protocol development. We thank Elliot DeVries for his artwork used in the figure. CHCC acknowledges U.S. NSF Polar Programs grant support (ANT1142158, OPP1645087) that enabled sampling of Southern Ocean waters and development of some of the described protocols.

References

1. Taberlet P, Coissac E, Hajibabaei M et al (2012) Environmental DNA. Mol Ecol 21: 1789–1793
2. Rees HC, Maddison BC, Middleditch DJ et al (2014) The detection of aquatic animal species using environmental DNA – a review of eDNA as a survey tool in ecology. J Appl Ecol 51: 1450–1459
3. Sassoubre LM, Yamahara KM, Gardner LD et al (2016) Quantification of environmental DNA (eDNA) shedding and decay rates for three marine fish. Environ Sci Technol 50: 10456–10464
4. Seymour M (2019) Rapid progression and future of environmental DNA research. Commun Biol 80:1–3
5. Thomsen PF, Willerslev E (2015) Environmental DNA – an emerging tool in conservation for monitoring past and present biodiversity. Biol Conserv 183:4–8
6. Foote AD, Thomsen PF, Sveegaard S (2012) Investigating the potential use of environmental DNA (eDNA) for genetic monitoring of marine mammals. PLoS One 7:e41781
7. Stoeckle MY, Soboleva L, Charlop-Powers Z (2017) Aquatic environmental DNA detects seasonal fish abundance and habitat preference in an urban estuary. PLoS One 12:e0175186
8. Knudsen SW, Ebert RB, Hesselsøe M et al (2019) Species-specific detection and quantification of environmental DNA from marine fishes in the Baltic Sea. J Exp Mar Biol Ecol 510:31–45
9. Kelly RP, Port JA, Yamahara KM et al (2014) Using environmental DNA to census marine fishes in a large mesocosm. PLoS One 9: e86175
10. Sawaya NA, Djuhuus A, Closek CJ et al (2018) Assessing eukaryotic biodiversity in the Florida Keys National Marine Sanctuary through environmental DNA metabarcoding. Ecol Evol 9: 1029–1040
11. Harper KJ, Goodwin KD, Harper LR et al (2020) Finding crush: environmental DNA analysis as a tool for tracking the Green Sea turtle *Chelonia mydas* in a marine estuary. Front Mar Sci 6:810
12. Laroche O, Kersten O, Smith CR et al (2020) Environmental DNA surveys detect distinct metazoan communities across abyssal plains and seamounts in the western Clarion Clipperton Zone. Mol Ecol 29:4588–4604
13. Djurhuus A, Pitz K, Sawaya NA et al (2018) Evaluation of marine zooplankton community structure through environmental DNA metabarcoding. Limnol Oceanogr Methods 16: 209–221
14. Berry TE, Saunders BJ, Coghlan ML et al (2019) Marine environmental DNA biomonitoring reveals seasonal patterns in biodiversity and identifies ecosystem responses to anomalous climatic events. PLoS Genet 15:e1007943
15. Djurhuus A, Closek CJ, Kelly RP et al (2020) Environmental DNA reveals seasonal shifts and potential interactions in a marine community. Nat Commun 11:1–9
16. Weltz K, Lyle JM, Ovenden J, Morgan JAT et al (2017) Application of environmental DNA to detect an endangered marine skate species in the wild. PLoS One 12:e0178124
17. Kim P, Kim D, Yoon TJ et al (2018) Early detection of marine invasive species, *Bugula neritina* (Bryozoa: Cheilostomatida), using species-specific primers and environmental DNA analysis in Korea. Mar Environ Res 139: 1–10
18. von Ammon U, Wood SA, Laroche O et al (2019) Linking environmental DNA and RNA for improved detection of the marine invasive fanworm *Sabella spallanzanii*. Front Mar Sci 6:621
19. Sigsgaard EE, Torquato F, Frøslev TG et al (2019) Using vertebrate environmental DNA from seawater in biomonitoring of marine habitats. Conserv Biol 34:697–710
20. Sigsgaard EE, Nielsen IB, Bach SS et al (2016) Population characteristics of a large whale shark aggregation inferred from seawater environmental DNA. Nat Ecol Evol 1:0004
21. Parsons KM, Everett M, Dahlheim M et al (2018) Water, water everywhere: environmental DNA can unlock population structure in elusive marine species. R Soc Open Sci 5: 180537
22. Bohmann K, Evans A, Gilbert MTP et al (2014) Environmental DNA for wildlife biology and biodiversity monitoring. Trends Ecol Evol 29:358–367
23. Goldberg CS, Turner CR, Deiner K et al (2016) Critical considerations for the application of environmental DNA methods to detect aquatic species. Methods Ecol Evol 7: 1299–1307

24. Tsuji S, Takahara T, Hideyuki D et al (2019) The detection of aquatic macroorganisms using environmental DNA analysis—a review of methods for collection, extraction, and detection. Environ DNA 1:99–108
25. Jeunen G-J, Knapp M, Spencer HG et al (2018) Species-level biodiversity assessment using marine environmental DNA metabarcoding requires protocol optimization and standardization. Ecol Evol 9:1323–1335
26. Lear G, Dickie I, Banks J et al (2018) Methods for the extraction, storage, amplification and sequencing of DNA from environmental samples. N Z J Ecol 4:1–51
27. Kusanke LM, Panteleit J, Stoll S et al (2020) Detection of the endangered European weather loach (*Misgurnus fossilis*) via water and sediment samples: testing multiple eDNA workflows. Ecol Evol 10:8331–8344
28. McGee KM, Robinson CV, Hajibabaei M (2019) Gaps in DNA-based biomonitoring across the globe. Front Ecol Evol 7:337
29. Andruszkiewicz EA, Starks HA, Chavez FP et al (2017) Biomonitoring of marine vertebrates in Monterey Bay using eDNA metabarcoding. PLoS One 12:e0176343
30. Bakker J, Wangensteen OS, Chapman DD et al (2017) Environmental DNA reveals tropical shark diversity in contrasting levels of anthropogenic impact. Sci Rep 7:16886
31. Cowart DA, Matabos M, Brandt MI et al (2020) Exploring environmental DNA (eDNA) to assess biodiversity of hard substratum faunal communities on the lucky strike vent field (mid-Atlantic ridge) and investigate recolonization dynamics after an induced disturbance. Front Mar Sci 6:783
32. Brandt MI, Pradillon F, Trouche B et al (2021) Evaluating sediment and water sampling methods for the estimation of deep-sea biodiversity using environmental DNA. Sci Rep 11:7856
33. Herbold CW, Pelikan C, Kuzyk O et al (2015) A flexible and economical barcoding approach for highly multiplexed amplicon sequencing of diverse target genes. Front Microbiol 6:731
34. Hebert PDN, Braukmann TWA, Prosser SWJ et al (2018) A sequel to sanger: amplicon sequencing that scales. BMC Genomics 19:219
35. Einsen JA (2007) Environmental shotgun sequencing: its potential and challenges for studying the hidden world of microbes. PLoS Biol 5:e82
36. Stat M, Huggett MJ, Bernasconi R et al (2017) Ecosystem biomonitoring with eDNA: metabarcoding across the tree of life in a tropical marine environment. Sci Rep 7:12240
37. Coissac E, Riaz T, Puillandre N (2012) Bioinformatic challenges for DNA metabarcoding of plants and animals. Mol Ecol 21:1834–1847
38. Hajibabaei M, Porter TM, Wright M et al (2019) COI metabarcoding primer choice affects richness and recovery of indicator taxa in freshwater systems. PLoS One 14:e0220953
39. Singer GAC, Shekarriz S, McCarthy A et al (2020) The utility of a metagenomics approach for marine biomonitoring. BioRxiv
40. Miya M, Sato Y, Fukunaga T et al (2015) MiFish, a set of universal PCR primers for metabarcoding environmental DNA from fishes: detection of more than 230 subtropical marine species. R Soc Open Sci 2:150088
41. Valentini A, Taberlet P, Miaud C et al (2016) Next-generation monitoring of aquatic biodiversity using environmental DNA metabarcoding. Mol Ecol 25:929–942
42. DiBattista JD, Coker DJ, Sinclair-Taylor TH (2017) Assessing the utility of eDNA as a tool to survey reef-fish communities in the Red Sea. Coral Reefs 36:1245–1252
43. Cowart DA, Murphy KR, Cheng CH (2018) Metagenomic sequencing of environmental DNA reveals marine faunal assemblages from the West Antarctic peninsula. Mar Genomics 37:148–160
44. Zilius M, Samuiloviene A, Stanislauskienė et al (2021) Depicting temporal, functional, and phylogenetic patterns in estuarine diazotrophic communities from environmental DNA and RNA. Microb Ecol 81:36–51
45. Lafferty KD, Garcia-Vedrenne AE, McLaughlin JP et al (2020) At Palmyra atoll, the fish-community environmental DNA signal changes across habitats but not with tides. J Fish Biol 98:415–425
46. Jamy M, Foster R, Barbera P et al (2019) Long-read metabarcoding of the eukaryotic rDNA operon to phylogenetically and taxonomically resolve environmental diversity. Mol Ecol Resour 20:429–443
47. van der Loos LM, Nijland R (2020) Biases in bulk: DNA metabarcoding of marine communities and the methodology involved. Mol Ecol 00:1–19
48. Pedersen MW, Overballe-Petersen S, Ermini L et al (2015) Ancient and modern environmental DNA. Philos Trans R Soc Lond B Biol Sci 370:20130383
49. Wilcox TM, McKelvey KS, Young MK et al (2013) Robust detection of rare species using environmental DNA: the importance of primer specificity. PLoS One 8:e59520

50. Stewart K, Ma H, Zheng J, Zhao J (2017) Using environmental DNA to assess population-wide spatiotemporal reserve use. Conserv Biol 31:1173–1182
51. Baker CS, Steel D, Nieukirk S et al (2018) Environmental DNA (eDNA) from the wake of the whales: droplet digital PCR for detection and species identification. Front Mar Sci 5:133
52. Ficetola GF, Miaud C, Pompanon F et al (2008) Species detection using environmental DNA from water samples. Biol Lett 4:423–425
53. Jerde CL, Mahon AR, Chadderton WL et al (2010) "Sight-unseen" detection of rare aquatic species using environmental DNA. Conserv Lett 4:150–157
54. Ma H, Stewart K, Lougheed S (2016) Characterization, optimization, and validation of environmental DNA (eDNA) markers to detect an endangered aquatic mammal. Conserv Genet Resour 8:561–568
55. LeBlanc F, Belliveau V, Watson E et al (2020) Environmental DNA (eDNA) detection of marine aquatic invasive species (AIS) in Eastern Canada using a targeted species-specific qPCR approach. Manag Biol Invasion 11:201–217
56. Simpson TJS, Dias PJ, Snow M et al (2017) Real-time PCR detection of *Didemnum perlucidum* (Monniot, 1983) and *Didemnum vexillum* (Kott, 2002) in an applied routine marine biosecurity context. Mol Ecol Resour 17: 433–453
57. Edgar RC (2010) Search and clustering orders of magnitude faster than BLAST. Bioinformatics 26:2460–2461
58. Caporaso JG, Kuczynski J, Stombaugh J et al (2010) QIIME allows analysis of high-throughput community sequencing data. Nat Methods 7:335–336
59. Rognes T, Flouri T, Nichols B et al (2016) VSEARCH: a versatile open source tool for metagenomics. PeerJ 4:e2584
60. Callahan BJ, McMurdie PJ, Rosen MJ et al (2016) DADA2: high-resolution sample inference from Illumina amplicon data. Nat Methods 13:581–583
61. Escudié F, Auer L, Bernard M et al (2018) FROGS: find, rapidly, OTUs with galaxy solution. Bioinformatics 34:1287–1294
62. R Core Team (2021) R: a language and environment for statistical computing. R Foundation for Statistical Computing, Vienna. https://www.R-project.org/
63. Turner CR, Uy KL, Everhart RC (2015) Fish environmental DNA is more concentrated in aquatic sediments than surface water. Biol Conserv 183:93–102
64. Buxton AS, Groombridge JJ, Griffiths RA (2018) Seasonal variation in environmental DNA detection in sediment and water samples. PLoS One 13:e0191737
65. Holman LE, de Bruyn M, Créer S et al (2019) Detection of introduced and resident marine species using environmental DNA metabarcoding of sediment and water. Sci Rep 9:11559
66. Brandt MI, Trouche B, Henry N et al (2020) An assessment of environmental metabarcoding protocols aiming at favoring contemporary biodiversity in inventories of deep-sea communities. Front Mar Sci 7:234
67. Deiner K, Lopez J, Bourne S et al (2018) Optimising the detection of marine taxonomic richness using environmental DNA metabarcoding: the effects of filter material, pore size and extraction method. Metabarcoding Metagenom 2:1–15
68. Robson HLA, Noble TH, Saunders RJ et al (2016) Fine-tuning for the tropics: application of eDNA technology for invasive fish detection in tropical freshwater ecosystems. Mol Ecol Resour 16:922–932
69. Turner CR, Barnes MA, Charles CYX et al (2014) Particle size distribution and optimal capture of aqueous macrobial eDNA. Methods Ecol Evol 5:676–684
70. Eichmiller JJ, Miller LM, Sorensen PW (2016) Optimizing techniques to capture and extract environmental DNA for detection and quantification of fish. Mol Ecol Resour 16:56–68
71. Renshaw MA, Olds BP, Jerde CL et al (2015) The room temperature preservation of filtered environmental DNA samples and assimilation into a phenol-chloroform- isoamyl alcohol DNA extraction. Mol Ecol Resour 15: 168–176
72. Wegleitner B, Jerde C, Tucker A et al (2015) Long duration, room temperature preservation of filtered eDNA samples. Conserv Genet Resour 7:789–791
73. Thomsen PF, Kielgast J, Iversen LL et al (2014) Detection of a diverse marine fish fauna using environmental DNA from seawater samples. PLoS One 7(8):e41732
74. Hinlo R, Gleeson D, Lintermans M et al (2017) Methods to maximise recovery of environmental DNA from water samples. PLoS One 12:e0179251
75. Guardiola M, Wangensteen OS, Taberlet P et al (2016) Spatio-temporal monitoring of deep-sea communities using metabarcoding of sediment DNA and RNA. PeerJ 4:e2807
76. Huber DH, Ugwuanyi IR, Malkaram SA et al (2018) Metagenome sequences of sediment from a recovering industrialized Appalachian

River in West Virginia. Genome Announc 6: e00350–e00318

77. Sellers GS, Di Muri C, Gómez A et al (2018) Mu-DNA: a modular universal DNA extraction method adaptable for a wide range of sample types. Metabarcoding Metagenom 2:1–11
78. Giguet-Covex C, Ficetola GF, Walsh K et al (2019) New insights on lake sediment DNA from the catchment: importance of taphonomic and analytical issues on the record quality. Sci Rep 9:14676
79. Wei N, Nakajima F, Tobino T (2018) Effects of treated sample weight and DNA marker length on sediment eDNA based detection of a benthic invertebrate. Ecol Indic 93:267–273
80. Spens J, Evans AR, Halfmaerten D et al (2017) Comparison of capture and storage methods for aqueous macrobial eDNA using an optimized extraction protocol: advantage of enclosed filter. Methods Ecol Evol 8:635–645
81. Majaneva M, Diserud OH, Eagle SHC et al (2018) Environmental DNA filtration techniques affect recovered biodiversity. Sci Rep 8: 4682
82. Wacker S, Fossøy F, Larsen BM (2019) Downstream transport and seasonal variation in freshwater pearl mussel (*Margaritifera margaritifera*) eDNA concentration. Environ DNA 1:64–73

Chapter 12

Metataxonomic Analysis of Bacterial Diversity Associated with Marine Organisms

Nadia Ruocco, Roberta Esposito, Valerio Zupo, and Maria Costantini

Abstract

In the present era marine biotechnologies are dominating the world of scientific research assisted by great advances in molecular biology techniques, and microbial community analysis provides useful tool to investigate their diversity and their potential for biotechnological applications. In fact, several marine organisms harbor diverse microbial associated communities, which play key roles for their host functioning and are rich sources of bioactive natural compounds. Here, we describe the fundamental steps of metataxonomic analysis of microbial communities associated with marine organisms.

Key words DNA, Heat-map, Krona plot, Microorganism, PCA analysis, Taxonomy BarPlot

1 Introduction

Marine organisms represent important sources of a wide range of bioactive natural compounds due to their vast biodiversity and molecular diversity of secondary metabolites synthetized. Several marine-derived natural products stem from associated microorganisms, such as the case of marine sponges [1]. In fact, sponges are considered a "gold mine" with respect to the diversity of their secondary metabolites [2], harboring several microorganisms, such as bacteria, cyanobacteria, archaea and unicellular eukaryotes. These microorganisms comprise up to 40% of the sponge volume and they play key roles in the evolution and ecological success of sponges [3–7]. Because of the presence of these associated microorganisms, sponges have an added value as potential providers of drugs against several diseases, such as cancer and inflammations [8–10].

Microorganisms are often difficult to culture in the laboratory but in the last years the high-throughput sequencing technologies transformed microbiology. In fact, marine microorganisms become

Nadia Ruocco and Roberta Esposito contributed equally to this work.

Cinzia Verde and Daniela Giordano (eds.), *Marine Genomics: Methods and Protocols*,
Methods in Molecular Biology, vol. 2498, https://doi.org/10.1007/978-1-0716-2313-8_12,

more accessible for functional analyses through cultivation-independent metagenomic approaches, thanks to the advances in genomics, metagenomics, proteomics, synthetic biology, and bioinformatics, so providing a great challenge in the discovery of novel bioactive compounds [11]. These molecular approaches allow to explore in depth the full genetic potential of microorganisms for drug discovery [12].

Here we report a metataxonomic analysis, describing the steps to perform these experiments used to characterize the assemblage of microorganisms living in a given environment [13]. In particular, metataxonomic analysis is a high-throughput process based on the sequencing of variable regions of the bacterial 16S rRNA gene [14, 15]. It is widely used for community-wide taxonomic classifications to characterize the entire microbiota and create a metataxonomic tree, showing the relationships between all sequences obtained.

2 Materials

2.1 DNA Extraction

1. DNeasy® PowerSoil® Pro Kit (QIAGEN) for metagenomic DNA extraction, according to the instruction manual of supplier. Material included:
   (a) PowerBead Pro Tube: allows the tissue to get disrupted.
   (b) Solution CD1.
   (c) Solution CD2 (*see* **Note 1**).
   (d) Solution CD3 (*see* **Note 2**).
   (e) Solution EA.
   (f) Solution C5 (*see* **Note 3**).
   (g) Solution C6 (*see* **Note 4**).

   Solutions are ready-to-use and stored at room temperature.
2. Filtered pipette tips.
3. Sterilized 1.5/2.0 mL tubes.
4. Microcentrifuge for 1.5/2.0 mL tubes.
5. Washed pipettes with 70% ethanol.
6. Appropriate boxes for liquid and solid wastes.
7. Tissue Lyzer.
8. NanoDrop spectrophotometer.

2.2 Preparation of 0.8% Agarose Gel

1. 0.8 g agarose.
2. 5× TAE buffer stock solution (40 mM Tris-acetate, 1 mM EDTA, pH 8.0):

(a) EDTA 0.5 M (pH 8.0): add 186.1 g of disodium EDTA to 800 mL of MilliQ H_2O. Stir vigorously on a magnetic stirrer. Adjust the pH to 8.0 with sodium hydroxide (NaOH; ~20 g of pellets), fill to a final volume of 1 L with MilliQ water, and sterilize by autoclaving. The disodium salt of EDTA will not go into solution until the pH of the solution is adjusted to ~8.0 by the addition of NaOH.

(b) Tris-base (molecular weight = 121.14 g/mol).

(c) 100% acetic acid.

3. Microwave for agarose melting.
4. Ethidium bromide (stock concentration of 10 mg/mL) (*see* **Note 5**).
5. Ready-to-load 1 Kb ladder (0.5 kb–10 kb).
6. 6× Orange Loading Dye for DNA.
7. Electrophoresis accessories (electrophoresis unit, gel tray, electrodes, well comb, power supply).
8. GelDoc Imaging System.

2.3 Sequencing of 16 rRNA on Illumina MiSeq Platform

1. Sample at final concentration of 30 ng/μL.
2. Thermocycler.
3. Polymerase Chain Reaction (PCR) of 16S rRNA with reagents (*see* **Note 6**).
4. 96-well 0.2 mL PCR plate.
5. Forward and Reverse primers targeting the hypervariable V3-V4 region of 16S rRNA [16], widely used for the taxonomic studies of microbial communities. The sequences of forward and reverse 16S rRNA primers as follows:

   Forward = 5′-TCGTCGGCAGCGTCAGATGTGTATAAGA GACAGCCTACGGGNGGCWGCAG-3′,

   Reverse = 5′-GTCTCGTGGGCTCGGAGATGTGTATAA GAGACAGGACTACHVGGGTATCTAATCC-3′.
6. PCR clean-up kit, according to the instruction manual of supplier.

2.4 Taxonomic Classification

1. Naive Bayesian Classifier, a simple "probabilistic classifier" based on applying Bayes' theorem with strong (naïve) independence assumptions between the features, following procedures reported https://scikit-learn.org/stable/modules/naive_bayes.html#multinomial-naive-bayes [17].
2. SILVA 132 database, a freely online database that provides comprehensive, quality checked and regularly updated datasets of aligned small (16S/18S) and large subunit (23S/28S)

rRNA sequences for Bacteria, Archaea, and Eukarya, available at https://www.arb-silva.de/documentation/release-132/ [18].

3. QIIME 2, a powerful, extensible and decentralized microbiome analysis package with a focus on data and analysis transparency. QIIME 2 enables researchers to start an analysis with raw DNA sequence data and finish with publication-quality figures and statistical results [19, 20]. The user guide for package downloading is available at https://docs.qiime2.org/2020.11/install/.
4. *Heatmapper* Software, a freely available web server (http://www.heatmapper.ca/) that allows users to interactively visualize their data in the form of heat maps through an easy-to-use graphical interface [21]. *Heatmapper* allows users to generate cluster and visualize expression-based heat maps from transcriptomic, proteomic and metabolomic experiments; pairwise distance maps; correlation maps; image overlay heatmaps; latitude and longitude and geopolitical (choropleth) heatmaps.

3 Methods

3.1 DNA Extraction According to the DNeasy® PowerSoil® Pro Kit (QIAGEN) Protocol

1. Excise up to 250 mg of tissue and put it in PowerBead Pro Tube.
2. Add 800 μL of Solution CD1 and briefly vortex.
3. Disrupt in Tissue Lyzer at 24 Hz for 2 min.
4. Centrifuge the PowerBead Pro Tube at 15,000 × *g* for 1 min (*see* **Note 7**).
5. Take the supernatant and add 200 μL of CD2 Solution.
6. Take the supernatant and add the Solutions CD2 and CD3 (*see* **Note 8**). After the addition of each solution proceed with a centrifugation step (15,000 × *g* for 1 min) and take the supernatant.
7. Load 650 μL of sample to a spin column (supplied by DNeasy® PowerSoil® Pro Kit) and centrifuge at 15,000 × *g* for 1 min.
8. Discard the flow-through and perform two wasting steps with Solutions EA and C5.
9. Add to the column 50–100 μL of Solution C6 and centrifuge for 1 min at full speed to elute DNA (*see* **Note 9**). The volume of Solution C6 depends on the weight of starting material. The yield of DNA may vary depending on the starting material (*see* **Note 10**).
10. Measure DNA quantity (ng/μL) and purity (A260/A280; A260/A230) by a NanoDrop spectrophotometer (*see* **Note 11**).

3.2 Evaluation of DNA Integrity on 0.8% Agarose Gel Electrophoresis

1. Preparation of 50× TAE stock: weight 242 g of Tris-base and dissolve under stirring in 700 mL of MilliQ water, add 57.1 mL of 100% acetic acid and 100 mm of 0.5 M EDTA (pH 8.0), fill until 1 L of final volume.
2. Dilute 5× TAE solution to 1X in 500 mL of MilliQ water.
3. Weight 0.8 g of agarose.
4. Add the agarose to 100 mL of 1× TAE buffer in glass bottle.
5. Microwave for 1–3 min until the agarose is completely dissolved.
6. Let the agarose solution cool down to about 50 °C for about 5 min.
7. Add 3 μL of ethidium bromide (final concentration of approximately 0.2–0.5 μg/mL) and pour the gel into the tray with the well comb in place.
8. Push away any bubbles from the well comb or toward the sides/edges of the gel with a pipette tip.
9. Let the gel to solidify (*see* **Note 12**).
10. Put the gel into the electrophoresis unit already filled of 1× TAE buffer.
11. Mix 200 ng of DNA sample to 1 μL of 6× Orange Loading Dye.
12. Load 10 μL of each sample and 5 μL of 1 Kb DNA ladder into separate wells.
13. Run the samples for 30 min at 80 V.
14. Visualize the high molecular weight DNA band under GelDoc Imaging System.

3.3 Sequencing on Illumina MiSeq Platform

1. Perform a PCR of 16S rRNA by using the following quantities of reagent per sample: 2.5 μL of microbial genomic DNA (5 ng/μL in 10 mM Tris pH 8.5), 5 μL of each Forward and Reverse primer and 12.5 μL of 2× KAPA HiFi HotStart ReadyMix to a final volume of 25 μL (*see* **Note 13**). Thermocycler conditions for PCR amplification: initial denaturation at 95 °C for 3 min, 25 cycles of 95 °C (30 s), 55 °C (30 s), 72 °C (30 s), final extension at 72 °C for 5 min, hold at 4 °C.
2. Proceed with a PCR clean-up to purify the V3-V4 amplicon from free primers and primer dimer species.
3. Apply an additional limited-cycle amplification step to add multiplexing indices and Illumina sequencing adapters by using a Nextera XT Index Kit (*see* **Note 14**). For more information, see the Illumina reference guide protocol [22].

4. Perform 16S sequencing on Illumina MiSeq Platform with 2×300 bp paired-end reads.
5. Use bioinformatic packages to filter, denoise, and remove chimeric sequences (*see* **Note 15**).
6. Produce high quality Amplicon Sequence Variants (ASV) datasets [23] (*see* **Note 16**).

3.4 Taxonomic Classification

1. Assign taxonomy to each ASV by "homemade" Naive Bayesian Classifier trained on V3-V4 16S sequences of SILVA 132 database.
2. Use QIIME 2 (Quantitative Insights Into Microbial Ecology) workflow for processing, analyzing, and visualizing microbiome data from ASV counts [19, 20]. Examples of QIIME 2 visualizations available are reported in Fig. 1 (Taxonomy BarPlot) and Fig. 2 (KronaPlot).

Fig. 1 Taxonomy BarPlot. Species diversity of the symbiotic communities living within eight marine samples are showed (Sample codes: 1–8). Each taxon is highlighted by a different color. Values are reported as percentage of relative frequency

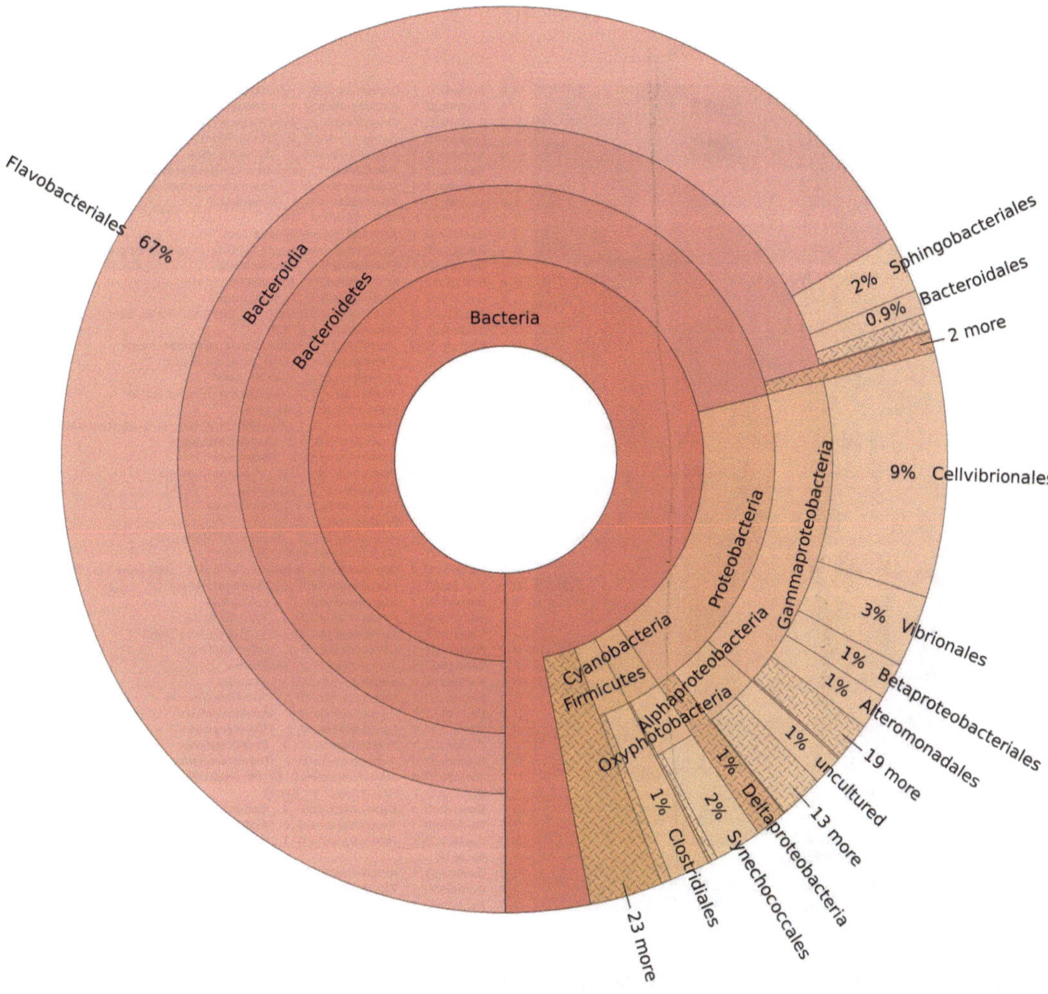

Fig. 2 Krona Plot. The relative abundance of the bacterial taxa associated to a marine sample are reported. Bacterial groups identified in trace levels are indicated as number of species plus "more"

3. To evaluate the difference of species composition among several specimens, generate heatmaps by *Heatmapper* Software using ASV counts of the most abundant taxa. To group samples by similar expression and compute distances between rows and columns select a Clustering method (None, Average, Centroid, Complete, or Single) and Distance Measurement algorithm (Euclidean, Pearson, Kendall's Tau, Spearman Rank Correlation, or Manhattan). In Fig. 3 the clustering of eight marine samples collected from different geographical sites is reported.
4. Perform a Principal Component Analysis (PCA) to reduce the complexity of sequencing data and reveal how much variance is accounted in the first few Principal Components (PC).

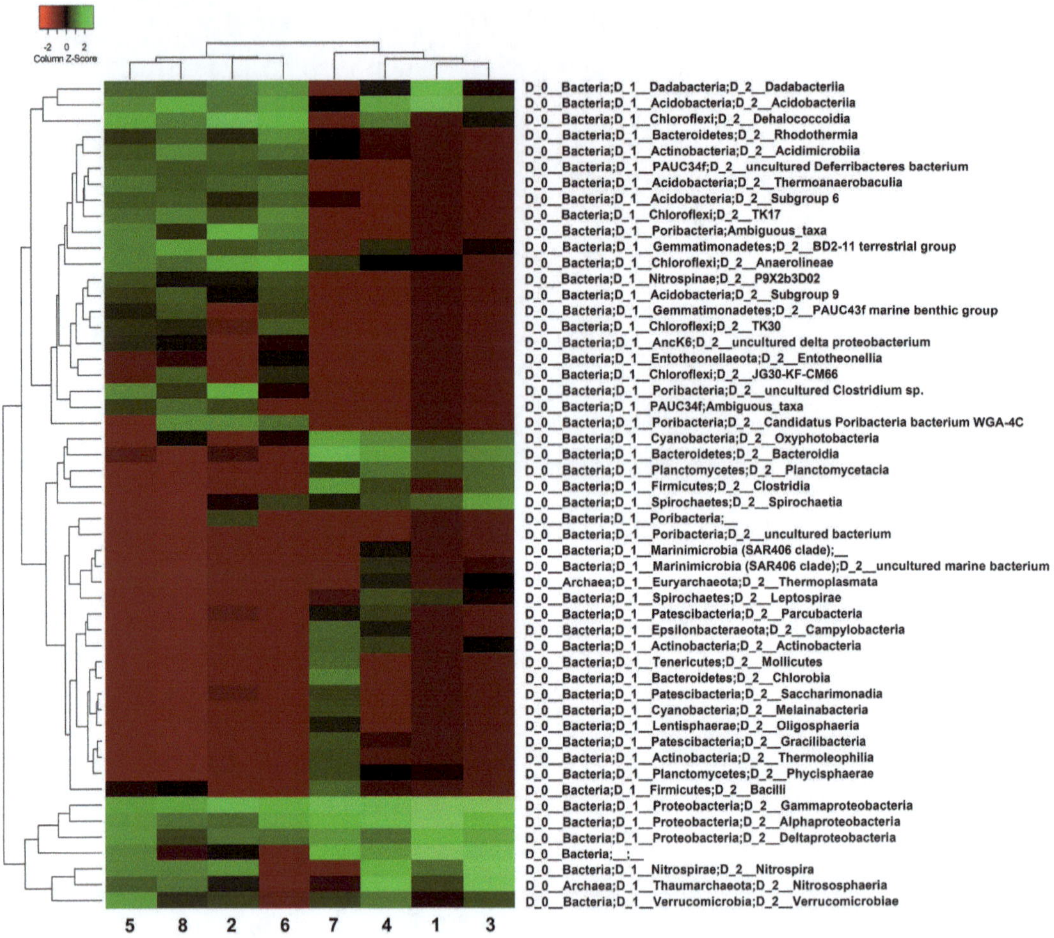

Fig. 3 Heatmap. The abundance of the most representative taxa is highlighted. Below are reported sample codes (1–8). Scaling is done by column and clustering is performed by using Average Linkage Method and Euclidean Distance Measurement. Amplicon Sequence Variants (ASV) values are normalized as Log10. Color code: green = high abundance, red = low abundance

PCA is an effective tool when the first few principal components account for most of the system variance. PCA is described in Fig. 4 which reports sequence data of eight marine samples collected from two different geographical sites, (a) highlighting how they cluster between the principal components PC1/PC2 (Fig. 4a) and PC2/PC3 (Fig. 4b); (b) comprising the highest amount of variability (76.5%).

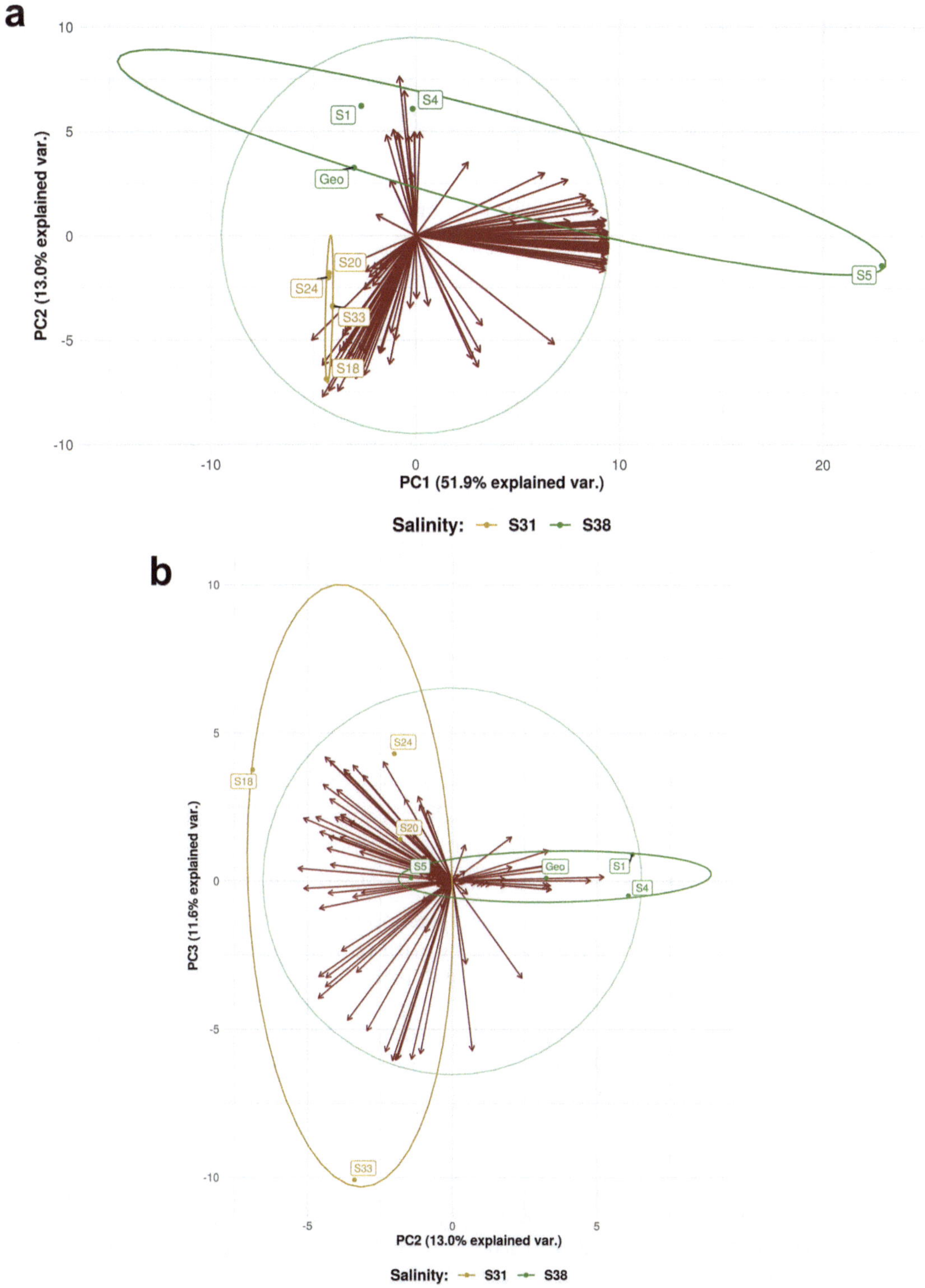

Fig. 4 PCA analysis - Biplot of individuals ($n = 8$) and explanatory variables ($n = 167$) of principal components PC1/PC2 (**a**) and PC2/PC3 (**b**) of metataxonomic data. The majority of species diversity is explained by the first three PCs (76.5%), with PC1 and PC2 having the highest contribution (PC1 = 51.9% and PC2 = 13%). Biplot

4 Notes

1. The removal of organic and inorganic matter that may reduce DNA purity is necessary to avoid inhibition of downstream applications.
2. At high salt concentrations, DNA binds tightly to the silica of Column filter membrane.
3. Solution C5 removes residual salt, humic acid, and other contaminants while allowing the DNA to bind the silica membrane.
4. We recommend storing the DNA frozen −30 °C to −15 °C or − 90 °C to −65 °C, as Solution C6 does not contain EDTA.
5. Ethidium bromide is a known mutagen. Wear a lab coat, eye protection, and gloves when working with this chemical.
6. 16 rRNA gene is a conserved molecular marker useful for phylogenetic studies in microbial ecology.
7. Perform all centrifugation steps at room temperature (15–25 °C).
8. If Solution CD3 precipitated, heat at 60 °C until precipitate dissolves.
9. To increase the DNA yield, incubate 5 min at room temperature with solution C6 prior to centrifugation.
10. Expect a yield range of 1–15 μg from 250 mg of tissue.
11. Use sterilized MilliQ water to clean and blank the instrument.
12. The gel will reach a complete solidification at 4 °C for 10–15 min or at room temperature for 20–30 min.
13. Store all reagents at −15 °C to −25 °C.
14. Through Nextera technology, DNA is simultaneously fragmented and tagged with sequencing adapters in a single-tube enzymatic reaction. Nextera XT supports ultra-low DNA input of only 1 ng. It supports a wide range of input samples, including small genomes, PCR amplicons greater than 300 bp, plasmids, microbial genomes, concatenated amplicons, and double-stranded cDNA.

Fig. 4 (continued) shows the PCA scores of the explanatory variables as vectors (in red) and individuals grouped for the site of collection which are characterized by a significant difference in water salinity measured as Practical Salinity Unit (PSU) (S38 = 38 PSU and S31 = 31 PSU). The circle represents the equilibrium of variables contribution. The weight of each variable is reflected by the magnitude of the corresponding eigenvectors (higher magnitude-higher importance). Vectors pointing toward similar (small angle) and opposite directions (0 to 180 degrees) indicate positively or negatively correlated variables, and vectors at approximately right angles (90 to 270 degrees) suggest a low correlation. The eight samples are indicated as S1, S4, S5, S18, S20, S24, S33, and Geo

15. Denoising pipeline is a bioinformatic approach to correct sequencing errors, thus improving taxonomic resolution. DADA2 is an open-source R package (https://github.com/benjjneb/dada2) for modeling and correcting Illumina-sequenced amplicon errors.
16. ASV is a term used to refer to single DNA sequences recovered from a high-throughput marker gene analysis. These amplicon reads are created following the removal of erroneous sequences generated during PCR and sequencing. This allows ASVs to distinguish sequence variation by a single nucleotide change. ASVs are utilized to classify groups of species based on DNA sequences, finding biological and environmental variation and to determine ecological patterns.

Acknowledgments

Nadia Ruocco was supported by a research grant "Antitumor Drugs and Vaccines from the Sea (ADViSE)" project (PG/2018/0494374). Roberta Esposito was supported by a PhD (PhD in Biology, University of Naples Federico II) fellowship funded by the Photosynthesis 2.0 project of the Stazione Zoologica Anton Dohrn. We also thank Michelle Sonnessa and Federico Andreani from Bio-Fab Research (Rome, Italy) for bibliographic material on metataxonomic analysis.

References

1. König GM, Kehraus S, Seibert SF et al (2006) Natural products from marine organisms and their associated microbes. Chembiochem 7: 229–238
2. Sipkema D, Franssen MCR, Osinga R et al (2005) Marine sponges as pharmacy. Mar Biotechnol 7:142–162
3. Quince C, Walker AW, Simpson JT et al (2017) Shotgun metagenomics, from sampling to analysis. Nat Biotechnol 35:833–844
4. Ruocco N, Costantini S, Palumbo F et al (2017) Marine sponges and bacteria as challenging sources of enzyme inhibitors for pharmacological applications. Mar Drugs 15(6): 173
5. Konstantinou D, Gerovasileiou V, Voultsiadou E et al (2018) Sponges-cyanobacteria associations: global diversity overview and new data from the Eastern Mediterranean. PLoS One 13:1–22
6. Konstantinou D, Voultsiadou E, Panteris E et al (2021) Revealing new sponge-associated cyanobacterial diversity: novel genera and species. Mol Phylogenet Evol 155:106991
7. Romano G, Ianora A, Costantini M et al (2016) Marine microorganisms as a promising and sustainable source of bioactive molecules. Mar Environ Res 128:58–69
8. Costantini S, Romano G, Rusolo F et al (2015) Anti-inflammatory effects of a methanol extract from the marine sponge *Geodia cydonium* on the human breast cancer MCF-7 cell line. Mediat Inflamm 2015:204975
9. Costantini S, Guerriero E, Teta R et al (2017) Evaluating the effects of an organic extract from the mediterranean sponge *Geodia cydonium* on human breast cancer cell lines. Int J Mol Sci 18:2112
10. Giordano D, Costantini M, Coppola D et al (2018) Biotechnological applications of bioactive peptides from marine sources. Adv Microb Physiol 73:171–220
11. Wilson MC, Piel J (2013) Metagenomic approaches for exploiting uncultivated bacteria

as a resource for novel biosynthetic enzymology. Chem Biol 20:636–647
12. Bashir Y, Pradeep Singh S, Kumar Konwar B (2014) Metagenomics: an application based perspective. Chinese J Biol 2014:1–7
13. Lederberg J, McCray AT (2001) "Ome sweet omics" - a genealogical treasury of words. Scientist 15:8
14. Marchesi JR, Ravel J (2015) The vocabulary of microbiome research: a proposal. Microbiome 3:1–3
15. Hilton SK, Castro-Nallar E, Pérez-Losada M et al (2016) Metataxonomic and metagenomic approaches vs. culture-based techniques for clinical pathology. Front Microbiol 7:484
16. Klindworth A, Pruesse E, Schweer T et al (2013) Evaluation of general 16S ribosomal RNA gene PCR primers for classical and next-generation sequencing-based diversity studies. Nucleic Acids Res 41:e1
17. Zhang H (2004) The optimality of Naive Bayes. Proc Seventeenth Int Florida Artif Intell Res Soc Conf FLAIRS 2004 2:562–567
18. Quast C, Pruesse E, Yilmaz P et al (2013) The SILVA ribosomal RNA gene database project: improved data processing and web-based tools. Nucleic Acids Res 41:590–596
19. Estaki M, Jiang L, Bokulich NA et al (2020) QIIME 2 enables comprehensive end-to-end analysis of diverse microbiome data and comparative studies with publicly available data. Curr Protoc Bioinformatics 70:1–46
20. Bolyen E, Rideout JR, Dillon MR et al (2019) Reproducible, interactive, scalable and extensible microbiome data science using QIIME 2. Nat Biotechnol 37:852–857
21. Babicki S, Arndt D, Marcu A et al (2016) Heatmapper: web-enabled heat mapping for all. Nucleic Acids Res 44:W147–W153
22. Illumina Inc. (2018) Nextera XT DNA Library Prep Kit. 1–24
23. Schloss PD (2021) Amplicon sequence variants artificially split bacterial genomes into separate clusters. bioRxiv

Chapter 13

From Sequences to Enzymes: Heterologous Expression of Genes from Marine Microbes

Angelica Severino, Alessandro Coppola, Monica Correggia, Costantino Vetriani, Donato Giovannelli, and Angelina Cordone

Abstract

Heterologous expression is an easy and broadly applicable experimental approach widely used to investigate protein functions without the need to genetically manipulate the original host. The approach is used to obtain large quantities of the desired protein, which can be further analyzed from a biochemical, structural, and functional perspective. The expression system consists of three main components: (1) a foreign DNA sequence coding for the protein of interest; (2) a suitable expression vector; (3) a suitable host (bacterial, yeast, or mammalian cells) which does not encode or express the protein of interest. Here, we show how to apply an *Escherichia coli*-based expression system to overexpress protein encoding genes from marine microbes.

Key words *E. coli*, Expression vector, Heterologous expression, Recombinant protein, Uncultured microorganism, Metagenome

1 Introduction

Heterologous expression refers to the expression of a gene into a host organism that does not naturally express the gene [1]. This process allows for the production and isolation of a protein of interest from difficult to grow or uncultured organisms to investigate the protein function and its biological implications [2]. Conventional genetic platforms established for heterologous expression consist of bacterial, yeast, algal, or mammalian cell systems. Bacterial heterologous hosts are the most desirable in terms of low-cost production and availability of genetic tools for gene cloning [3]. Heterologous expression systems are widely used for large-scale production of proteins with significant biotechnological applications. One of the most famous examples is *Taq* DNA Polymerase, a well-known and widely used enzyme, initially isolated from its native host, *Thermus aquaticus*, in 1976 [4], and then cloned and

Cinzia Verde and Daniela Giordano (eds.), *Marine Genomics: Methods and Protocols*,
Methods in Molecular Biology, vol. 2498, https://doi.org/10.1007/978-1-0716-2313-8_13,

overexpressed in *Escherichia coli* [5]. Heterologous expression offers additional advantages besides the potential for high yields, including heterologous expression of genes obtained from uncultured microorganisms [6], metagenomes or Metagenome Assembled Genomes (MAGs) [7] obtained from diverse samples (e.g., marine water, sediments, fluids), as well as expression of ancestral reconstructed genes obtained from comparative genomic analysis (as reported in Chapter 5). This has the potential to unlock information from the large fraction of genes encoding hypothetical proteins [8]. In marine microbiology, heterologous expression approaches can be coupled to comparative genomics methods to study the functions and the evolutionary history of genes of interest, providing insight into their biochemical, structural, and functional properties.

Heterologous expression methods are well established in molecular biology, and it is achieved by cloning the DNA fragment, previously isolated from the organism or metagenome of interest, into a suitable expression vector and introducing the recombinant plasmid into a host that will provide the biochemical apparatus to generate and express the "foreign" protein. *E. coli* is the most popular host used to express recombinant proteins for several reasons: (1) it is easy to handle; (2) it grows rapidly; (3) it is easy to transform; (4) it does not have specific nutritional needs, and (5) a large number of expression vectors are available for *E. coli* [3]. These features make *E. coli*-based heterologous expression a feasible and versatile tool to probe marine functional diversity.

Here we illustrate how to apply a heterologous expression method to express a protein coding gene from marine microbes. Briefly, the target DNA is cloned downstream and in frame with a sequence that encodes an N-terminal polyhistidine tag (His-tag), under the control of a strong bacteriophage T7 promoter. The His-tag acts as a metal-binding domain essential for the subsequent protein purification step. *E. coli* BL21 (DE3) is an ideal host for recombinant protein expression of T7 regulated genes because it contains the DE3 bacteriophage lambda lysogen (λDE3) that carries the T7 RNA polymerase gene under the *lacUV5* control. Using Isopropyl-β-D-1-thiogalactopyranoside (IPTG), an allolactose analog, it is possible to induce the expression of T7 RNA polymerase to express the gene cloned under the T7 promoter's control [3]. The method requires several individual steps: (1) cloning and propagation of the recombinant expression vector, (2) expression and extraction of the recombinant protein, (3) detection of the recombinant protein by SDS-PAGE and Western blotting analysis.

2 Materials

2.1 Making *Escherichia coli* DH5α and BL21 (DE3) Chemically Competent Cells

1. Sterile loops, 2 mL sterile tubes, and 100 mL flasks.
2. A basket filled with ice.
3. Cuvette UV/VIS.
4. Tryptone–yeast (TY) liquid medium (1 L final volume): 10 g tryptone, 5 g yeast extract, 8 g NaCl. Add into a cylinder containing at least 800 mL of deionized water and a small stirring magnet at the bottom. Mix until the components are completely dissolved and add water up to 1 L. Autoclave before use.
5. Tryptone–yeast (TY) agar plates: add 1.8% agar (w/v) to TY liquid medium and autoclave before use (*see* **Note 1**).
6. A fresh stock of 50 mM $CaCl_2$ solution (500 mL final volume): add 2.77 g of $CaCl_2$ into a beaker containing 400 mL of water. Let it dissolve completely and add water up to 500 mL. Autoclave before use (*see* **Note 2**).
7. 100% glycerol (*see* **Note 3**).
8. *E. coli* DH5α glycerol stock.
9. *E. coli* BL21 (λDE3) glycerol stock.

2.2 Molecular Cloning Step

1. 0.2 mL and 0.5 mL sterile tubes.
2. Template genomic or metagenomic DNA to be used to amplify the target protein encoding gene.
3. Taq DNA polymerase with proofreading activity with the appropriate buffer.
4. Forward and reverse primers targeting the 5′ and 3′ of the gene of interest (*see* **Note 4**).
5. 10 mM dNTP.
6. Sterile deionized H_2O (ddH_2O).
7. Thermocycler.
8. pRSET-A expression vector.
9. Restriction endonucleases with appropriate buffers.
10. T4 ligase enzyme with appropriate buffer.
11. A tube rack.
12. Gel extraction kit, according to the instruction manual of the supplier.
13. PCR purification kit, according to the instruction manual of the supplier.

2.3 Bacterial Transformation

1. 90 mm TY agar plates supplemented with ampicillin (100 μg/mL) (*see* **Note 5**).
2. TY liquid medium (*see* **Note 3**).
3. *E. coli* DH5α or BL21 (λDE3) competent cells.
4. Plasmid DNA or ligation mix (product of Subheading 2.2).
5. 1.5 mL sterile tubes.
6. A basket filled with ice.
7. A tube rack.

2.4 Agarose Gel Electrophoresis

1. Comb, gel tray and electrophoresis apparatus.
2. 1% Agarose gel, made in 1× Tris–acetic acid–EDTA (TAE) buffer and casted to size according to the gel tray used.
3. 10 mg/mL ethidium bromide solution, molecular grade.
4. 50× TAE: 2 M Tris base, 1 M acetic acid, 0.5 M EDTA, pH 8.6. To prepare 500 mL 50× TAE, weigh 121 g of Tris base and dissolve in approximately 300 mL of deionized water. Add 28.6 mL of 100% acetic acid and 50 mL of 0.5 M EDTA (*see* **Note 6**).
5. 6× Loading buffer: 0.25% bromophenol blue, 0.25% xylene cyanol, 30% glycerol, water.
6. UV light transilluminator.

2.5 DNA Plasmid Extraction

1. TY liquid medium (*see* **Note 3**).
2. A basket filled with ice.
3. Ampicillin stock solution (100 mg/mL). To prepare ampicillin 100 mg/mL stock, weigh 1 g of ampicillin (keep the powder stock stored at 4 °C). Transfer into a sterile 15 mL falcon having care not to miss any granule of the ampicillin powder (help yourself with a washing flask). Add water up to 10 mL. Vortex to help to dissolve ampicillin. Filter the solution with a 0.22 μM filter and aliquot into sterile 2 mL tubes. Store at −20 °C.
4. Solution 1: 50 mM glucose, 10 mM EDTA, 25 mM Tris–HCl, pH 8.0. Prepare Tris–HCl 1 M solution and adjust the pH to 8. Weigh 0.90 g of glucose and add to a beaker containing 50 mL of water. Add 2.5 mL of Tris–HCl 1 M, 2 mL of 0.5 M EDTA. Add water up to 100 mL and mix until dissolved. Adjust the pH to 8 if needed (*see* **Note 3**).
5. Solution 2: 0.2 M NaOH, 1% SDS. Mix 10 mL of 10% SDS and 5 mL of 4 M NaOH into 8.5 mL of distilled water. Add deionized water up to 100 mL (*see* **Note 3**).

6. 3 M Sodium acetate pH 5.2 (100 mL final volume): Add 24.6 g of sodium acetate to 80 mL of water. Adjust pH if needed and add water up to 100 mL (*see* **Note 3**).
7. Phenol–chloroform–isoamyl alcohol (25:24:1, v:v).
8. EtOH 100%, EtOH 70% (*see* **Notes** 7 and **10**): To prepare 100 mL of EtOH 70%, add 30 mL of EtOH to 70 mL of deionized water. Store at −20 °C.
9. RNase A stock solution (10 mg/mL). To prepare RNase A 10 mg/mL stock, weigh 10 mg of RNase A (keep the powder stock stored at 4 °C). Transfer into a 2 mL sterile tube having care not to miss any granule of the RNase powder (help yourself with a washing flask and a rack to stabilize the sterile tube). Add water up to 1 mL. Vortex to help to dissolve the powder. Store at −20 °C.
10. Sterile 100% glycerol.
11. 1.5 mL, 2 mL, and 15 mL sterile tubes.

2.6 Protein Expression and Extraction

1. *E. coli* BL21 (DE3) competent cells.
2. Recombinant plasmid (product of DNA plasmid extraction, Subheading 2.5).
3. TY agar plate supplemented with ampicillin (*see* **Note 6**).
4. TY liquid medium.
5. Ampicillin stock solution (100 mg/mL) (*see* Subheading 2.5, **item 3**).
6. Sterile loops.
7. 2 mL, 1.5 mL, 50 mL sterile tubes, and flasks.
8. Cuvette for UV/VIS.
9. 100 mM IPTG stock solution. To prepare a stock of IPTG 100 mM, weigh 0.24 g of IPTG (keep the powder stock stored at −20 °C). Transfer into a sterile 15 mL falcon tube having care not to miss any granule of the powder (help yourself with a washing flask). Add deionized water up to 10 mL. Vortex to help to dissolve the powder. Filter the solution with a 0.22 μM filter and aliquot into sterile 2 mL tubes. Store at −20 °C.
10. Phosphate buffered saline (PBS)–Triton X-100 1% solution (1 L final volume): 0.1 M NaCl, 15 mM Na_2HPO_4, 4.5 mM NaH_2PO_4, 1% Triton X-100, pH 7.3.
11. Dry ice–ethanol bath or liquid nitrogen.
12. 37 °C water bath.

2.7 Bradford Assay

1. 2 mL tubes.
2. Cuvette for UV/VIS.
3. Pasteur pipettes.

4. 0.1 mg/mL bovine serum albumin (BSA) stock solution.
5. Bradford reagent (1 L final volume): 100 mg Coomassie Brilliant Blue G250, 50 mL 95% ethanol, 100 mL 85% (w/v) phosphoric acid, add ddH_2O up to 1 L.
6. Deionized H_2O (ddH_2O).
7. Spectrophotometer UV/VIS.

2.8 SDS–Polyacrylamide Electrophoresis (SDS-PAGE)

1. Glass plates, comb, spacer, casting gel apparatus.
2. Resolving gel (12.5%) (10 mL final volume): 4 mL 30% acrylamide, 2.6 mL 1 M Tris–HCl pH 8.8, 100 μL 10% SDS, 100 μL 10% APS, ddH_2O up to 10 mL, 5 μL tetramethylethylenediamine (TEMED) (*see* **Note 8**).
3. Stacking gel (5%) (4 mL final volume): 670 μL 30% acrylamide, 500 μL 1 M Tris–HCl pH 6.8, 40 μL 10% SDS, 40 μL 10% APS, ddH_2O up to 4 mL, 4 μL TEMED (*see* **Note 8**).
4. 2-propanol 100%.
5. Deionized H_2O (ddH_2O).
6. Clean absorbent paper.
7. Electrophoresis apparatus.
8. 10× Tris–glycine–SDS (TGS) running buffer: 0.25 M Tris base, 1.92 M glycine, 1% SDS, pH 8.3 (*see* **Note 9**).
9. 5× Loading sample buffer (10 mL final volume): 1 mL 1 M Tris–HCl pH 6.8, 1.4 mL 10% SDS, 2 mL of 20% glycerol, 2.5 mL of 25% β-mercaptoethanol and 1% Bromophenol Blue. Add water up to 10 mL and aliquot into clean 2 mL tubes (*see* **Note 10**).
10. Coomassie staining solution: 0.2% Coomassie Brilliant-blue R, 25% 2-propanol and 10% acetic acid (*see* **Note 11**).
11. Destaining solution: 10% acetic acid, 30% denatured ethanol. Prepare 500 mL of destain solution adding 50 mL of acetic acid and 150 mL denatured ethanol to 300 mL of ddH_2O (*see* **Note 12**).

2.9 Immunoblotting and Detection

1. A spatula and clamps.
2. Nitrocellulose membrane 6 cm × 9 cm.
3. 6 absorbent filter paper 6 cm × 9 cm for a single gel.
4. Plastic containers.
5. 10× Tris–glycine (*see* **Note 13**): 0.25 M Tris base pH 8.3, 1.92 M glycine.
6. 100% methanol.
7. Ponceau S.

8. 10× PBS: 80 mM $NaHPO_4$; 20 mM $NaH_2PO_4 \times 2\ H_2O$; 100 mM NaCl, pH 7.2.
9. 5% milk-blocking solution: add 2.5 g skimmed milk to 50 mL of 1× PBS (*see* **Note 14**).
10. Anti-His tag antibody.
11. Chemiluminescent substrate.
12. Deionized H_2O.

3 Methods

Carry out all the procedures under sterile conditions, unless otherwise specified.

3.1 Cloning and Propagation of the Recombinant Expression Vector in Escherichia coli

Prepare *E. coli* DH5α and BL21 (DE3) competent cells following the standard calcium chloride method [9] prior to cloning procedures:

1. Streak *E. coli* from a glycerol stock or a liquid culture on a TY agar plate with a sterile loop and incubate the plate overnight at 37 °C.
2. Inoculate a single colony of *E. coli* in 5 mL of TY liquid medium and incubate overnight at 37 °C with vigorous shaking.
3. Add 1 mL of the overnight *E. coli* culture to 99 mL of fresh TY liquid medium and incubate at 37 °C with shaking until OD_{600} nm reaches 0.5.
4. Place the culture on ice for 30 min then centrifuge at 4 °C at 4000 × *g* for 10 min and discard the supernatant.
5. Gently resuspend the pellet with 100 mL ice-cold 50 mM $CaCl_2$, then centrifuge at 4 °C at 4000 × *g* for 10 min and discard the supernatant.
6. Resuspend the pellet with 50 mL of ice-cold 50 mM $CaCl_2$ and incubate for 15 min on ice, then centrifuge at 4 °C at 4000 × *g* for 10 min and pour off the supernatant.
7. Lastly, resuspend the pellet with 8.5 mL of ice-cold 50 mM $CaCl_2$ and 1.5 mL of Glycerol (15% final concentration) and place on ice for 1 h.
8. Aliquot 400 μL of the competent cells into sterile 2 mL tubes. The competent cells are ready to be used or can be stored at −80 °C (*see* **Note 15**).

Once *E. coli* competent cells are ready, subject your target DNA to PCR amplification using the following method:

1. In a 0.2 mL sterile tube combine 10–50 ng of DNA target, 0.4 μM of appropriate primers forward and reverse, 0.3 mM dNTP, 1× buffer (supplied by the manufacturer of the DNA polymerase), 1 U of Taq DNA polymerase with proofreading activities. Prepare a negative control mix by adding the same reagents except for the DNA target in another 0.2 mL sterile tube.
2. Add the right amount of sterile water to reach 50 μL final volume per reaction and gently pipette up and down.
3. Place the PCR samples into the thermocycler and start the PCR amplification program as follows: (a) initial denaturation step at 94 °C for 5 min; (b) denature 15 s at 94 °C; (c) anneals primers at 56 °C for 30 s; (d) extend the DNA at 68 °C for 90 s; (e) repeat **steps b–d** for 30 times; (f) final extension at 68 °C for 5 min.
4. In the meantime, prepare 1% agarose gel by mixing 1 g agarose with 100 mL of TAE 1× buffer previously prepared (*see* **Note 16**).
5. Melt agarose by heating in a microwave oven until the solution boils and then leave the gel solution to cool at room temperature. Chill until the gel reaches approximately 50 °C.
6. Gently pour the gel into a gel tray, add 0.5 μg/mL of ethidium bromide and evenly mix into the agarose gel with the help of a pipette tip.
7. Add the comb to one side of the gel and remove any bubbles in the solution.
8. Let the gel harden for approximately 20 min until it becomes milky and opaque. Remove the comb and soak the gel into an electrophoresis tank with 1× TAE running buffer.
9. Prepare the samples by adding 6× loading buffer to the DNA samples and the DNA ladder.
10. Load 5 μL of the DNA target into the wells and close the lid of the electrophoresis chamber. Set the run at 120 V (usually 30 min to 1 h).
11. Visualize the gel under UV light to reveal the DNA bands.
12. Purify the PCR products with a PCR purification kit, according to the instruction manual of the supplier and proceed with the cloning step.

Prior to proceeding with the cloning step, digest both the DNA insert and the expression plasmid with the selected restriction enzymes applying the following digestion procedure:

1. Add the following reagents in a sterile 1.5 mL tube: 1 μg of DNA target, 10 units of the selected restriction endonuclease,

1× restriction enzyme Buffer (supplied by the manufacturer of the restriction endonuclease), sterile water to reach the 50 μL final volume. Gently mix by pipetting up and down.

2. Incubate the tube at 37 °C for at least 1 h.
3. Load 5 μL of the digested DNA on 1% agarose gel, together with the undigested DNA (100 ng) and run the gel by starting the electrophoresis process for 20 min at 200 V.
4. Purify the digested DNA with a gel extraction kit, according to the instruction manual of the supplier.
5. Set up the ligation mixture reaction in a sterile tube using a 1:5 molar ratio of vector:insert DNA following the equation (*see* **Note 17**):

$$
\begin{aligned}
&(\text{ng of vector} \\
&\quad \times \text{kb size of insert} \\
&\quad \div \text{kb size of vector}) \\
&\quad \times (\text{insert} \\
&\quad \div \text{vector molar ratio})
\end{aligned}
$$

6. Add the following reagents in a 20 μL final volume: water, 1× ligation buffer (supplied by the manufacturer of the T4 ligase), DNA vector, DNA insert, T4 DNA ligase (1 unit). Mix gently (T tube). Carry out a negative control mix containing all the reagents except the DNA insert (C tube).
7. Incubate both ligation mixtures for 12 h at 16 °C.
8. Add 200 μL of *E. coli* competent cells into the T tube gently pipette up and down. Do the same with the C tube.
9. Incubate on ice for 30 min.
10. Heat-shock both tubes at 42 °C for 2 min and then transfer at room temperature for 5 min.
11. Add 800 μL of prewarmed TY liquid medium and incubate at 37 °C for 1 h.
12. Spread 1:10 and 1:100 dilutions from the T and C tubes on TY agar plates containing 100 μg/mL ampicillin and incubate at 37 °C for 12–15 h.
13. Inoculate 10 colonies obtained from the transformation of *E. coli* cells with T tube and 1 colony obtained from the transformation of *E. coli* cells with C tube (control colony), in 5 mL TY liquid medium supplemented with 100 μg/mL ampicillin and incubate at 37 °C for 12 h with energetic shaking for the subsequent plasmid DNA extraction.

14. Centrifuge bacterial cultures at 10,000 × *g* for 5 min and discard the supernatant.
15. Resuspend the pellet in 100 μL of Solution 1 and incubate for 5 min at room temperature.
16. Add 200 μL of Solution 2 and quickly place it on ice.
17. Add 150 μL of 3 M sodium acetate and place on ice for 5 min.
18. Centrifuge at 10,000 × *g* for 5 min and collect 400 μL of supernatant into 2 mL sterile tubes.
19. Add 1 volume of phenol–chloroform–isoamyl alcohol solution (25:24:1, v/v) and centrifuge at 10,000 × *g* for 5 min.
20. Transfer the aqueous phases in a new 2 mL tube and add 3 volumes of Ethanol (EtOH) 100% (v/v) and 1/10 volume of 3 M sodium acetate.
21. Centrifuge at 10,000 × *g* for 5 min and wash two more times using EtOH at 70% final concentration.
22. Discard the supernatant and let samples dry at room temperature.
23. Resuspend the pellet in 50 μL of sterile water containing 20 μg of RNase (10 mg/mL) and incubate at 37 °C for 30 min.
24. Load 5 μL of the isolated plasmid DNA on 1% agarose gel, together with 5 μL of plasmid DNA extracted from the control colony and run the gel by starting the electrophoresis process for 20 min at 200 V.
25. Visualize the gel under UV light to reveal the DNA bands.
26. Digest the selected plasmid DNA with one or two one cutter restriction endonucleases based on the sequence of the recombinant plasmid, to verify the success of the cloning procedure and load the digested DNA on 1% agarose gel for 20 min at 100 V.
27. Prepare a glycerol stock of the strain containing the recombinant plasmid, by adding glycerol at 30% final concentration to the bacterial culture collected in **step 13**. Store it at −80 °C (Fig. 1).
28. Use 10–50 ng of the recombinant plasmid to transform *E. coli* BL21 (DE3) competent cells and to select the recombinant strain on TY agar plate supplemented with 100 μg/mL ampicillin, as reported above.

3.2 Expression and Extraction of Recombinant Protein

1. Inoculate a single BL21 (DE3) recombinant colony in 5 mL TY liquid medium containing ampicillin 100 μg/mL using a sterile loop and incubate the culture at 37 °C for 12 h.
2. Add 1 mL of the culture to 49 mL of fresh TY liquid medium and incubate at 37 °C with shaking until OD_{600} nm reaches 0.4.

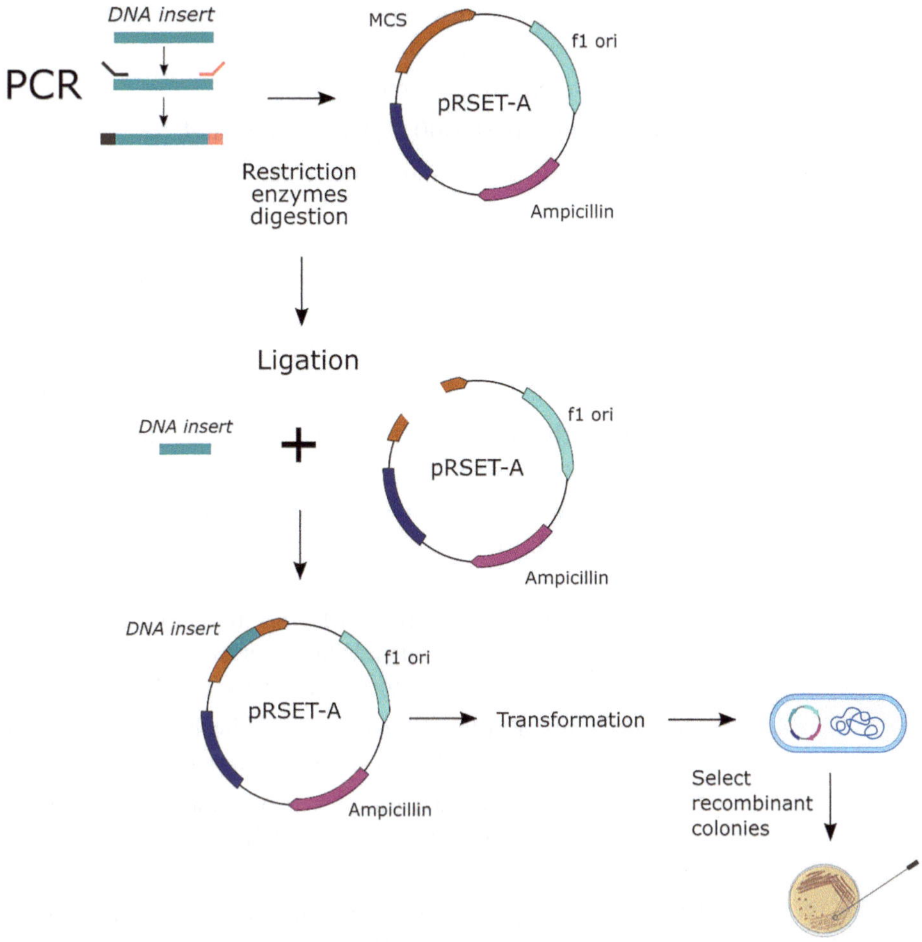

Fig. 1 A simplified scheme of the cloning strategy applied to obtain the recombinant plasmid. Green and red lines indicate forward and reverse primers used for the PCR amplification step. The amplified products (PCR insert) and the expression vector (pRSET-A) are subjected to digestion with the appropriate restriction enzymes, ligated thanks to the action of the T4 ligase enzyme, and cloned into *E. coli* DH5a host cells. The recombinant plasmid is then used to transform *E. coli* BL21 (DE3) competent cells. *Ampicillin* ampicillin resistance gene, *MCS* multi cloning site

3. Split the cultures into 2 sterile flasks containing 25 mL of culture each one.
4. Add IPTG to a final concentration of 1 mM only to one flask (labeled induced flask) and incubate both flasks at 37 °C with vigorous shaking.
5. After 6 h induction, collect the cultures from both flasks and centrifuge at 10,000 × *g* for 10 min. Discard the supernatant.
6. Resuspend the pellets in 4 volumes of fresh and cold PBS–Triton X-100 1% and aliquot the samples into clean 2 mL tubes.

7. Freeze in a dry ice-ethanol bath for 10 min. Immediately, thaw at 37 °C water bath for 10 min.
8. Repeat **step** 7 for 5 times.
9. Centrifuge at 10,000 × *g* for 15 min at 4 °C.
10. Collect the supernatants into clean 2 mL tubes and resuspend the pellets with the same PBS–Triton X-100 1% volumes used in **step 6**.
11. Quantify the protein concentration of the soluble (supernatant) and insoluble (pellet) fractions using the Bradford assay. Warm up the spectrophotometer before use.
12. Prepare a 0.1 mg/mL BSA stock solution to use as standard.
13. Dilute the Bradford reagent fivefold in deionized H_2O to minimize analysis bias.
14. Create the blank sample adding 1 mL Bradford reagent into a clean 2 mL tube.
15. Prepare the standard samples into 2 mL clean tubes using increasing concentrations of BSA to elaborate the standard curve (i.e., 2, 3, 6, 8 μg). When pipetting, be careful to release the samples at the bottom of the tube (*see* **Note 18**).
16. Add 1 mL Bradford reagent to all the BSA samples, mix well and incubate for 20 min at room temperature.
17. Dilute the unknown samples if necessary in deionized H_2O (generally 10- to 100-fold) to obtain approximately 5–50 μg of protein.
18. Set up a triplicate of each unknown sample using increasing volumes (i.e., 20, 30, 40 μL) to get approximately increasing protein concentration.
19. Add 1 mL Bradford reagent to all the unknown samples, mix well and incubate for 20 min.
20. With the help of Pasteur pipettes, transfer each sample into clean UV/VIS cuvettes to read the absorbance.
21. Use the blank sample as the zero to calibrate the spectrophotometer. Measure the absorbance of each sample (BSA standards and unknown samples) at 595 nm.
22. Calculate the concentration of the unknown samples based on the linear equation of the standard curve.
23. Load 15 μg of each soluble fraction or 15 μL of each pellet fraction on SDS-PAGE gel and run the electrophoresis process as follows.

3.3 Recombinant Protein Detection by SDS-PAGE

The first step of SDS-PAGE consists of casting the polyacrylamide gel in two distinct layers: the stacking gel (5%) and the resolving gel (12.5%). The polyacrylamide gel is composed of 30% acrylamide–bis acrylamide 29:1.

1. Assemble the glass plates and the spacer in the gel casting apparatus. Accurately mix the resolving gel mixture before pouring it into the glass plates. Make sure to reach a level below approximately 2 cm from the shorter plates.
2. Add a layer of 2-propanol 100% with the help of a pipette to avoid bubbles formation. This step also prevents the gel from drying out.
3. Let the resolving gel polymerize for 20 min at room temperature.
4. Drain the 2-propanol 100%, and add ddH_2O to wash out any 2-propanol or gel residual. Wipe any residual water trapped into the glass plates with the help of clean absorbent paper.
5. Mix the stacking gel components and carefully pour the gel above the resolving gel to fill the glass plates. Insert the comb at the top of the gel.
6. Stand at room temperature for at least 20 min. Once the gel has polymerized, remove the comb and wash the wells with ddH_2O.
7. Place the casted gel into the electrophoresis apparatus and fill it with TGS 1× running buffer.
8. Accurately prepare the samples resuspending in 1× loading sample buffer.
9. Incubate the samples at 100 °C for 5 min. Add a protein marker into the first well and then load 15 μg or 15 μL of the samples into the wells.
10. Run the gel at 120 V until the samples reach the bottom of the gel. Stop the run and pick up the gel. Carefully remove the glass plate and place the gel into a plastic container. Wash the gel with ddH_2O.
11. Drain the water and add approximately 20 mL of Coomassie staining solution. Incubate 1 h at room temperature with shaking.
12. Remove the staining solution and add 5 mL of destaining solution for 1 min. Discard the destain solution and add 20 mL of destaining solution. Place at room temperature with shaking until completely destained. Wash with ddH_2O and visualize (Fig. 2).

3.4 Recombinant Protein Detection by Western Blotting

To detect the recombinant fusion protein by Western blot analysis, we use an anti-His antibody against the N-terminal His-tag. Separate the proteins by SDS-PAGE as reported above and then proceed with the electrotransfer. Be careful to avoid the staining step with Coomassie staining solution when performing Western blotting analysis.

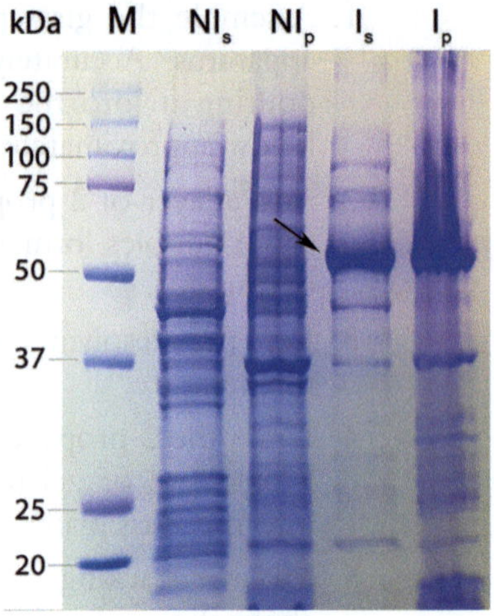

Fig. 2 An example of SDS-PAGE analysis of soluble and insoluble proteins extracted from induced and uninduced cultures. The image represents an SDS-PAGE analysis of the expression of a novel putative collagenase from a marine microorganism *Sulfurovum riftiae* [10], class *Epsilonproteobacteria* (recently proposed to be reclassified as a new phylum *Campylobacterota*) [11]. The arrow indicates a strong expression of the protein in the induced cultures. *M*: Protein marker (Precision Plus Protein Dual Color Standards, Bio-Rad); NI_s indicates soluble proteins extracted from the uninduced BL21 (DE3) strain carrying the recombinant plasmid; NI_p indicates insoluble proteins extracted from the uninduced BL21 (DE3) strain carrying the recombinant plasmid; I_s indicates soluble proteins extracted from the induced BL21 (DE3) strain carrying the recombinant plasmid; I_p indicates insoluble proteins extracted from the induced BL21 (DE3) strain carrying the recombinant plasmid

1. Once the electrophoresis is completed, collect the gel with the help of a spatula and remove the stacking gel.
2. Place the gel and the nitrocellulose membrane in separate plastic containers with the 1× Tris–glycine, 20% methanol transfer buffer (*see* **Note 19**). Incubate for 30 min at 4 °C.
3. Assemble the transfer sandwich with the help of clean clamps and using absorbent filter papers. Rinse 3 filter paper into the transfer buffer and stack them on top of each other. Make sure the nitrocellulose membrane is on the cathode and the gel on the anode. Stack the other 3 filter paper rinsed into the transfer buffer onto the sandwich.
4. Eliminate any potential air bubbles trapped into the sandwich with the help of a roller.
5. Run the electrotransfer at 15 V for 30 min.
6. Shut the transfer machinery and pick up the membrane with the help of clean clamps. Place it into a plastic container.

7. To reveal the transferred bands, stain the membrane with Ponceau S dye for a couple of minutes.
8. Once the protein bands display, destain thoroughly with 1× PBS for further immunorevelation steps.
9. Incubate the membrane with 1× PBS–milk 5% blocking solution for 30 min on a shaker.
10. Add 1:1000 HRP conjugate anti-His tag antibody to the blocking 5% solution and incubate for 1 h.
11. Rinse the blot 3–4 times with 1× PBS for 5 min.
12. Reveal using the chemiluminescent substrate according to the manufacturer instructions.
13. Visualize the chemiluminescent signal using a CCD camera-based imager.

4 Notes

1. To prepare 400 mL of TY agar plate, weigh 7.4 g of agar and transfer it into a clean bottle. Adjust the volume to 400 mL using fresh TY liquid medium and autoclave. Let it cool at room temperature before pouring onto petri dishes in sterile conditions.
2. We find that preparing this fresh each time will implement the results.
3. We find best results when using it sterile.
4. Primers are designed by using the web-based program *Primer3*. The forward primer includes 15–18 nucleotides complementary to the sequence of interest starting from its start codon and a selected restriction site at the 5′ end. The reverse primer includes 15–18 nucleotides complementary to the sequence of interest starting from the stop codon and a selected restriction site at the 5′ end of this primer. The restriction sites are chosen amongst that present in the multicloning site (MCS) of the vector and should guarantee that the DNA fragment is inserted *in frame* with the start codon of the expression vector.
5. Before pouring onto petri dishes, add 400 μL ampicillin into a 400 mL TY agar plate. We maintain a ratio of 1:1000 when adding ampicillin to medium or suspension. This means that for 1 mL of medium, 1 μL of 100 μg/mL ampicillin is added.
6. Tris base will take a while to dissolve completely; thus, we find it convenient to use it as the first ingredient. Store stock solution at room temperature.

7. Always use cold freeze EtOH. When preparing or handling EtOH wear gloves and a mask to prevent from inspiring any fumes.
8. In a 15 mL conical tube add the acrylamide, Tris–HCl and SDS. Carefully mix and then add APS. Reach the final volume by adding ddH_2O and mix well. Add TEMED at last, evenly mix and pour it into the glass plates directly from the tube. This will lessen bubbles forming at the top of the gel. Always add TEMED as the last ingredient because this will initiate the polymerization reaction.
9. To prepare 100 mL stock of TGS 10×, weigh 3 g of Tris base and 14.4 g of glycine. Transfer into a bottle and add 10 mL of 10% SDS. Add deionized water up to 100 mL.

   To prepare 1 L of TGS 1×, add 100 mL of TGS 10× to 900 mL of ddH_2O previously spilt into a cylinder. Seal the cylinder with parafilm and carefully mix upside down to evenly blend the ingredients.
10. Always store at −20 °C. When the solution is being used, place it on ice. We find it proper not to let the solutions sit at room temperature.
11. To prepare 400 mL of stain solution, add 0.8 g of Brilliant Blue R to 100 mL of 2-propanol and let the Coomassie powder dissolve completely. Add 40 mL of acetic acid and up to the final volume with ddH_2O.
12. Always wear gloves when handling acetic acid or ethanol and prepare the solution in a fume hood to avoid any fumes inhalation.
13. Always prepare a large stock of 10× Tris–glycine (as reported in **Note 9** except for SDS) to store at room temperature.
14. Prepare 50 mL of 1× PBS–milk 5% blocking solution, weigh 2.5 g of skimmed milk and transfer into a falcon tube containing 50 mL of 1× PBS. Vortex the solution may help dissolve faster.
15. It seems to be common that competent cells may have low-efficiency transformation. We find that this issue can be sorted out by changing the molarity of $CaCl_2$ or using fresh $CaCl_2$, or resuspending more gently.
16. Prepare TAE 1× adding 20 mL TAE 50× to 980 mL ddH_2O evenly mixing.
17. Frequently this ligation reaction step does not work as intended causing low or no recombinant colonies downstream the transformation step. Thus, troubleshoot modifying the vector:insert molar ratio for the insert as well as scaling down ng of the vector.

18. When performing the Bradford assay, put the samples first into the tubes (BSA standard or unknown) paying attention to release the sample completely and then add the Bradford reagent. It is recommended to use the same equipment throughout the assay to minimize biases that frequently can occur when handling with little volumes.
19. Prepare 1× Tris–glycine–20% methanol in a beaker adding 10 mL 10× Tris–glycine, 20 mL methanol 100%, and 70 mL ddH_2O. The gel and the nitrocellulose membrane must be thoroughly covered with the transfer buffer solution.

Acknowledgments

This work was partially supported by the European Research Council (ERC) under the European Union's Horizon 2020 research and innovation programme (grant agreement No. 948972), NSF grant IOS 19-51690, and NASA grant 80NSSC21K0485.

References

1. Binder MD, Hirokawa N, Windhorst U (2009) Heterologous expression. In: Binder MD, Hirokawa N, Windhorst U (eds) Encyclopedia of neuroscience. Springer, Heidelberg, pp 1830–1830
2. Frommer WB, Ninnemann O (1995) Heterologous expression of genes in bacterial, fungal, animal, and plant cells. Annu Rev Plant Physiol Plant Mol Biol 46:419–444
3. Rosano GL, Ceccarelli EA (2014) Recombinant protein expression in *Escherichia coli*: advances and challenges. Front Microbiol 5: 172
4. Chien A, Edgar DB, Trela JM (1976) Deoxyribonucleic acid polymerase from the extreme thermophile *Thermus aquaticus*. J Bacteriol 127:1550–1557
5. Lawyer FC, Stoffel S, Saiki RK et al (1989) Isolation, characterization, and expression in *Escherichia coli* of the DNA polymerase gene from *Thermus aquaticus*. J Biol Chem 264: 6427–6437
6. Hofer U (2018) The majority is uncultured. Nat Rev Microbiol 16:716–717
7. Malmstrom RR, Eloe-Fadrosh EA (2019) Advancing genome-resolved metagenomics beyond the shotgun. mSystems 4:e00118-19
8. Mehlin C, Boni E, Buckner FS et al (2006) Heterologous expression of proteins from Plasmodium falciparum: results from 1000 genes. Mol Biochem Parasitol 148:144–160
9. Sambrook J, Russell DW (2006) The condensed protocols from molecular cloning: a laboratory manual. CSHL Press
10. Giovannelli D, Chung M, Staley J et al (2016) *Sulfurovum riftiae* sp. nov., a mesophilic, thiosulfate-oxidizing, nitrate-reducing chemolithotrophic Epsilonproteobacterium isolated from the tube of *Riftia pachyptila*, a deep-sea hydrothermal vent tube worm. Int J Syst Evol Microbiol 66(7):2697–2701
11. Waite DW, Chuvochina MS, Hugenholtz P (2019) Road map of the Phylum Campylobacterota. In: Bergey's manual of systematics of archaea and bacteria. American Cancer Society, pp 1–11

Chapter 14

Expression of Recombinant Cold-Adapted (Hemo)Globins from Marine Bacteria

Daniela Giordano and Cinzia Verde

Abstract

The production of recombinant proteins in bacteria made possible to obtain large quantities of proteins essential for basic and applied research. *Escherichia coli* remains one of the organisms of choice for recombinant proteins because of its ability to grow at high density and availability of a vast catalog of cloning vectors and mutant host strains. Here, we describe the protocols for the expression of cold-adapted (hemo)globins in *Escherichia coli*.

Key words Bacterial hemoprotein, Cold adaptation, Recombinant-protein expression, *Escherichia coli*, Expression plasmid, UV–visible absorption spectrum

1 Introduction

The production of recombinant proteins in bacterial cell was revolutionary for molecular biology and biochemistry and made possible to obtain large quantities of proteins essential for basic and applied research. Native and recombinant proteins are fundamental for pharmaceutical and agricultural industry [1]. The protocols for protein expression are simple and may be summarized as follows: (1) selection of gene; (2) cloning in the expression vector; (3) transformation in a suitable host; (4) induction; (5) purification and biophysical/biochemical characterization. However, there are many issues often found along the pipeline, for example, poor growth of the host, formation of inclusion body and protein denaturation that need attention. Proteins which are produced as inclusion bodies are sometime inactive and require refolding although many recombinant proteins still retain biological activity when expressed as inclusion bodies. In addition, proteins with many disulfide bonds are often denatured after expression and their refolding is extremely time consuming. Selecting the perfect protocol is impossible a priori, and many experimental conditions often

Cinzia Verde and Daniela Giordano (eds.), *Marine Genomics: Methods and Protocols*,
Methods in Molecular Biology, vol. 2498, https://doi.org/10.1007/978-1-0716-2313-8_14,

need to be tested before obtaining the desired protein since protein expression is sometimes protein specific. In this chapter we describe the methods and protocols that have been successful for expressing cold-adapted globins and focus on the advantages of using *Escherichia coli* as host organism for protein expression, for example, fast growth kinetics and easy transformation with DNA of interest. Proteins overexpressed in *E. coli* account approximately for 90% of the structures present in the Protein Data Bank (PDB) [2]. The BL21(DE3) and its derivatives are by far the most popular strains for protein expression. Competent cells of *E. coli* BL21(DE3) strain, routinely used for expressing recombinant genes, are suitable for transformation and high level protein expression using a T7 RNA polymerase. The T7 promoter systems and the pET vectors are widespread in recombinant protein expression protocols. Many inducible promoters have been developed recently and among these, the T7 or lac-based promoter is the most popular and can be effectively induced by the addition of isopropyl-β-Δ-thiogalactopyranoside (IPTG). The IPTG inducible production system may not be desirable only when a very large scale production is required since its cost is high. The T7 expression strains are lysogens of bacteriophage DE3 [3] and carry achromosomal copy of the T7 RNA gene under control of the lacUV5 promoter. In BL21(DE3) cells, the lacUV5 promoter drives the expression of the T7 RNA polymerase. By using this system, the gene of interest is cloned behind the promoter which is recognized by the T7 RNA polymerase which is produced from the lysogenic prophage DE3. In the case of formation of inclusion bodies, for cold-adapted proteins it is possible to use *E. coli* Arctic Express Cells, a BL21(DE3) strain optimized for growth at low temperatures thanks to the expression of two cold-adapted chaperonins from the psychrophilic bacterium *Oleispira antarctica*. The expression at low temperatures allows for the production of functional cold-adapted enzymes, which are typically poorly stable at higher temperatures [4, 5]. Such strains are suitable for production of protein from target genes cloned in appropriate T7 expression vectors (pET), using IPTG as an inducer.

The pET systems, first described in the 1980s, have increasingly become popular for their versatile application. To date there are 103 vectors in the pET series. pET28a is the most used plasmid and it is described in more than 40,000 published articles [6]. It contains the typical genetic elements of a pET vector with T7 promoter and T7 terminator into the pBR322 backbone, as well as the translation initiation region of Shine–Dalgarno sequence. Between T7 promoter and T7 terminator there is the multiple cloning site containing restriction enzyme sites to be used for engineering the plasmid and inserting the gene of interest. Within the restriction sites there is the coding sequence for a polyhistidine purification tag (His6) and a thrombin protease recognition site so that the recombinant protein produced can be easily purified using affinity chromatography.

We express the recombinant hemoglobin encoded by the *PSHAa0030* gene from marine Antarctic bacterium *Pseudoalteromonas haloplanktis* TAC125 [7] using *E. coli* BL21(DE3) in the presence of precursor of heme as prosthetic group [8–11]. Full-length globin cDNA was cloned in the expression vector pET-28a under T7 promoter and with kanamycin as antibiotic resistance. Sequences corresponding to the restriction enzyme sites (*NcoI* and *BamHI*) were introduced in the forward and reverse primers, respectively. The expression vector pET-28a carrying the respective plasmid construct (pET28a-*globin*) was subsequently transformed into *E. coli* BL21(DE3) and the bacterial hemoglobin expressed in the presence of the precursor of heme as prosthetic group. The complete sequence of the construct was checked by dideoxy sequencing, verifying that the construct was correctly positioned for expression.

2 Materials

All solutions are prepared by using purifying deionized water to attain a sensitivity of 18 MΩ-cm at 25 °C. Analytical grade reagents are sterilized by using the autoclave and stored according to the manufactory instructions.

2.1 Transformation in the E. coli Strain BL21(DE3)

1. Competent cells of *E. coli* BL21(DE3) strain (*see* **Note 1**).
2. Expression vector pET-28a carrying the plasmid construct of interest (pET28a-*globin*), previously checked by sequencing (*see* **Note 2**).
3. Plastic petri dishes (9 cm diameter).
4. Luria–Bertani (LB) agar (1 L): weigh 10 g of tryptone, 5 g of yeast extract, 5 g of NaCl, 15 g of agar and transfer to a 1 L cylinder. Add water to a volume of 900 mL. Autoclave. Cool to 55 °C, and then add antibiotic (50 μg/mL kanamycin, final concentration), and pour into petri dishes (~25 mL/9 cm diameter for plate).
5. Stock solution of 50 mg/mL kanamycin: weigh 50 mg of powder of kanamycin resuspended in 1 mL of sterile deionized water, vortex and then filter through a 0.22 μm filter. Leave one aliquot at 4 °C for current use and store the remaining aliquots at −20 °C.

2.2 Globin Expression

1. 15 mL sterile tubes with aerated cap (suitable for inoculation).
2. Erlenmeyer flasks (1 L) wrapped with aluminum foil and scotch autoclave tape.
3. Microbiological hood.
4. LB medium (1 L): weigh 10 g of tryptone, 5 g of yeast extract, and 5 g of NaCl and transfer to a 1 L cylinder. Add water to a volume of 900 mL. Mix and make up to 1 L with water. Autoclave the broth (*see* **Notes 3–5**).
5. Stock solution of 1 M δ-aminolevulinic acid hydrochloride (precursor of heme group): weigh 167.5 mg of powder resuspended in 1 mL of sterile deionized water, vortex and then filter through a 0.22 μm filter. Leave one aliquot at 4 °C for current use and store the remaining aliquots at −20 °C (*see* **Note 6**).
6. Stock solution of 1 M IPTG: weigh 238 mg of powder in 1 mL resuspended in 1 mL of sterile deionized water, vortex and then filter through a 0.22 μm filter. Leave one aliquot at 4 °C for current use and store the remaining aliquots at −20 °C (*see* **Note** 7).
7. Incubator (37–25 °C) with an orbital shaking (capable of 150 rpm motion).
8. 50 mL plastic conical tubes.
9. Refrigerated centrifuge with rotor for 50 mL conical tubes (2000–12,000 × *g*).
10. Lysis buffer: 50 mM Tris-HCl, pH 7.6, 1.0 mM EDTA, pH 8.0, 100 mM NaCl, 5% glycerol, 1.0 mM phenylmethylsulfonyl fluoride (PMSF) and protease-inhibitor cocktail (*see* **Notes 8–10**).
11. Refrigerated ultracentrifuge with rotor (75,000 × *g*).
12. Polypropylene tubes for ultracentrifugation.
13. French press for cell rupture (*see* **Note 11**).
14. Spectrophotometer UV-Visible (250–700 nm).
15. Quartz and plastic cuvettes with 1 cm of optical path.

3 Methods

3.1 Transformation in the E. coli Strain BL21(DE3)

1. Thaw one aliquot (50–100 μL) of competent cells of *E. coli* BL21(DE3) strain on ice (*see* **Note 1**).
2. Gently mix the competent cells and add 50–100 ng of the expression vector pET-28a carrying the plasmid construct of interest (pET28a-*globin*), according to the instruction manual of supplier (*see* **Note 2**).

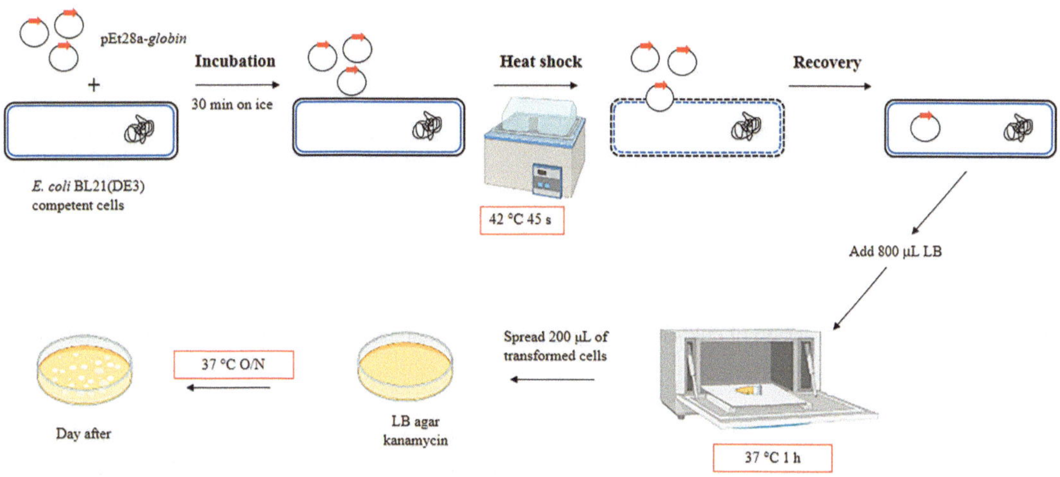

Fig. 1 Chemical transformation in the *E. coli* strain BL21(DE3). The main steps of bacterial transformation are indicated

3. Follow the transformation protocol, according to the instruction manual of supplier. Incubate the mix for 30 min on ice and heat shock for 45 s at 42 °C. Incubate on ice for 2 min, add 800 μL of LB medium and incubate at 37 °C while shaking for 1 h.
4. After growth, spread 200 μL of the cells transformed with pET28a-*globin* using a sterile spreader onto prewarmed LB agar plates that contain the appropriate antibiotic (50 μg/mL kanamycin, final concentration).
5. Incubate the plates overnight at 37 °C.
6. The day after the positive colonies carrying the plasmid construct (pET28a-*globin*) will be visible on plate (Fig. 1).

3.2 Globin Expression

1. Inoculate a positive single colony carrying the plasmid construct (pET28a-*globin*) in 3 mL of LB medium supplemented with 3 μL of 50 mg/mL kanamycin (50 μg/mL, final concentration) in a 15 mL sterile tube with aerated cap (*see* **Note 2**).
2. Grow the cells overnight in an incubator with an orbital shaking at 150 rpm at 37 °C.
3. After 16 h, measure the optical density (OD) of grown culture at 600 nm in a plastic cuvette using a spectrophotometer and calculate the volume to add to 200 mL of LB medium to start the growth at 0.08 OD (divide 0.08 for the read absorbance and multiply for 200 mL) (*see* **Note 12**).
4. Working under a microbiological hood, transfer the grown culture (*see* **step 3**) into a 1 L Erlenmeyer flask containing 200 mL LB medium (previously autoclaved) containing 200 μL of 50 mg/mL kanamycin (50 μg/mL, final

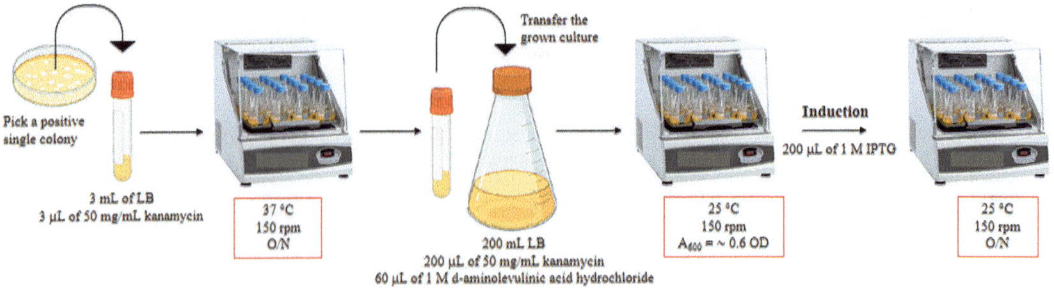

Fig. 2 Recombinant expression of cold-adapted globin. The main steps of expression of protein of interest are indicated

concentration) and 60 μL of 1 M δ-aminolevulinic acid hydrochloride (0.3 mM, final concentration) (*see* **Notes 3–6**).

5. Allow to grow in an incubator with an orbital shaking at 150 rpm at 25 °C until A_{600} reached ~0.6 OD (almost 4–5 h) (*see* **Note 13**).
6. Induce the culture by the addition of 200 μL of 1 M IPTG (1 mM, final concentration) (*see* **Note 7**).
7. Continue the expression for 16 h at 25 °C (overnight) at 150 rpm (Fig. 2) (*see* **Note 14**).
8. Recover each 50 mL of grown culture in 50 mL conical tubes and harvest the cells at 2000 × *g* at 4 °C for 20 min.
9. Discard the supernatant and freeze each pellet.
10. After 1–2 nights of freezing, thaw the frozen cells and resuspend each pellet (50 mL) in 3 mL of lysis buffer (*see* **Notes 8–10**).
11. Vortex each pellet solution after lysis.
12. Broke the resuspended cells in a French press until the supernatant is reddish and clear (*see* **Notes 11** and **15**).
13. Centrifuge at 12,000 × *g* at 4 °C for 10 min to remove the cell debris.
14. Transfer the supernatant in tubes for ultracentrifugation, balance them perfectly and ultracentrifuge at 75,000 × *g* for 1 h to be sure to remove all cell debris.
15. Transfer the clear and reddish/brown supernatant to new tubes.
16. Check the expression of the globin, properly folded with the heme group inside the distal cavity, by registering an electronic absorption spectrum in the UV-visible range (250–700 nm) of the lysate, comparing it with the lysate obtained from the expression of empty vector (Fig. 3). Before registering the spectrum of lysate, register the blank spectrum (lysis buffer).

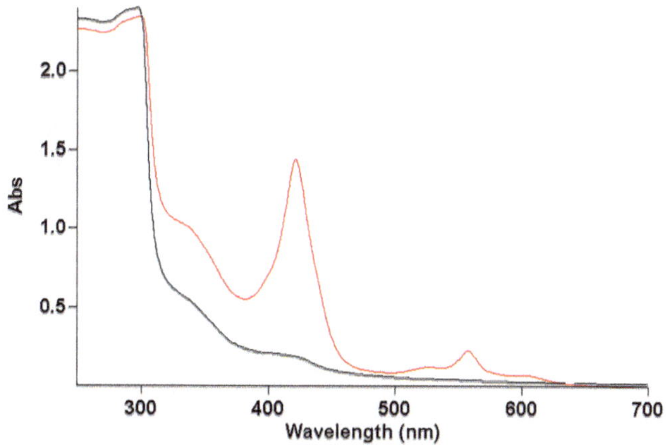

Fig. 3 UV–visible absorption spectra of lysates of recombinant globin and empty vector. Absorbance (Abs) as function of wavelength (nm). The spectrum of globin (red line) shows the typical shape of ferrous heme (maxima of absorbance of Soret peak at 420 nm and peaks at 558 and 527 nm) with respect of spectrum of lysate of empty vector (black line)

4 Notes

1. The best choice was to use *E. coli* BL21(DE3) cells to obtain high-level expression of this recombinant globin. However, *E. coli* BL21(DE3)pLysS cells are usually preferred for the expression of potentially toxic proteins, as occurs in other globins [12, 13]. In fact, these hosts carry a plasmid (pLysS) that encodes T7 lysozyme, a natural inhibitor of T7 RNA polymerase, to suppress basal expression of polymerase before induction. However, some globins can precipitated as inclusion bodies. In these cases, it is possible to denature inclusion bodies and refold the native globin with an excess of hemin [12]. As an alternative approach, for cold-adapted proteins it is possible to use *E. coli* Arctic Express BL21(DE3) cells (Agilent, 230192).
2. It is a good practice to do the transformation of *E. coli* BL21 (DE3) cells by carrying the empty vector pET-28a for comparison of the expression with the vector carrying the construct of interest.
3. As an alternative medium, Terrific broth (TB) can be also used [12, 13]. However, LB medium was chosen for the expression of this cold-adapted globin to avoid strong oxidative conditions.
4. Prepare 8 Erlenmeyer flasks (1 L) containing 200 mL LB medium for a total of 1.6 L of bacterial culture. The small-scale (200 mL) cultures in shake flasks are necessary to improve the quality of expression. In fact, production of the protein

under strong oxygenation conditions may generate different forms of the heme [9].

5. Take into account that the best ratio between volume of medium and volume of flask is 1:5 (200 mL LB medium–1 L of Erlenmeyer flask) to ensure enough aerated space for the culture.
6. δ-aminolevulinic acid hydrochloride (Sigma-Aldrich, A3785) is an intermediate in heme biosynthesis and its addition is necessary for the expression of this protein to augment the correct folding of prosthetic group in the protein as reported for other globins [12, 13]. In other globins this addition is not necessary [14].
7. The concentration of IPTG to be used depends on the protein and has to be tested. For this globin the final concentration is 1 mM IPTG; 0.4 mM IPTG may be enough for other globins [12, 13].
8. Prepare lysis Buffer 1–2 h before using it. The PMSF is unstable in aqueous solution and has a short half-life. The best thing is to weigh the powder and resuspend it in the lysis buffer on a stirring plate for 1–2 h until the powder has completely dissolved.
9. Prepare the protease-inhibitor cocktail (Sigma-Aldrich, P8465) fresh: 1 mL of the cocktail solution is prepared for the inhibition of the protease activity in 20 mL of cell lysate from 4 g of pellet cells; for 100 mL of lysis buffer dissolve 215 mg of the lyophilized powder per 1 mL DMSO, vortex and then add 4 mL of deionized water. This mixture of specific protease inhibitors of serine, cysteine, aspartic metalloprotease, and aminopeptidases is necessary to degrade endogenous enzymes of *E. coli* present in the extract probably responsible at least in part of an unexpected proteolytic cleavage occurring in solution [8].
10. Do not use DTT or β-mercaptoethanol in the lysis buffer with bacterial globins, because they show to form high-affinity complexes with hydrogen sulfide in their ferric state. The presence of a ferric sulfide adduct with a typical visible absorption spectrum peaked at 427 nm was also reproducibly observed in hemoglobins from *Bacillus subtilis* and from *Thermobifida fusca* preparations immediately after cell disruption, thus indicating that these proteins were at least partially saturated with sulfide within the *E. coli* expression vector and were endowed with a high affinity for this ligand [15].
11. The French press is the best cell disruption method for this protein because it does not denature the globin with respect to the sonicator.

12. Be sure to measure OD within absorbance 1. If the reading exceeds 1, you have to dilute your sample (200 μL of grown culture in 800 μL of LB medium) and after the reading you have to take into account the dilution factor of 5.
13. It is important to induce at $A_{600} \sim 0.6$ OD and not more because we observed a stronger stress condition with an unusual peak absorbance at 615 nm when the induction occurs at higher optical density. This depends on the globin. For other globins, the best thing was to induce at 0.8 or 1.0 OD [12, 14].
14. After induction, the time of expression depends on the globin. In other globins, 3–5 h are enough to obtain the maximum yield of expression [16].
15. For some globins is necessary to add 1% lysozyme to the lysis buffer to efficiently break the cells [14].

Acknowledgments

This research was financially supported by the Italian National Programme for Antarctic Research (PNRA) of the Ministry of University and Research (MUR). We wish to thank Valentina Brasiello and Chiara Nobile for technical support and assistance.

References

1. Rosano GL, Ceccarelli EA (2014) Recombinant protein expression in *Escherichia coli*: advances and challenges. Front Microbiol 5: 172
2. Berman HM, Westbrook J, Feng Z et al (2000) The Protein Data Bank. Nucleic Acids Res 28: 235–242
3. Daegelen P, Studier FW, Lenski RE et al (2009) Tracing ancestors and relatives of *Escherichia coli* B, and the derivation of B strains REL606 and BL21(DE3). J Mol Biol 394:634–643
4. Ferrer M, Chernikova TN, Yakimov MM et al (2003) Chaperonins govern growth of *Escherichia coli* at low temperatures. Nat Biotechnol 21:1266–1267
5. Ferrer M, Chernikova TN, Timmis KN et al (2004) Expression of a temperature-sensitive esterase in a novel chaperone-based *Escherichia coli* strain. Appl Environ Microbiol 70: 4499–4504
6. Shilling PJ, Mirzadeh K, Cumming AJ et al (2020) Improved designs for pET expression plasmids increase protein production yield in *Escherichia coli*. Commun Biol 3:214
7. Médigue C, Krin E, Pascal G et al (2005) Coping with cold: the genome of the versatile marine Antarctica bacterium *Pseudoalteromonas haloplanktis* TAC125. Genome Res 15: 1325–1335
8. Giordano D, Parrilli E, Dettaï A et al (2007) The truncated hemoglobins in the Antarctic psychrophilic bacterium *Pseudoalteromonas haloplanktis* TAC125. Gene 398:69–77
9. Howes BD, Giordano D, Boechi L et al (2011) The peculiar heme pocket of the 2/2 hemoglobin of cold-adapted *Pseudoalteromonas haloplanktis* TAC125. J Biol Inorg Chem 16: 299–311
10. Giordano D, Russo R, Ciaccio C et al (2011) Ligand- and proton-linked conformational changes of the ferrous 2/2 hemoglobin of *Pseudoalteromonas haloplanktis* TAC125. IUBMB Life 63:566–573
11. Giordano D, Pesce A, Boechi L et al (2015) Structural flexibility of the heme cavity in the cold-adapted truncated hemoglobin from the Antarctic marine bacterium *Pseudoalteromonas haloplanktis* TAC125. FEBS J 282:2948–2965

12. Dewilde S, Mees K, Kiger L et al (2008) Expression, purification, and crystallization of neuro- and cytoglobin. Methods Enzymol 436:341–357
13. Giordano D, Boron I, Abbruzzetti S et al (2012) Biophysical characterisation of neuroglobin of the icefish, a natural knockout for hemoglobin and myoglobin. Comparison with human neuroglobin. PLoS One 7:e44508
14. Cuypers B, Vermeylen S, Hammerschmid D et al (2017) Antarctic fish versus human cytoglobins—the same but yet so different. J Inorg Biochem 173:66–78
15. Nicoletti FP, Comandini A, Bonamore A et al (2010) Sulfide binding properties of truncated hemoglobins. Biochemistry 49:2269–2278
16. Coppola D, Giordano D, Milazzo L et al (2018) Coexistence of multiple globin genes conferring protection against nitrosative stress to the Antarctic bacterium *Pseudoalteromonas haloplanktis* TAC125. Nitric Oxide 73:39–51

Chapter 15

Isolation of UV-Resistant Marine Bacteria by UV-C Assays

Daniela Coppola, Cinzia Verde, and Daniela Giordano

Abstract

Marine organisms have developed physiological and biochemical strategies to survive under the exposure of UV-B radiation. In particular, Antarctic marine bacteria, exposed to extremes of temperature, UV and ice, have adapted to cope with UV radiation by producing photoprotective molecules. Here, we describe (1) the sampling strategy to collect marine samples (surface water/ice and sediment samples) and (2) the selection strategy to isolate in these samples only UV-resistant marine bacteria.

Key words UV radiation, UV-C assay, Marine bacterium, Photoprotection, Bioactive compound, Isolation strategy

1 Introduction

Marine biotechnology, known also as blue biotechnology, has a considerable potential thanks to the huge phylogenetic diversity of marine organisms and the untapped biotechnological potential for the discovery of novel bioactive molecules. The increasing number of the sequenced bacterial genomes and the advent of metagenomics are opened new avenues to the metabolic diversity of the oceans, greatly facilitating the discovery of new products of marine biotechnology.

Advancements in omics technologies have facilitated the discovery of novel molecules and avoided problems associated with the repeated isolation of already known compounds [1], also in inaccessible extreme environments. However, the use of microorganisms to produce natural products may be considered an efficient process for the high yields of biomass that can be obtained and for the simplicity to extract the bioactive natural compound. In this context, microbial biosynthesis processes constitute a sustainable and environmentally friendly alternative to produce biomolecules of technological interest [2, 3].

Cinzia Verde and Daniela Giordano (eds.), *Marine Genomics: Methods and Protocols*,
Methods in Molecular Biology, vol. 2498, https://doi.org/10.1007/978-1-0716-2313-8_15,

Marine organisms represent a promising source of natural products with bioactivities useful for the treatment and prevention of human diseases and for cosmeceutical, nutraceutical, and biotechnological applications [4–7]. Among bioactive molecules, natural compounds to cope with solar UV-B radiation are of particular interest in marine species. In fact, organisms living exposed to sunlight have developed defense mechanisms to survive under the exposure of UV-B radiation, including avoidance mechanisms, synthesis of UV-absorbing substances, enzymatic and non-enzymatic quenching of reactive oxygen species (ROS), and the activation of DNA repair pathways [8]. In Antarctic marine organisms, living in one of the ocean's most inhospitable scenarios, with extremes of temperature, UV and ice, photoprotective compounds are fundamental to mitigate the effects of solar UV-B radiation [9]. UV-resistant Antarctic marine bacteria may count either on a non-enzymatic antioxidant defense based on the synthesis of carotenoid pigments [10–13] and an efficient enzymatic system to cope with ROS [14].

In this chapter, we describe the sampling strategy used to collect marine samples (surface water/ice and sediment samples) exposed naturally to UV radiation and the selection strategy to isolate UV-resistant marine bacteria. By applying this protocol, selected marine bacteria will be identified by 16S rRNA sequencing and their extracts screened for anti-UV activity on human cells for further applications.

2 Materials

2.1 Collection of Water/Ice Samples and Surface Sediments

1. 50 mL plastic sterile tubes with cap.
2. 10 L plastic sterile carboy with handle(s) and cap.
3. Autoclave.
4. Microbiological hood or Bunsen burner.
5. Filtration apparatus for the collected water: peristaltic pump, sterile filter holder, sterile filters 0.45 μm.
6. Sterile sea water: sea water previously collected and then filter through a 0.22 μm filter to be used as a medium for samples. Keep at 4 °C.
7. 2 mL sterile cryovials.
8. Sterile 40% glycerol solution: dilute 40 mL of 100% glycerol in 60 mL of water, and sterilize it in autoclave.
9. Liquid nitrogen.

2.2 Isolation of UV-Resistant Marine Bacteria from Surface Water/Ice/Sediments

1. Water/ice/sediment samples previously collected and stored at 4 °C and/or −20 °C/−80 °C.
2. Autoclave.
3. Sterile sea water: sea water previously collected and then filter through a 0.22 μm filter. Keep at 4 °C.
4. 15 mL sterile tubes with aerated cap (suitable for inoculation).
5. 2 mL sterile safe-lock tubes.
6. Plastic petri dishes (9 cm diameter).
7. Sterile serological pipettes.
8. 2 mL cryovials.
9. Sterile L-shape cell spreader.
10. Balance.
11. Magnetic stirrer.
12. Magnetic stir bar.
13. Marine Broth (MB): weigh 40.20 g of medium and transfer to a 1 L cylinder. Add water to a volume of 1 L. Autoclave (*see* **Note 1**).
14. Marine Agar (MA): weigh 15.0 g of agar for 1 L of MB. Autoclave. Cool to 55 °C, and then pour into petri dishes (~25 mL for 9 cm diameter plate).
15. Microbiological hood or Bunsen burner.
16. Bio-Link crosslinker (UV-C, 254 nm), germicidal lamp of 80 W (*see* **Note 2**).
17. Incubator (4–25 °C) with an orbital shaking (capable of 150 rpm motion).

10. Glycerol solution (40%): dilute 40 mL of glycerol 100% in 60 mL of water, and sterilize it in autoclave.
11. Liquid nitrogen.

2.3 UV-C Assay

1. Autoclave.
2. Microbiological hood and Bunsen burner.
3. Spectrophotometer.
4. Plastic cuvettes with 1 cm of optical path.
5. 15 mL and 50 mL sterile tubes with aerated cap (suitable for inoculation).
6. Sterile serological pipettes.
7. Plastic petri dishes (9 cm diameter).
8. 200 μL sterile pipette tip or loop.
9. Sterile L-shape cell spreader.
10. Balance.

11. Magnetic stirrer.
12. Magnetic stir bar.
13. *Escherichia coli* MG1655, UV-C sensitive to be used as negative control.
14. *Deinococcus radiodurans*, UV-C resistant to be used as positive control.
15. Luria–Bertani (LB) broth: weigh 10.0 g of tryptone, 5.0 g of yeast extract, 10.0 g of sodium chloride, and transfer to 1 L cylinder. Add water to a volume of 1 L and autoclave.
16. LB agar (1 L): weigh 15.0 g of agar and transfer to 1 L cylinder. Autoclave. Cool to 55 °C and then pour into petri dishes (~25 mL/9 cm diameter for plate).
17. Tryptone yeast glucose (TYG) broth: weigh 3.0 g of tryptone, 3.0 g of yeast extract, 3.0 g of glucose, 1.0 g of K_2HPO_4, and transfer to 1 L cylinder. Add water to a volume of 1 L and autoclave.
18. TYG Agar (1 L): weigh 15.0 g of agar and transfer to 1 L TYG broth. Autoclave. Cool to 55 °C, and then pour into petri dishes (~25 mL/9 cm diameter for plate).
19. Bio-Link crosslinker (UV-C, 254 nm), germicidal lamp of 80 W (*see* **Note 2**).
20. Incubator (4–37 °C) with an orbital shaking (capable of 150 rpm motion).

3 Methods

3.1 Collection of Surface Water/Ice/Sediments

You can collect different types of samples (water, ice, sediments) for the isolation of UV-resistant bacteria (*see* **Notes 3** and **4**).

1. Collect surface water in different sampling stations (in triplicate) in 50 mL sterile tubes with cap and store at 4 °C and/or −20 °C until further use (*see* **Note 5**).
2. Collect surface water (up to 10 L) in different sampling stations (in triplicate) in sterile 10 L plastic carboy with handle(s) and cap. Working under a microbiological hood or near Bunsen burner flame, filter the collected liters of water under microbiological hood through a 0.45 μm filter inserted in a filter holder, by using a peristaltic pump. Incubate the filter for one night in 2 mL of sterile sea water and transfer 1 mL of this seawater containing the resuspended bacterial cells in a 2 mL cryovials. Add 20% glycerol (therefore 1 mL of 40% glycerol previously sterilized), gently mix, and freezes it in liquid nitrogen. Store at −80 °C (*see* **Notes 6** and 7). Use 1 mL of leftover

sample to make also 10 aliquots of sample stored in 20% glycerol to thaw for each experiment.

3. Collect surface ice in different stations (in triplicate) in 50 mL sterile tubes with cap and store at 4 °C and/or −20 °C (*see* **Note 5**).
4. Collect surface sediments along the coast (in triplicate) in 50 mL sterile tubes with cap and store at 4 °C and/or −20 °C (*see* **Note 5**).

3.2 Primary Screening for UV-Resistant Marine Bacteria from Water/Ice Samples

1. Take water/ice samples from 4 °C or thaw them (it is better to thaw 100 μL aliquots, *see* **Note 5**), if stored at −20 °C or −80 °C, and gently mix the samples.
2. Working under a microbiological hood or near Bunsen burner flame, transfer from each sample (Subheading 3.1, **steps 1–3**) 100 μL into a 2 mL sterile tube containing 0.9 mL of MB.
3. Allow to grow in an incubator with an orbital shaking at 150 rpm at 15 °C for at least 1 week (*see* **Notes 8** and **9**).
4. After growth, spread 100 μL of the bacterial cells using a sterile L-shape cell spreader onto an MA plate for the negative control (CTN 1:1) and other 100 μL of the bacterial cells onto an MA plate to be used for the UV-C assay (UV-C 1:1). Only the latter plate (UV-C 1:1) will be exposed to UV-C irradiation for 45 s (corresponding to 0.135 J/cm^2) inside the Bio-Link crosslinker (*see* **Notes 10** and **11**).
5. From the same bacterial cells (Subheading 3.2, **step 3**), proceed with dilutions up to 10^6 (*see* **Notes 12** and **13**). Therefore, take other 100 μL of the bacterial cells and dilute them in 0.9 mL of MB to obtain the dilution 1:10. Gently mix the sample and spread 100 μL of the bacterial cells using a sterile L-shape cell spreader onto an MA plate for the negative control (CTN 1:10) and other 100 μL of the bacterial cells onto an MA plate to be used for the UV-C assay (UV-C 1:10). Only the latter plate (UV-C 1:10) will be exposed to UV-C irradiation for 45 s (corresponding to 0.135 J/cm^2) inside the Bio-Link crosslinker (*see* **Notes 10** and **11**).
6. From the dilution 1:10, take 100 μL of the bacterial cells and dilute them in 0.9 mL of MB to obtain the dilution 1:100. Gently mix the sample and spread 100 μL of the bacterial cells onto an MA plate for the negative control (CTN 1:100) and other 100 μL of the bacterial cells onto an MA plate to be used for the UV-C assay (UV-C 1:100). Only the latter plate (UV-C 1:100) will be exposed to UV-C irradiation for 45 s (corresponding to 0.135 J/cm^2) inside the Bio-Link crosslinker (*see* **Notes 10** and **11**).

7. From the dilution 1:100, take 100 μL of the bacterial cells and dilute them in 0.9 mL of MB to obtain the dilution 1:1000. Gently mix the sample and spread 100 μL of the bacterial cells onto an MA plate for the negative control (CTN 1:1000) and other 100 μL of the bacterial cells onto an MA plate to be used for the UV-C assay (UV-C 1:1000). Only the latter plate (UV-C 1:1000) will be exposed to UV-C irradiation for 45 s (corresponding to 0.135 J/cm^2) inside the Bio-Link crosslinker (*see* **Notes 10** and **11**).
8. Continue dilutions until 1:10^6 and from each dilution spread 100 μL of the bacterial cells onto an MA plate for the negative control (CTN) and other 100 μL of the bacterial cells onto an MA plate to be used for the UV-C assay and exposed to UV-C irradiation for 45 s (corresponding to 0.135 J/cm^2) inside the Bio-Link crosslinker (*see* **Notes 10–13**), (Fig. 1a).
9. Incubate all plates at 15 °C for at least 1 week (*see* **Notes 8** and **14**).
10. After 1 week (or more) only the UV-resistant colonies will be visible on plates.

3.3 Primary Screening for UV-Resistant Marine Bacteria from Surface Sediments

1. Take sediment samples (Subheading 3.1, **step 4**) from 4 °C or thaw them (it is better to thaw 100 μL aliquots, *see* **Note 5**), if stored at −20 °C, and gently mix the samples.
2. Working under a microbiological hood or near Bunsen burner flame, from each sample transfer 1 g of sediments into a 15 mL sterile tube containing 9 mL of sterile sea water.
3. Incubate the samples in an incubator with an orbital shaking at 150 rpm at 15 °C (*see* **Note 8**) for one night.
4. The day after spread 100 μL of the sediment solution using a sterile L-shape cell spreader onto an MA plate for the negative control (CTN 1:1) and other 100 μL of the sediment solution onto an MA plate to be used for the UV-C assay (UV-C 1:1). Only the latter plate (UV-C 1:1) will be exposed to UV-C irradiation for 45 s (corresponding to 0.135 J/cm^2) inside the Bio-Link crosslinker (*see* **Notes 10** and **11**).
5. From the same sediment sample obtained from one-night growth (Subheading 3.3, **step 3**), take 1 mL of the sediment solution and dilute it in a 9 mL of MB to obtain the dilution 1: 10. Gently mix the samples and spread 100 μL of the sediment solution 1:10 using a sterile L-shape cell spreader onto an MA plate for the negative control (CTN 1:10) and other 100 μL of the sediment solution 1:10 onto an MA plate to be used for the UV-C assay (UV-C 1:10). Only the latter plate (UV-C 1:10) will be exposed to UV-C irradiation for 45 s (corresponding to 0.135 J/cm^2) inside the Bio-Link crosslinker (*see* **Notes 10** and **11**).

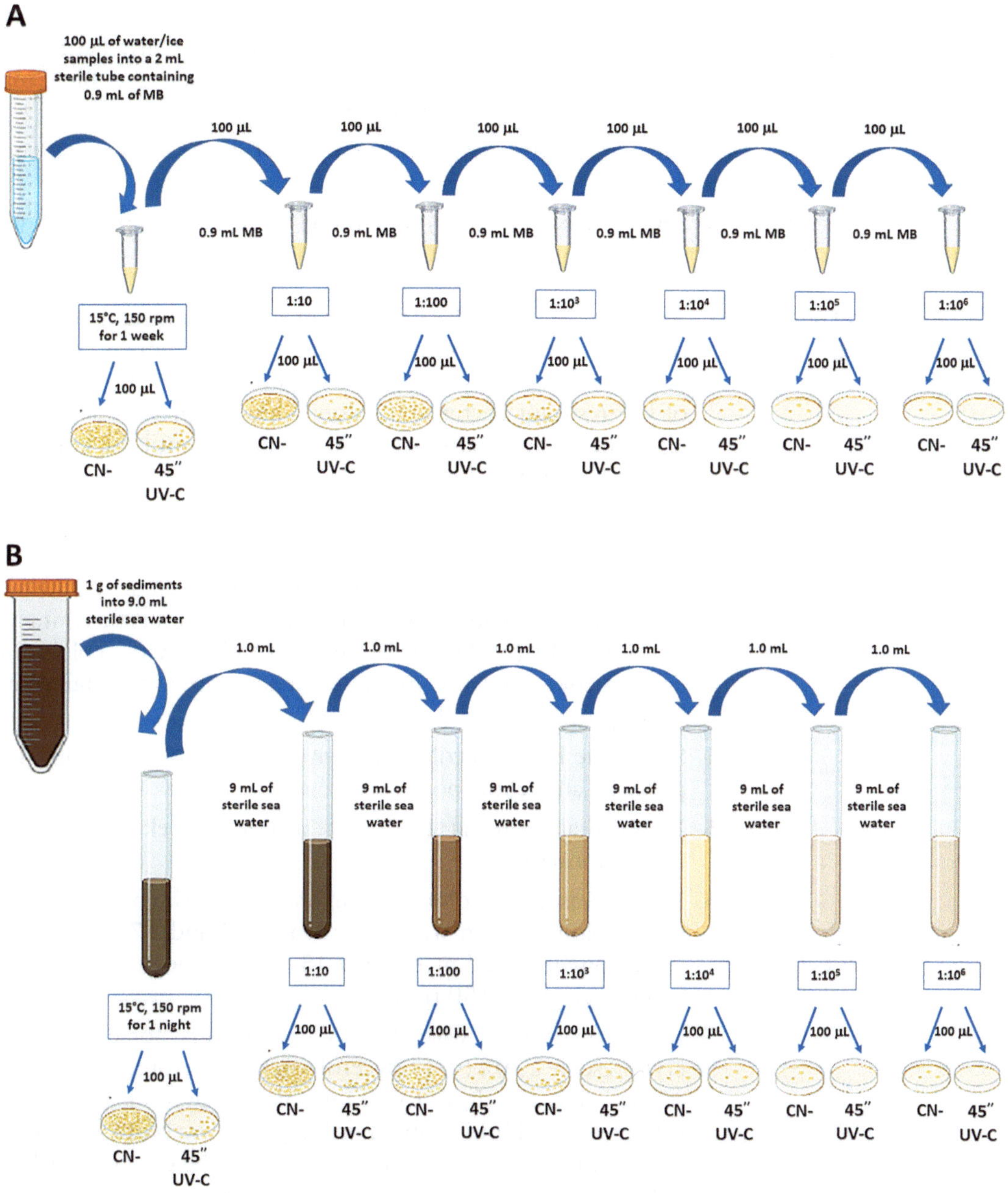

Fig. 1 Serial dilution procedure for the isolation of marine bacteria from water/ice samples (**a**) and surface sediments (**b**). Main steps of serial dilution procedure are indicated

6. Prepare all dilutions from 1:100 to $1{:}10^6$ (as Subheading 3.2, **steps** 7 and **8**) transferring 1 mL of the sediment solution previously diluted into a 9 mL of MB to obtain the next dilution. Gently mix the samples and spread 100 μL of the obtained sediment solution onto an MA plate for the negative control (CTN) and other 100 μL of the same sediment solution

onto an MA plate to be used for the UV-C assay (UV-C). Only the latter plate (UV-C) will be exposed to UV-C irradiation for 45 s (corresponding to 0.135 J/cm^2) inside the Bio-Link crosslinker (*see* **Notes 10** and **11**) (Fig. 1b).

7. Incubate all plates at 15 °C for at least 1 week (*see* **Notes 8** and **14**).
8. After 1 week (or more) only the UV-resistant colonies will be visible on plates.

3.4 Isolation of UV-Resistant Marine Bacteria

1. After 1 week at 15 °C, check for grown UV-resistant colonies obtained from the primary screening (Subheadings 3.2 and 3.3). If colonies are present, you can proceed with the isolation of UV-resistant bacteria. If colonies are not present, leave the plates in the incubator at 15 °C for longer until the colonies appear (*see* **Note 14**).
2. Add 3 ml of MB into 15 ml inoculum tubes by using sterile serological pipettes.
3. Pick well-isolated colonies from the petri dish by a 200 μL sterile pipette tip and drops it into the inoculum tube containing 3 mL MB (*see* **Notes 15** and **16**) and grow the colony in liquid medium at 15 °C for at least 5 days with orbital agitation (150 rpm) (*see* **Note 9**).
4. Streak the bacterial colony grown in the liquid media on MA plates by an aseptic loop.
5. Incubate all plates at 15 °C for at least 1 week, until single colony formation.
6. Pick well-isolated colonies from the petri dish (not touching other colonies) as in **step 3** (*see* **Notes 16** and **17**), and grow the colony in liquid medium at 15 °C for at least 5 days with orbital agitation (150 rpm).
7. Before proceeding with the UV experiments, for each strain cryopreserve a 20% (v/v) glycerol stock. Therefore, add 500 μL of the bacterial culture to 500 μL of 40% glycerol (previously sterilized) (*see* **Note 6**) in a 2 mL cryovial (*see* **Note** 7), gently mix, and freeze it in liquid nitrogen. Store the glycerol stock at −80 °C.

3.5 UV-C Assay

1. Measure the optical density at 600 nm (OD_{600}) of bacterial cultures isolated in Subheading 3.4, **step 6**, in a plastic cuvette using a spectrophotometer (*see* **Notes 18** and **19**).
2. After measuring the ODs, add an adequate amount of cell culture into 50 mL inoculum tubes containing 10 mL (*see* **Note 20**) of appropriate liquid medium to start from 0.08 OD (*see* **Note 21**).

3. At the same time, grow the cultures of *E. coli* MG1655 and *D. radiodurans*, UV-C resistant strains that will be used as negative and positive controls, respectively. Use LB at 37 °C for *E. coli* and TYG at 30 °C for *D. radiodurans*. Then, proceed as in **steps 1** and **2**.
4. Grow bacterial cultures in incubator with orbital shaking at 150 rpm at 15 °C (UV-resistant isolated marine bacteria), 37 °C (negative control) and 30 °C (positive control) for a few hours until reaching around 0.3 OD [11] (*see* **Note 22**).
5. 100 μL of each UV-resistant bacterium was plated by a sterile L-shape cell spreader on MA plates (in triplicate). Each plate will be exposed to UV-C irradiation (*see* **Note 10**) for 0, 1, 2, 3 min (corresponding to 0, 0.180, 0.360, 0.540 J/cm^2, respectively) (*see* **Note 23**), and then incubated at 15 °C for 1 week (Fig. 2).
6. 100 μL of negative (*E. coli* MG1655) and positive (*D. radiodurans*) control were plated by a sterile L-shape cell spreader on LB or TYG agar plates, respectively (in triplicate). Each plate will be exposed to UV-C irradiation and then incubated at 37 °C for 1–2 days for *E. coli* and 30 °C for 3–4 days for *D. radiodurans* (*see* **Notes 10** and **24**).
7. After 1 week at 15 °C, take MA plates left in the incubator and counts the number of colony forming units (CFUs), to select the most UV-resistant bacteria (*see* **Note 25**).
8. Compare the number of CFUs obtained from isolated marine bacteria with those of positive and negative controls, to evaluate the difference of degree of resistance between them.

4 Notes

1. MB is one of the most used medium for the growth of marine bacteria. This saline medium appears with a colour of light yellow with a natural turbidity. To check its contamination, you should let it settle for hours and then see the saline sediment at the bottom of the bottle and the solution above completely clear. Be careful because there may be differences between MB of different brands. We recommend to use always the same brand to standardize the experiment. For example, in our experiments, marine bacterial growths were optimized with MB, Conda cat number 1217.00. However, in addition to MB, many other different media can be used for the isolation of marine bacteria [15, 16].
2. UV-C lamp Stratagene (now Vilber). Similar devices can be used. Consider that by using different germicidal lamps, the UV-C exposure times should be modified appropriately.

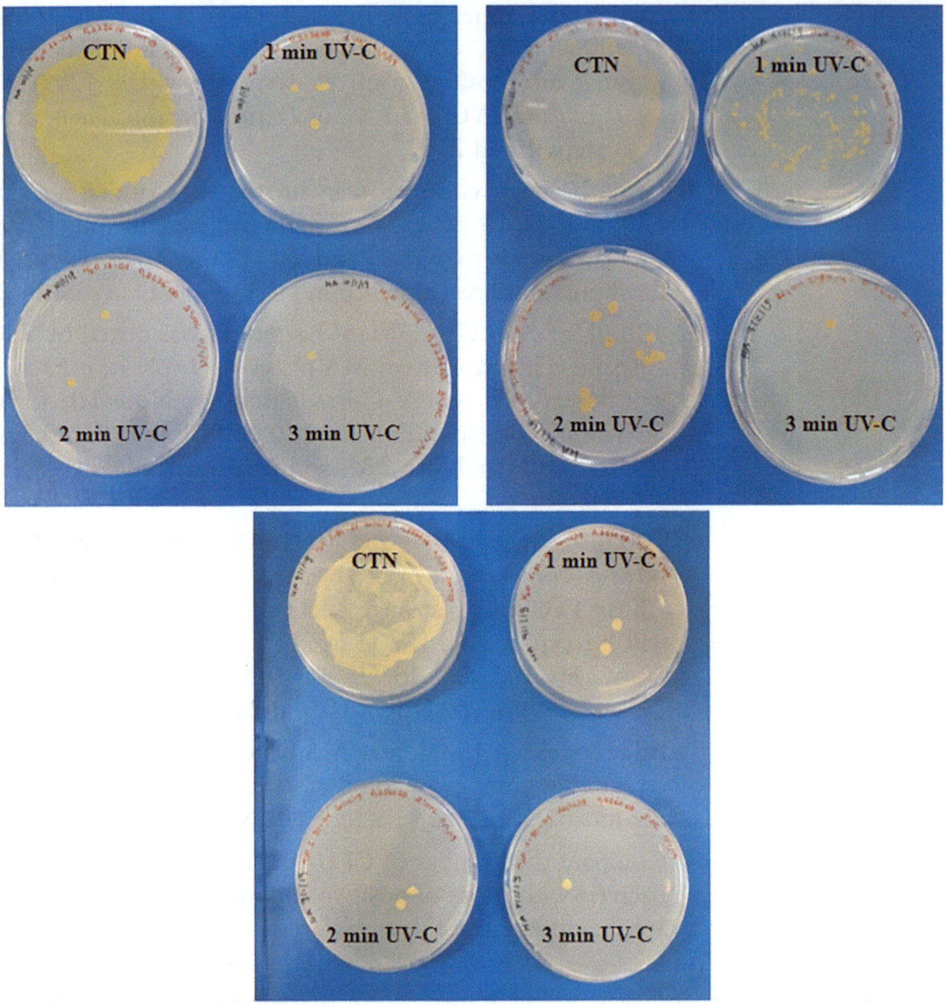

Fig. 2 Examples of Antarctic marine bacteria exposed to UV-C irradiation for 0 min (CTN), 1 min (0.180 J/cm^2), 2 min (0.360 J/cm^2), 3 min (0.540 J/cm^2). The number of colonies decreases going from the CTN (not exposed to UV-C) to the plate exposed to 3 min of UV-C

3. Searching UV-resistant marine bacteria, it is a good choice to collect surface samples because more exposed to UV radiation.
4. To isolate bacteria from water, two different methods are possible: the collected water can be used directly for bacteria isolation (Subheading 3.1, **step 1**) or it can be filtered to concentrate the number of bacteria, and the filters then used for isolation (Subheading 3.1, **step 2**). It is recommended that you try both methods in order to isolate more bacteria of interest.
5. You can store your water/ice samples at 4 °C if you plan to use them as soon as possible. Alternatively, you can store them at −20 °C if you plan to use them after a while (at least

2–3 months). Note that freezing and thawing samples causes the death of most bacteria. For this, it is better to make also 100 μL aliquots of the samples before freezing them, so as to thaw only the amount of sample needed for the experiment, preserving the unused sample.

6. You can also use 100% glycerol directly, but pipetting viscous liquids like glycerol is difficult. It is recommended to use a very slow aspiration speed, wait for a few seconds until the glycerol stops, and then use a very slow dispensing rate. The glycerol stock is stable for years, as long as it is frozen. Therefore, use it always on ice. Be careful not to let the glycerol stock thawed.
7. Do not use snap top tubes because they can open unexpectedly at −80 °C.
8. Note that the bacterial growth temperature depends on the collected samples. The optimal growth temperature range of psychrophiles is between 5 and 20 °C [17]. Higher temperatures (>20 °C) induce cellular stress [18]. Therefore, we used 15 °C in our experiments as good compromise between the optimal growth temperature and the growth rate, but other growth temperatures can be tested in order to obtain a greater number of isolates. However, in the case of nonpsychrophilic samples higher temperatures are generally recommended.
9. If the liquid medium is still clear, leave the colony at 15 °C until the medium becomes cloudy (cloudy media indicates bacterial growth). Be careful with the MB as it can be quite cloudy due to the salts present. In any case, in the presence of bacterial growth, the turbidity will be much greater than that in liquid medium where bacterial growth is not present.
10. When you expose plated bacteria to the UV-C irradiation, remove the lid from the petri dish so that rays are not shielded causing false positives.
11. Forty-five seconds was chosen as the UV-exposure time because the negative control *E. coli* MG1655 is resistant up to 45 s.
12. Dilutions are necessary to isolate dominant marine bacteria found in lower dilution series from those less abundant/dominant in higher dilution series [19].
13. For water/ice samples collected in a tube of 50 mL (Subheading 3.1, **steps 1** and **3**), it is enough to prepare dilutions until 1:100 or 1:1000 because these samples are not concentrated. In contrast, for samples obtained from filtered water and sediments it is worth doing dilution up to 10^6.
14. Check the plates also after 1 to 2-months because some bacteria can grow very slowly. Therefore, even if colonies are present on the agar plates and you proceed with the isolation of the

colonies grown in a week, then you can leave the plate at 15 °C to observe if new colonies grow with longer incubation times (e.g., 2 weeks and/or 1 month). If new colonies are present, repeat the whole experiment on the new colonies.

15. You could also pick the colony with an aseptic loop and swirl the colony into the inoculum tube.
16. It is important never to touch the tip or loop surface with your hands, in order to keep sterility. When picking the colonies, never touch other colonies or agar not occupied by your colony, to be sure you have selected a single colony.
17. Isolates should be streaked at least three times before being considered pure.
18. If you do not have the bacterial culture already available, use the glycerol stock stored at −80 °C to retrieve the isolated UV-resistant bacteria. In this case, use a 200 μL sterile pipette tip or loop to scrape off a portion from the top of the frozen glycerol stock, and drops it into the 15 mL inoculum tube containing 3 mL MB. Grow the colony in liquid medium at 15 °C for at least 5 days with orbital agitation (150 rpm), then measures the optical density at 600 nm (OD_{600}).
19. Be sure to measure OD within absorbance 1. If the reading exceeds 1, you have to dilute your sample (e.g., 200 μL of grown culture in 800 μL of medium, and after the reading you have to take into account the dilution factor of 5).
20. Take into account that the best ratio between volume of medium and volume of tube is 1:5 (10 mL medium:50 mL tube) to ensure enough aerated space for the culture.
21. To calculate the volume to add to 10 mL of medium to start the growth at 0.08 OD use the formula: (0.08 OD/measured OD) × 10 mL.
22. It is very important do not exceed 0.3–0.35 OD because the more cells they grow, the more resistant they will be, causing a false result. Be careful because each strain has a different time for the growth, some strain can reach 0.3 OD after 3–4 h and others after 6–7 h (or more).
23. Some strains very resistant were exposed to UV-C irradiation also for 4, 5, 6 min (corresponding to 0.720, 0.900, 1.08 J/cm^2, respectively) to be compared with the positive control *D. radiodurans*.
24. *E. coli* MG1655 as negative control was exposed to UV-C irradiation for 0 and 1 min (corresponding to 0 and 0.180 J/cm^2, respectively) whereas the positive control (*D. radiodurans*), being UV-resistant, was exposed to UV-C irradiation for 0, 1, 2, 3, 4, 5, 6 min (corresponding to

0, 0.180, 0.360, 0.540, 0.720, 0.900, 1.08 J/cm^2, respectively).

25. Check the plates also after 1–2 months because some bacteria can grow very slowly, especially after UV-C irradiation.

Acknowledgments

This research was financially supported by the Italian National Programme for Antarctic Research (PNRA) of the Ministry of University and Research (MUR), 2016/AZ1.06-Project PNRA16_00043. We wish to thank Giovanni Del Monaco and Francesca Segreti for technical support and assistance.

References

1. Sheridan C (2012) Recasting natural product research. Nat Biotechnol 30:385–387
2. Williamson NR, Fineran PC, Leeper FJ et al (2006) The biosynthesis and regulation of bacterial prodiginines. Nat Rev Microbiol 4: 887–899
3. Rehm BH (2010) Bacterial polymers: biosynthesis, modifications and applications. Nat Rev Microbiol 8:578–592
4. Jaspars M, De Pascale D, Andersen JH et al (2016) The marine biodiscovery pipeline and ocean medicines of tomorrow. J Mar Biol 96: 151–158
5. Bruno S, Coppola D, di Prisco G et al (2019) Enzymes from marine polar regions and their biotechnological applications. Mar Drugs 17: 544
6. Núñez-Pons L, Shilling A, Verde C et al (2020) Marine terpenoids from polar latitudes and their potential applications in biotechnology. Mar Drugs 18(8):401
7. Giordano D (2020) Bioactive molecules from extreme environments. Mar Drugs 18(12):640
8. Banaszak AT (2003) Photoprotective physiological and biochemical responses of aquatic organisms. In: Helbling EW, Zagarese H (eds) UV effects in aquatic organisms and ecosystems. Royal Society of Chemistry, Cambridge
9. Núñez-Pons L, Avila C, Romano G et al (2018) UV-protective compounds in marine organisms from the Southern Ocean. Mar Drugs 16(9):336
10. Correa-Llantén DN, Amenábar MJ, Blamey JM (2012) Antioxidant capacity of novel pigments from an Antarctic bacterium. J Microbiol 50:374–379
11. Órdenes-Aenishanslins N, Anziani-Ostuni G, Vargas-Reyes M et al (2016) Pigments from UV-resistant Antarctic bacteria as photosensitizers in dye sensitized solar cells. J Photochem Photobiol B 162:707–714
12. Silva TR, Tavares RSN, Canela-Garayoa R et al (2019) Chemical characterization and biotechnological applicability of pigments isolated from Antarctic bacteria. Mar Biotechnol 3: 416–429
13. Silva TRE, Júnior SLCF, de Queiroz AC et al (2021) Pigments from Antarctic bacteria and their biotechnological applications. Crit Rev Biotechnol 23:1–28
14. Coppola D, Giordano D, Tinajero-Trejo M et al (2013) Antarctic bacterial hemoglobin and its role in the protection against nitrogen reactive species. Biochim Biophys Acta 1834: 1923–1931
15. Joint I, Mühling M, Querellou J (2010) Culturing marine bacteria—an essential prerequisite for biodiscovery. Microb Biotechnol 3(5): 564–575
16. Sipkema D, Schippers K, Maalcke WJ et al (2011) Multiple approaches to enhance the cultivability of bacteria associated with the marine sponge Haliclona (gellius) sp. Appl Environ Microbiol 77(6):2130–2140
17. Feller G, Gerday C (2003) Psychrophilic enzymes: hot topics in cold adaptation. Nat Rev Microbiol 1:200–208
18. Gerday C, Aittaleb M, Bentahir M et al (2000) Cold-adapted enzymes: from fundamentals to biotechnology. Trends Biotechnol 18(3): 103–107
19. Schut F, de Vries EJ, Gottschal JC et al (1993) Isolation of typical marine bacteria by dilution culture: growth, maintenance, and characteristics of isolates under laboratory conditions. Appl Environ Microbiol 59(7):2150–2160

Chapter 16

Fractionation Protocol of Marine Metabolites

Genoveffa Nuzzo, Emiliano Manzo, Carmela Gallo, Giuliana d'Ippolito, and Angelo Fontana

Abstract

Marine organisms are well known for their capability to produce chemically unique and biological relevant small molecules. However, marine extracts are complex mixtures of compounds with a huge presence of salts. Thus, biological screening and chemical analysis of marine extracts pose specific technical constraints and require adequate sample preparation. We have developed an automated solid-phase extraction (SPE)-based method to desalt marine extracts and recover metabolites, adapt to be integrated in platform of high-throughput screening. The procedure uses a poly(styrene-divynylbenzene)-based support and a stepwise organic solvent elution herein described.

Key words Marine natural product, Secondary metabolites, Solid-phase extraction, Chromatographic fractionation

1 Introduction

The study of natural secondary metabolites is important to a broad range of fields, including drug discovery, ecology, and chemical biology [1]. Due to the chemical difficulties to purify minority metabolites, most natural products being published today bear structural similarity to previously published compounds [1], whereas it is always more challenging identified new scaffolds. However, marine organisms continue to be a rich source natural products with no structural precedents [2], suggesting that development of innovative fractionation methods can be useful to isolate new compounds with unique chemical and biological properties.

Marine extracts are mixture of large number of metabolites containing high concentration of salt, which poses specific technical constraints (e.g., overweight, overlapped effect, synergic reaction) [3]. Thanks to its versatility, solid-phase extraction (SPE) has become a common strategy for purification and enrichment of natural products [4]. With the aim to favor the recovery of minor

Cinzia Verde and Daniela Giordano (eds.), *Marine Genomics: Methods and Protocols*,
Methods in Molecular Biology, vol. 2498, https://doi.org/10.1007/978-1-0716-2313-8_16,

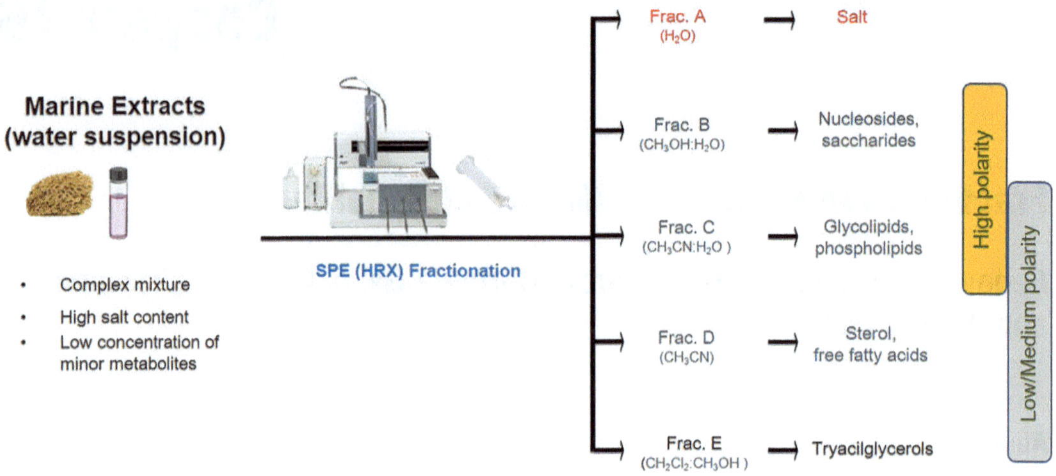

Fig. 1 Schematic procedure of fractionation by solid-phase extraction of marine extracts

compounds, we have developed an SPE fractionation method [5] based on modern polymeric Chromabond® HR-X columns. This support is composed of hydrophobic polystyrene-divinylbenzene (PS/DVB) resin in spherical particles with highest reproducibility and excellent recovery rates, especially for the enrichment of pharmaceutical/active ingredients [6].

PS-DVB, due to its pH stability and the hydrophobic surfaces, has been largely used as chromatographic support for purification of organic molecules from different sources [7–9] and has been long used by several authors to desalt marine extracts [10–13].

Applied to marine extracts, our fractionation protocol allows to remove salts, eluted into initial fraction (A), and a predictable distribution of metabolites in the subsequent four fractions B–E (Fig. 1). Based on the decreasing polarity sequence of the eluents, primary metabolites like sugars and amino acids were eluted in fractions A and B, polar lipids including glycolipids and most phospholipids in C, sterols and free fatty acids in D and, finally, triglycerides and other neutral lipids in E. The protocol was set up on an open column, but following development and fine-tuning, it was transferred onto a completely automated fractionation system (GX-271 ASPEC Gilson apparatus, Fig. 2). The use of a branded analytical platform for loading and eluting the sample on prepacked cartridges has required only an adjustment of the solvent volumes, as specified below. Thus, the analytical procedure has been adapted to be integrated into traditional procedures of high-throughput bioassay guided fractionation [14] and it has been applied for the analysis of minor marine metabolites [15–17].

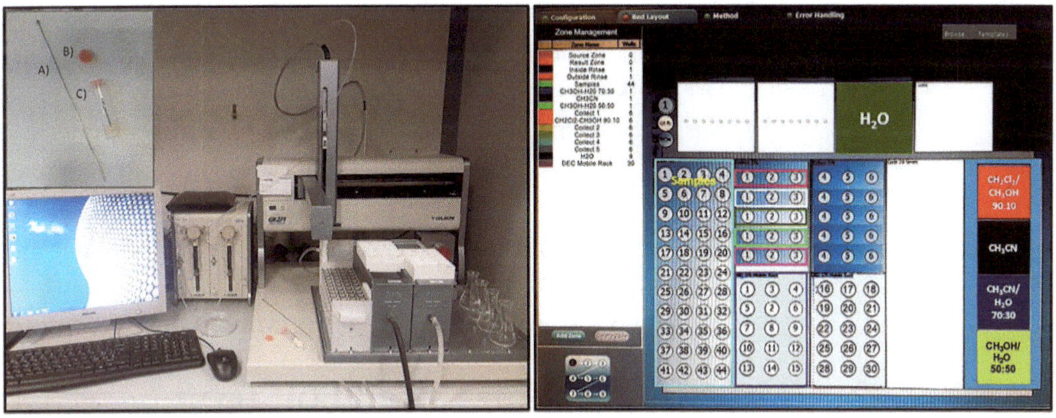

Fig. 2 On the left the image of GX-271 ASPEC Gilson liquid handling apparatus: the insert shows (**a**) needle probe, (**b**) sealing caps 6 mL, and (**c**) prepacked Chromabond HRX column; on the right the image of the Bed layout of GX-271 ASPEC Gilson liquid handling apparatus showing in detail how to accommodate solvents, samples, columns, and glass tubes

2 Materials

1. HR-X cartridge of 6 mL/500 mg (*see* **Note 1**).
2. 15 mL conical tube.
3. Borosilicate glass tubes.
4. Evaporating flasks.
5. Glass screw-top vials.
6. CH_3CN high-performance liquid chromatography (HPLC) grade (≥99%).
7. CH_3OH HPLC grade (≥99%).
8. CH_2Cl_2 HPLC grade (≥99%).
9. Milli-Q deionized water.
10. GX-271 ASPEC Gilson apparatus (*see* **Note 2**).
11. TRILUTION® LC Software (*see* **Note 3**).
12. Rotary evaporation.
13. Argon gas.
14. Dry extract (*see* **Note 4**).
15. Collector.

3 Methods

3.1 HR-X Fractionation

1. Prepare 20 mL of each eluent: H_2O 100%, CH_3OH–H_2O 1:1 (v/v), CH_3CN–H_2O 7:3 (v/v), CH_3CN 100%, CH_2Cl_2–CH_3OH 9:1 (v/v) (*see* **Note 5**).

2. Activate the HR-X cartridge with 3 mL of CH_3OH.
3. Equilibrate the column with 6 mL of Milli-Q deionized water (*see* **Note 6**).
4. Suspend 30 mg of dry extract in 1 mL of distilled water (*see* **Note 7**).
5. Load the suspension on the activated HR-X column.
6. Wash the vial of the extract with 2 mL of distilled water and transfer onto column.
7. Elute and discard salt with 15 mL of Milli-Q deionized water (fraction A, Fig. 1).
8. Elute and collect 18 mL of $CH_3OH–H_2O$ 1:1 (v/v) (fraction B, Fig. 1, *see* **Note 8**).
9. Elute and collect 18 mL of $CH_3CN–H_2O$ 7:3 (v/v) (fraction C, Fig. 1, *see* **Note 8**).
10. Elute and collect 18 mL of 100% CH_3CN (fraction D, Fig. 1, *see* **Note 9**).
11. Elute and collect 18 mL of $CH_2Cl_2–CH_3OH$ 9:1 (v/v) (fraction E, Fig. 1, *see* **Note 9**).
12. Remove solvent by rotary evaporation (*see* **Note 10**).
13. Dissolve dry fractions and transfer into vials with a minimal volume of solvent (*see* **Note 11**).
14. Remove solvent from the organic fractions and store under argon at −20 °C until analysis.

3.2 Automated HRX Fractionation on GX-271 ASPEC Gilson Apparatus

1. Turn on GX-271 ASPEC Gilson apparatus (*see* **Note 2**).
2. Open TRILUTION® LC Software (*see* **Note 3**).
3. Set bed layout positions according to Fig. 1 (*see* **Note 12**).
4. Place HR-X column and glass tubes in the well "DEC Mobile 376—Rack no. 1" (*see* **Note 13**).
5. Prepare solvents (*see* Subheading 3.1, **step 1**).
6. Place 5 glass tubes in the wells of the "Collector" from 1 to 5.
7. Suspend 30 mg of dry extract in 1 mL distilled water (*see* **Notes** 7 and **14**).
8. Transfer the suspension to a 15 mL conical tube.
9. Place the sample to load in the well "Sample 1."
10. Set liquid handling operations as follows.
    (a) Conditioning 5 mL CH_3OH.
    (b) Conditioning 6 mL H_2O.
    (c) Conditioning 6 mL H_2O.
    (d) Load and collect 1 mL (*see* **Note 14**).

(e) Dispense 2 mL of H_2O.

(f) Mix 1 mL of sample.

(g) Load and collect the above 2 mL from the sample onto the column.

(h) Fractionate 6 mL H_2O (to collect in well 1 of Collector 1).

(i) Fractionate 9 mL CH_3OH–H_2O 1:1 (v/v) (to collect in well 1 of Collector 2).

(j) Fractionate 9 mL CH_3CN–H_2O 7:3 (v/v) (to collect in well 1 of Collector 3).

(k) Fractionate 9 mL 100% CH_3CN (to collect in well 1 of Collector 4).

(l) Fractionate 9 mL CH_2Cl_2–CH_3OH 9:1 (v/v) (to collect in well 1 of Collector 5).

11. Transfer all the fractions to an evaporating flask.
12. Remove solvent by rotary evaporation (*see* **Note 10**).
13. Dissolve dry fractions and transfer into vials with a minimal volume of solvent (*see* **Note 11**).
14. Remove solvent from the organic fractions and store under argon at −20 °C until analysis.

4 Notes

1. Polystyrene-divinylbenzene (PS/DVB) copolymer is a robust material used as the base of SPE columns by different brand. Different commercial types of PS-DVB absorbent can be used. However, pore sizes, surface areas, degrees of cross-linking, and particle sizes can affect chemical interaction between the analyte and stationary phase resulting in differential selectivity. Chromabond® HR-X offers the advantage to have narrow pores (260 Å) and regular, little particles (around 65 μm) that increase the surface area. Therefore, to have the same results we suggest to use Chromabond®, on which the method has been developed.
2. GX-271 ASPEC Gilson apparatus is a single-probe X/Y/Z liquid handler that paired with a VERITY® 4060 Single Syringe Pump or VERITY® 4260 Dual Syringe Pump can be used to automated liquid handling and SPE procedures that use positive pressure elution for cartridge-based applications.
3. TRILUTION® LC software is a software package for controlling high-pressure liquid chromatography systems and automated liquid handling instruments.

4. Dry extract can be obtained by organic solvent extraction (generally methanol) of marine organisms or biomass, as a pellet of microorganisms (e.g., diatoms) or invertebrate (sponge, tunicate, molluscs, echinoderm, etc.).
5. The indicated volume (20 mL) is referred to only one SPE fractionation.
6. To remove completely the organic solvent, carry out two equilibration steps with 3 mL each.
7. Samples rich of salt can be suspended in water easily. However, we always suggest to sonicate samples before loading. For highly lipophilic extracts, precipitates or aggregates of lipid drops may be observed, resulting in a low recovery of the organic fractions. To overcome this issue, adsorb the extract onto silica gel: dissolve the extract with the best solubilizing solvent and add 60 mg of silica (1:2 w/w) and dry by rotary evaporation. Finally, load manually the silica containing the extracts onto the HRX-column after the equilibration step.
8. Fraction B and C are mainly composed of polar compounds, like nucleosides/saccharides and glycolipids/phospholipids eluted with CH_3OH–H_2O 1:1 and CH_3CN–H_2O 7:3 respectively (Fig. 1).
9. In fraction D and E are mainly collected compounds with medium and low polarity, as sterol and free fatty acids eluted with 100% CH_3CN or triacylglycerol with CH_2Cl_2–CH_3OH 9:1 (Fig. 1).
10. To avoid degradation of very labile compounds, it is recommended to use bath temperature never higher than 28 °C.
11. Due to the chemical diversity of the four organic fractions, transfer the more polar fractions B and C with CH_3OH, whereas for fractions D and E it needed a mixture of CH_3OH–CH_2Cl_2 2:1 (v/v).
12. The Bed Layout tab window is a graphical interface used to create a bed layout, one of the building blocks of a method (for more details on "How to Create a Bed Layout" refer to the manual downloadable at the following link https://it.gilson.com/pub/media/docs/TRILUTIONLH_UG_LT302006-01.pdf).
13. The GX-271 ASPEC can be configured with a variety of rack types and sizes. "DEC Mobile 376" is the code of that available for 6 mL SPE cartridges and 10 mL tubes in the collection block.
14. Generally, 1 mL of sample solution is drawn and loaded onto the column. However, if more solvent is needed to suspend the extract, remember to draw the same volume.

Acknowledgments

The authors acknowledge the financial support of the project "Antitumor Drugs and Vaccines from the Sea (ADViSE)" (CUP B43D18000240007–SURF 17061BP000000011) funded by POR Campania FESR 2014–2020 "Technology Platform for Therapeutic Strategies against Cancer"—Action 1.1.2 and 1.2.2.

References

1. Pye CR, Bertin MJ, Lokey RS et al (2017) Retrospective analysis of natural products provides insights for future discovery trends. Proc Natl Acad Sci U S A 114:5601–5606
2. Carroll AR, Copp BR, Davis RA et al (2021) Marine natural products. Nat Prod Rep 38: 362–413
3. Houssen WE, Jaspars M (2006) Isolation of marine natural products. In: Sarker SD, Latif Z, Gray AI (eds) Natural products isolation. Methods in biotechnology, vol 20. Humana Press, Totowa, New Jersey
4. Wilson BAP, Thornburg CC, Henrich CJ et al (2020) Creating and screening natural product libraries. Nat Prod Rep 37:893–918
5. Cutignano A, Nuzzo G, Ianora A et al (2015) Development and application of a novel SPE-method for bioassay-guided fractionation of marine extracts. Mar Drugs 13:5736–5749
6. Faraji M, Yamini Y, Gholami M (2019) Recent advances and trends in applications of solid-phase extraction techniques in food and environmental analysis. Chromatographia 82: 1207–1249
7. Lundanes E, Reubsaet L, Greibrokk T (2014) Chromatography: basic principles, sample preparations and related methods, 1st edn. Wiley-VCH Verlag, Weinheim, Germany
8. Huck CW, Bonn GW (2005) Poly(styrene-divynylbenzene) based media for liquid chromatography. Chem Eng Technol 28: 1457–1472
9. Chambers TK (1996) Liquid chromatographic retention studies using polystyrene-divinylbenzene stationary phases in reversed-phase and normal-phase eluents. Dissertation, Iowa State University, Ames, IA, USA
10. Minale L, Pizza C, Riccio R, Zollo F (1982) Steroidalglycosides from starfishes. Pure Appl Chem 54:1935–1950
11. West LM, Northcote PT (2000) Peloruside a: a potent cytotoxic macrolide isolated from the New Zealand marine sponge *Mycale* sp. J Org Chem 65:445–449
12. West LW, Faulkner DJ (2006) Hexaprenoid Hydroquinones from the sponge *Haliclona* (aka *Adocia*) sp. J Nat Prod 69:1001–1004
13. Bugni TS, Harper MK, McCulloch MWB et al (2008) Fractionated marine invertebrate extract libraries for drug discovery. Molecules 13:1372–1383
14. Gallo C, Barra G, Saponaro M et al (2020) A new bioassay platform Design for the Discovery of small molecules with anticancer immunotherapeutic activity. Mar Drugs 18:604
15. Nuzzo G, Gomes BA, Gallo C et al (2019) Potent cytotoxic analogs of Amphidinolides from the Atlantic Octocoral *Stragulum bicolor.* Mar Drugs 17:58
16. Gallo C, d'Ippolito G, Nuzzo G et al (2017) Autoinhibitory sterol sulfates mediate programmed cell death in a bloom-forming marine diatom. Nat Commun 8:1292
17. Gallo C, Nuzzo G, d'Ippolito G et al (2018) Sterol sulfates and sulfotransferases in marine diatoms. In: Moore BS (ed) Methods in enzymology, vol 605. Academic Press, Cambridge, Massachusetts, pp 101–138

Chapter 17

Detection and Quantification of Small Noncoding RNAs in Marine Diatoms

Alessandra Rogato and Angela Falciatore

Abstract

Endogenous small noncoding RNAs (sRNAs) are a large family of essential regulators of gene expression in both eukaryotes and prokaryotes. Various types of sRNAs with different size and mapping to different genome locations have been recently identified in diatoms, a successful group of phytoplankton in the marine environment. However, their biogenesis and regulatory function are still largely unknown and unexplored in these microalgae, also due to the lack of methods for their experimental analysis. Herein, we present a point-by-point description of the protocols for detection and quantification of sRNAs by Northern-blot analysis and quantitative stem-loop RT-PCR, established in the diatom molecular model specie *Phaeodactylum tricornutum*.

Key words Diatom, Small noncoding RNA, Northern blot, Stem-loop qPCR

1 Introduction

In the organisms of the three domains of life, a plethora of small regulatory non coding RNAs (sRNAs) of around 20–32 nucleotide acts as major regulators of gene expression [1]. The most extensively studied sRNAs are the microRNAs (miRNAs) and small interfering RNAs (siRNAs). Both exert gene silencing effects via RNA-induced silencing complex (RISC) pathways [2].

In various organisms, next-generation sequencing (NGS) has provided an unprecedented opportunity to identify and quantify diverse kinds of sRNAs derived from different genomic location and regulating different cellular pathways. These sRNAs differ in their mode of biogenesis and in their function [3]. However, in order to exclude artefactual biases due to the library preparations or sequencing, and to get function insights of novel sRNAs categories identified in less established model organisms, the predicted sRNAs need to be experimentally validated.

Cinzia Verde and Daniela Giordano (eds.), *Marine Genomics: Methods and Protocols*,
Methods in Molecular Biology, vol. 2498, https://doi.org/10.1007/978-1-0716-2313-8_17,

Diatoms are prominent marine unicellular algae [4]. Although recent diatom genome analyses have highlighted the presence of unique combinations of genes, recruited during their history as secondary endosymbionts, as well as by horizontal gene transfer [5], very little is known about the mechanisms of regulation of gene expression in these algae and on the processes controlling adaptive responses to changing environment. In 2014, Rogato and colleagues [6] and Lopez-Gollom and coworkers [7] identified a plethora of sRNAs in three diatom species for which complete genome sequences have been assembled, annotated and published: *Phaeodactylum tricornutum* [5], *Thalassiosira pseudonana* [8], and *Fragilariopsis cylindrus* [9]. By the high-throughput sequencing of short RNA fragments extracted from diatoms grown under different conditions of light and nutrients, different sRNA populations of different size (19–35 nucleotide) have been described [6, 7]. These sRNAs map different genome localization such as tRNAs, transposable elements (TEs), intergenic region and small nuclear RNA (snRNAs). Canonical miRNAs were not found while TEs marked by DNA methylation represent the majority source of sRNAs. A comparative analysis of genome composition in diatoms revealed a strong variation in transposable element dynamics, suggesting that TEs play a key role in genome evolution and gene function [10]. Rogato and coworkers suggested that the TE-derived sRNAs, differentially expressed in the different tested growth conditions, may affect TE transcripts by sequence complementarity, contributing to the regulation of stress responses. Furthermore, other sRNAs mapping tRNAs, intergenic regions, and snRNAs were highly expressed and have been experimentally detected and validated. These sRNAs are likely implicated in the regulation of still uncharacterized genetic processes.

Here, we describe in detail the optimized protocols for the detection and quantification of sRNAs expression by Northern-blot analysis and quantitative stem-loop RT-PCR [11] in the molecular model specie *P. tricornutum* exposed to different stress conditions. These protocols represent a useful support to further characterize the biogenesis of these sRNAs, and understand their molecular tasks and functions in diatoms and other phytoplanktonic organisms [6].

2 Materials

2.1 Diatom Culture and Medium

1. Cell culture flasks with vented filter cap.
2. Wild-type *P. tricornutum* culture.
3. F/2 medium containing: deionized sterile water, 20 g/L sea salts, 1× F/2 solution [12], without silicate.

4. Artificial Sea Water (ASW): 410 mM NaCl, 20 mM $MgSO_4{\cdot}7H_2O$, 20 mM $MgCl_2{\cdot}6H_2O$, 10 mM KCl, 3 mM KNO_3, 9.9 mM $CaCl_2$, 0.03 mM Na_2EDTA, 3 mM glycylglycine, 1.9 μM Fe-NH_4-citrate, 8.1 μM H_3BO_3, 1.04 μM $ZnSO_4{\cdot}7H_2O$, 0.41 μM $CuSO_4{\cdot}5H_2O$, 0.49 μM Na_2MoO_4, 0.42 μM $CoCl_2{\cdot}6H_2O$, 0.5 μM $MnCl_2{\cdot}4H_2O$, 6.5 μM Na_2tartrate·$2H_2O$, 21.6 nM O_2Se, 33.6 μM KBr, 3.75 μM $SrCl_2{\cdot}6H_2O$, 10 μM NaF, 0.65 μM $NiSO_4{\cdot}6H_2O$, 9.8 nM Na_3VO_4, 0.4 nM $CrK(SO_4)_2{\cdot}12H_2O$, 8.25 mM Tris, pH was adjusted before autoclaving to 8.0 by adding HCl.

   After sterilization add: 0.3 mM K_2HPO_4, Vitamins: 0.1 μg/L thiamine hydrochloride, 50 μg/L B12, 50 μg/L biotin, [13], without silicate.

2.2 RNA Extraction

1. RNase Surface Decontaminant.
2. Ice-cold phosphate buffer saline 1× (PBS): 137 mM NaCl, 2.7 mM KCl, 8 mM Na_2HPO_4, and 2 mM KH_2PO_4.
3. TRIzol Reagent (*see* **Note 1**).
4. Chloroform, molecular biology grade.
5. Isopropanol, molecular biology grade.
6. 75% ethanol, RNase free.
7. Absolute ethanol, molecular biology grade.
8. DEPC-treated water.
9. 50 mL conical tubes.
10. Refrigerated centrifuge.
11. Centrifuge tubes (2 and 1.5 mL).
12. Micropipettes.
13. Sterile filter pipette tips.
14. Vortex mixer.
15. Spectrophotometer.
16. Heat block.
17. Agarose gel: molecular biology Agarose and 0.5× TAE buffer (Tris-acetate-EDTA) RNase-free obtained from a 50× TAE buffer stock containing: 50 mM EDTA sodium salt, 2 M Tris–HCl (adjust pH to 8.0) and 1 M acetic acid.
18. RNA Gel-loading dye: 50% glycerol, 50 mM Tris–HCl (adjust pH to 7.7), 5 mM EDTA, 0.03% Bromophenol Blue.
19. UV Transilluminator.

2.3 RNA Precipitation to Enrich the Small RNA Fraction (Not Mandatory)

1. Total RNA.
2. 5–10% Polyethylene glycol (PEG) 200.
3. 0.5 M NaCl.
4. Absolute ethanol, molecular biology grade.

5. RNase-free water.
6. Microfuge tubes.
7. High-speed microcentrifuge.

2.4 Northern Blot Analysis of Low Molecular Weight RNAs

1. Acrylamide Gel: 6 M urea; 40% acrylamide 19:1; 10× TBE Buffer (Tris–Borate–EDTA) RNase-free stock containing 0.02 M EDTA sodium salt, 1 M Tris–HCl (adjust pH to 8.0) and 1 M boric acid; ammonium persulfate (APS); TEMED (*N*′-tetramethylethylenediamine).
2. N^+ membranes.
3. Transfer semidry system.
4. UV Crosslinker.
5. Hybridization oven.
6. Glass hybridization tubes.
7. Rectangular Pyrex dishes.
8. Orbital shaker.
9. Plexiglas shield.
10. Geiger counter.
11. Long metal forceps.
12. X-ray cassettes.
13. Phosphorimager and image eraser device.
14. Low-range RNA ladder.

2.5 Isotope Labeling of Oligo DNA Probes and Washing Buffer (See Note 2)

1. γ-^{32}P-ATP.
2. Reverse complement oligonucleotide (20 to 32 nt) to the predicted sRNA, diluted to working concentration (10 μM).
3. T4 polynucleotide kinase (T4 PNK).
4. 10× T4 Polynucleotide Kinase reaction Buffer.
5. Ready-to-use hybridization buffer for Northern (*see* **Note 3**).
6. Microfuge tubes.
7. Vortex.
8. High-speed microcentrifuge.
9. Sterile distilled water.
10. Wash Buffer: 2× SSC (Saline-sodium citrate) buffer, RNase-free obtained from a 20× SSC stock containing: 3 M sodium chloride, 0.3 M sodium citrate, (adjust pH to 7), and sodium dodecyl sulfate (SDS).
11. U6 sRNAs loading control oligo probe sequence: GCTAATCTTCTCTGTATC.

2.6 Reverse Transcriptase Reactions

1. Thermocycler.
2. Microfuge tubes.
3. PCR tubes.
4. Micropipettes.
5. Sterile filter pipette tips.
6. DNase (stored at −20 °C).
7. 10× DNase Buffer (stored at −20 °C).
8. Reverse Transcriptase (RT) (stored at −20 °C).
9. 10× Reverse Transcriptase Buffer (stored at −20 °C).
10. 2 mM dNTPs (obtained from a stock of 10 mM) (stored at −20 °C).
11. RT-loop Primer 1 μM (obtained from a stock of 10 μM) (stored at −20 °C).
12. DTT (DL-Dithiothreitol) to stabilize enzymes (stored at −20 °C).
13. Recombinant Ribonuclease Inhibitor RNaseOUT.

2.7 Stem-Loop qPCR

1. Micropipettes.
2. Sterile filter pipette tips.
3. Real-time PCR thermal cycler.
4. Microcentrifuge.
5. PCR tubes.
6. Plates: 96- or 384-well plates.
7. Plate seals.
8. SYBR Green mix.
9. cDNA template.
10. 10 μM forward and reverse primers.
11. PCR grade water.
12. Oligos for Stem Loop qPCR.

    Universal Reverse PCR primer: AGTGCAGGGTCCGAGGTATTC

    SnoR85 sequence: GTGCATTCAAAAGCCCTTACATGT

3 Methods

3.1 Diatom Cultures Conditions

1. Start a *P. tricornutum* culture, growing it at 18 °C under 12 h:12 h day–night cycles. For northern blot analysis shown in Fig. 1, cells were grown under different conditions [6] Low Light (LL) (30 μmol photons m^{-2} s^{-1}), Normal Light (NL) (80 μmol photons m^{-2} s^{-1}); High Light

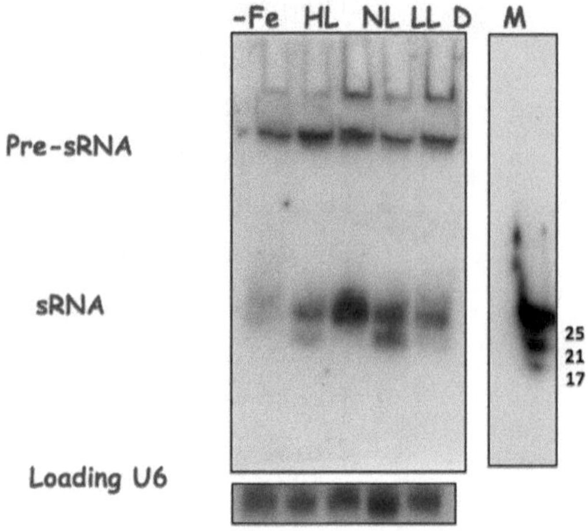

Fig. 1 Low-molecular weight Northern blot analysis for the validation of the size and expression of sRNAs derived from the U2 snRNAs of *P. tricornutum* (chr5: 696.774–696.999). Total RNAs was extracted from cell cultures grown under different conditions: High Light (HL), Normal Light (NL), Low Light (LL), Dark (D), and iron starvation (-Fe). sRNAs mapping the U2–3′ were detected using a 24 nt long oligonucleotide complementary to the sRNA (*see* **Note 14**). The U6 sRNAs were used as loading control. Two sRNAs of diverse sizes were detected, whose intensity varies in the different tested growth condition. The higher molecular weight band corresponds to the U2 sRNA precursor (Pre-sRNA), whose intensity remains unaltered across the samples. M, microRNA Marker. (Modified from [6])

(500 μmol photons m^{-2} s^{-1}) for 5 h (HL), dark for 60 h (D) and under iron starvation (-Fe), either in F/2 media or ASW medium (*see* Subheading 2).

2. Prepare F/2 liquid medium and for the iron starvation, growth cultures of *P. tricornutum* at low light, in ASW medium, containing 5 nmol/l Fe-NH_4-citrate as final concentration.
3. When the cells have reached the exponential phase of growth, collect 100 mL of culture. Per each condition, collect cell pellets by centrifugating the cultures in two conical tubes of 50 mL at 1300 × *g* at 4 °C for 10 min. Discard large part of the supernatant.
4. Transfer the cell pellet into 2 mL tubes, and spin again at 12,000 × *g* at 4 °C for 3 min.
5. Remove the residual supernatant and wash quickly the cell pellet with 1 mL of ice-cold PBS 1×, form PBS Buffer 10× RNase-free stock, to remove salts.
6. Spin at 12,000 × *g* at 4 °C for 3 min.

7. Discard supernatant and freeze the cells with liquid N_2, then proceed with the extraction or store the cell pellet at −80 °C for months.

3.2 RNA Extraction (See Note 4)

1. Lyse cells with 1.5 mL TRIzol Reagent by repetitive pipetting.
2. Incubate the homogenized sample for 5 min at room temperature to permit the complete dissociation of nucleoprotein complexes.
3. Centrifuge to remove cell debris.
4. Transfer the supernatant to new 2 mL tube.
5. *Phase separation*: Add 0.3 mL of chloroform per 1.5 mL of TRIzol Reagent. Cap sample tubes securely. Vortex samples vigorously for 15 s and incubate them at room temperature for 15 min. Centrifuge the samples at no more than 12,000 × *g* for 15 min at 4 °C. Following centrifugation, the mixture separates into lower red, phenol–chloroform phase, an interphase, and a colorless upper aqueous phase. RNA remains exclusively in the aqueous phase. Transfer upper aqueous phase carefully without disturbing the interphase into fresh 1.5 mL tube (*see* **Note 5**).
6. *RNA Precipitation:* Precipitate the RNA from the aqueous phase by adding isopropyl alcohol. Use 0.75 mL of isopropyl alcohol per 1.5 mL of TRIzol Reagent used for the initial homogenization. Cap the tubes and invert several times to mix it. Incubate samples at Room Temperature (RT) for 10 min and centrifuge at 12,000 × *g* for 10 min at 4 °C. Remove the supernatant completely. The RNA precipitate forms a gel-like pellet on the side and bottom of the tube.
7. *RNA wash:* Wash the RNA pellet once with 75% ethanol, adding 1.5 mL of 75% ethanol per 1.5 mL of TRIzol Reagent used for the initial homogenization (*see* **Note 6**). Mix the samples by inverting the tubes and centrifuge at 12,000 × *g* for 5 min at 4 °C. Repeat above washing procedure once. Remove all leftover ethanol.
8. *Redissolving RNA:* Air-dry or vacuum-dry RNA pellet for 5–10 min. Dissolve RNA in RNase-free water by passing solution a few times through a pipette tip.
9. *Spectrophotometric Analysis*: Take OD at 260 nm and 280 nm to determine sample concentration and purity. The A260:A280 ratio should be above 1.6.

3.3 Northern Blot Analysis

Perform Northern blot analysis of low molecular weight RNA, with 20 μg of total RNA (Fig. 1).

1. *Sample preparation*: Start with 20 μg (up to 40 μg for low expressed sRNAs) of total RNA and fill it up with H_2O to get the same volume each tube. Then add 1 vol of formaldehyde

100%, denaturate at 95 °C for 5 min, and put in ice for another 5 min. Spin down and add ¼ volume of gel-loading dye.

2. Separate the RNA extracted from cells on 14% acrylamide gel, by using a vertical gel electrophoresis system. Use combs of 1.5 mm and also corresponding 1.5 mm glass.
3. Remove urea in each well and prerun at 100 V for 30 min.
4. Remove again urea in each well and load the samples.
5. Migrate the samples for 4 h at 80 V in 0.5× TBE.
6. Transfer the RNA on a N^+ membrane using a semidry system for 1 h at 80 V—300 mA.
7. *Probe preparation:* 9 μL water (RNase-free), 3 μL of oligonucleotide complement to the sRNA at 10 μM, denaturate at 95 °C for 5 min and put in ice for another 5 min, add 2 μL of 10× PNK buffer, 5 μL of γ-^{32}P-ATP (50 μCU), then add 1 μL T4 PNK and put at 37 °C for 45–60 min.
8. After transfer, put the membrane on 2× SSC for 10 min.
9. Cross-link in a UV Stratalinker. Turn on the UV Stratalinker. Place the membrane with RNA side up into the chamber. Close the door. Press Auto Cross Link. Press Start.
10. Prehybridization with 20 mL of Hybridization buffer for at least 20 min at 42 °C.
11. Load the probe into the hybridization buffer and hybridize the membrane over night with oligonucleotide complementary to the target sRNA sequence. The temperatures for the hybridization depending on the T_m of the probe (*see* **Notes 7** and **8**).
12. *Washing:* wash 2 times for 20 min with 2× SSC, 0.2% SDS at 40 °C. If the radioactivity is still high performed another washing 2× SSC, 1% SDS for 10 min (*see* **Note 9**).
13. Remove blot from hybridization tube, wrap securely with Saran wrap, and expose to phosphorimaging screen or X-ray screen.
14. Scan the screen after 5–10 h exposure and then expose overnight or several days for signal optimization.
15. As reference loading control hybridized with radiolabeled probes recognizing the U6 snRNA gene. U6 snRNA is a spliceosome core component (Fig. 1).

3.4 Stem-Loop Reversed Transcription Reaction

Because of their small size and possible low expression levels, detection of the small RNAs by traditional methods is technically demanding. The stem-loop real-time reverse transcriptase polymerase chain reaction (qRT-PCR) can be used to quantify sRNA expression in diatoms, with a protocol adapted from [11]. The primers for these reactions required the design of a stem-loop RT primer, and a set of primers complementary to the sRNA for the subsequent quantitative PCR amplification. The stem-loop RT

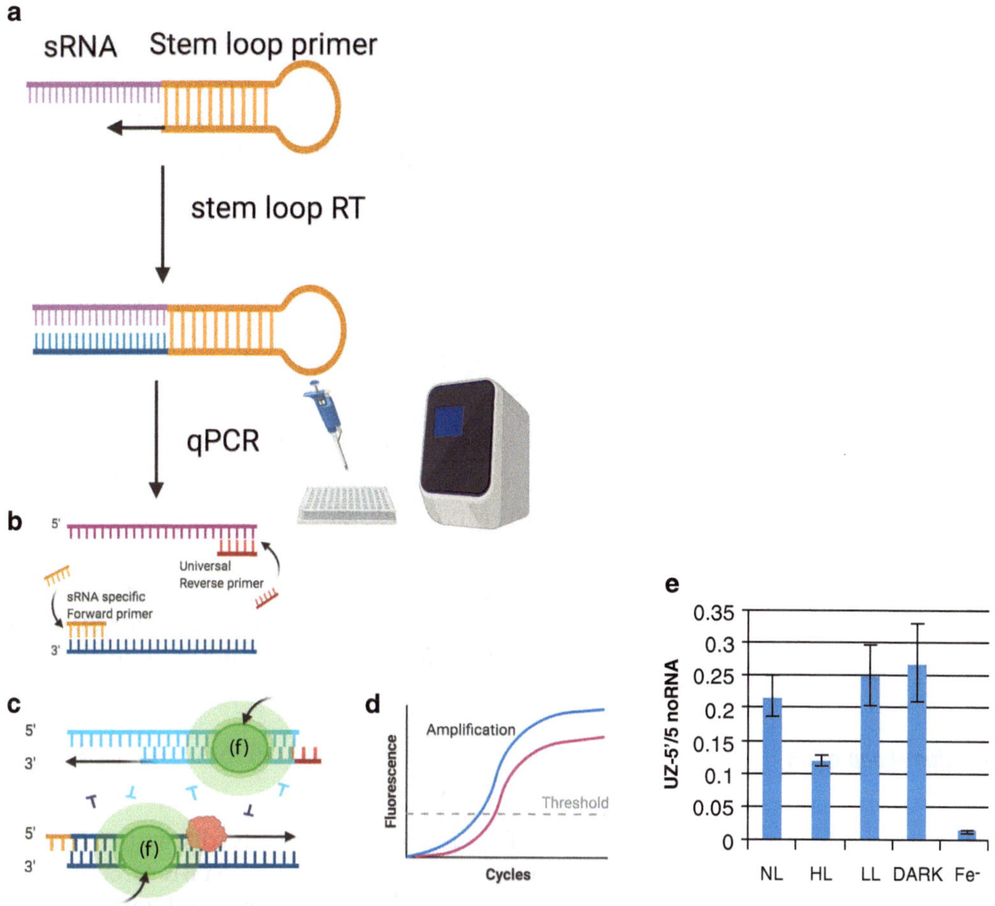

Fig. 2 Schematic showing the stem-loop RT-PCR method for the relative transcript levels of low expressed sRNAs in *P. tricornutum*. (**a**) A stem-loop RT primer binds to the 3′ portion of the sRNA, initiating reverse transcription in the presence of RT. (**b**) The RT product is amplified using an sRNA-specific forward primer and the universal reverse that is reverse complement of the last six nucleotides at the 3′ end of the sRNA (**c**) The quantification is achieved by qPCR through SYBR Green (F in green circle) incorporation during amplification. (**d**) The quantification is achieved by measuring the fluorescence of the distinct reactions. The fluorescence increases at the level of the amplification products. (**e**) For the analysis performed in *P. tricornutum*, we have successfully used this approach to quantify a sRNAs called U2–5′, mapping the 5′-region of the chr5: 696.774–696.999 (modified from Fig. 3c in [6]), not detected by Northern blot (*see* **Note 15**), contrary to the sRNAs U2–3′ shown in Fig. 1. The expression values of the U2–5′ sRNA was normalized to the snoRNA (Created with BioRender.com)

primers bind the 3′ portion of sRNA molecules, initiating reverse transcription of the sRNA and provide more specificity and sensitivity than linear primers owing to the base stacking and spatial constraints of the stem-loop structure.

The stem-loop RT primer is used to reverse transcribe the sRNA molecule using total RNA isolated from *P. tricornutum*. The qPCR is carried out with a forward primer of 21 nt specific for the sRNA target sequence and a universal reverse primer complementary to the sequence of the loop portion of an RT stem-loop primer (*see* **Note 10**) (Fig. 2).

1. *Starting materials:* Total RNA obtained by TRIzol extraction (*see* Subheading 3.2), and suspended in RNase-free water.
2. *DNase treatment:* Prepare a mix as follows: 12 μg of RNA (*x* μL); DNase Buffer 10× (1 μL); RNase OUT (0.1 μL); DNase (2 μL) H_2O up to 10 μL. Incubate at 37 °C for 50 min. Add 2 μL Stop solution and incubate at 65 °C for 10 min, followed by 5 min on ice. The RT reactions contain 1 μg/μL of total RNA.
3. *Stem-loop reverse transcriptase reactions*: Prepare Mix A as follows: 2 mM dNTPs (2 μL); 1 μM RT-loop primer (1 μL); H_2O 10.4 μL. Heat at 70 °C for 5 min and then place the samples on ice for at least 5 min.
4. Prepare Mix B as follows: RT Buffer; RT enzyme; 0.1 M DTT; RNaseOUT.
5. Assemble MixA+MixB and add 1 μL of DNase-treated RNA (1 μg). Final volume 20 μL.
6. Perform the Stem-loop reverse transcription reaction as follows: 16 °C for 30 min, 30 °C for 30 s, 42 °C for 30 s, 50 °C for 1 s for 60 cycles 85 °C for 5 min and holding at 4 °C (*see* **Note 11**).

3.5 Stem-Loop qPCR

Twenty microliters of the obtained cDNAs is used for the qRT-PCR, performed as described here.

1. Prepare the SYBR Green master mix as follows (*see* **Note 12**): volume per single reaction 10 μL. 2× SYBR Green 1× Final 5 μL; (2.5 μM) Specific Forward Primer + Universal Reverse Primer 2 μL; PCR grade H_2O 2 μL. Mix gently and spin down the solution.
2. Carefully aliquot 9 μL of master mix into each plate well, add 1 μL of specific cDNA (*see* **Note 13**).
3. Mix reactions and spin.
4. Seal the PCR plate and label (according to instrument requirements).
5. Run samples as per instrument manufacturer's recommendations using the following protocol: initial denaturation 95 °C for 10 min; 45 cycles 95 °C for 15 s, 60 °C for 20 s, 72 °C for 40 s. Perform a melting curve at the end of the program to check the amplification efficiency by steps of 1 °C (from 95 °C to 50 °C).

Run all samples in triplicate. Analyze data using the $2^{-\Delta\Delta Ct}$ method for relative expression [14], with the small nucleolar RNA SnoR85 as endogenous reference because it is not susceptible to silencing factors.

4 Notes

1. Always use gloves and eye protection. TRIzol may cause severe damage, use it in a chemical hood. Avoid contact with skin or clothing, and avoid breathing vapor.
2. Always use gloves, eye protection, and Plexiglas shield. Avoid contact with skin or clothing.
3. This material was provided by Sigma-Aldrich. Sigma's PerfectHyb Plus is a ready-to-use hybridization buffer, up to 20× more sensitive than conventional buffers. In particular, it is recommended to use this buffer for low expressed sRNAs. A small amount of precipitate that has formed in the solution can be dissolved at 30–35 °C.
4. Work in an RNase-free environment, wash bench-top with MilliQ H_2O and 60% ethanol, spread RNase Surface Decontaminant.
5. Measure the volume of the aqueous phase. The volume of the aqueous phase is about 60% of the volume of TRIzol Reagent used for homogenization.
6. The RNA precipitate can be stored in 75% ethanol for 1 week at 4 °C, or at least 1 year at −20 °C.
7. The hybridization buffer containing the probe can be stored at −20 °C for 2 weeks and reused.
8. Include the appropriate controls, by using the U6 probe.
9. Use 20 mL of washing solution.
10. The stem loop RT primer is specific to the individual sRNA, with a six-nucleotide extension at the 3′ end that is reverse complement of the last six nucleotides at the 3′ end of the sRNA. Forward primers are specific to the sRNA sequence but exclude the last six nucleotide at the 3′ end of the sRNA. A 5′ extension of 5–7 nucleotides is added to each primer to increase the melting temperature; these sequences are chosen randomly and are GC-rich.
11. For this step, prepare 1 sample extra to cover pipetting errors.
12. The 20 μL of cDNA obtained can be used for the stem-loop qPCR immediately or stored at least 1 week at 4 °C.
13. As control, include negative RT 1 μL and RNase-free water 1 μL samples.
14. U2 sRNAs oligo probe sequence: GTACACGGTCAAGGCGGTGAGTG.
15. Oligos for Stem Loop qPCR, used to validate and quantify the predicted sRNA U2–5′ (*see* Fig. 2 and [3]):

U2–5′ mature sequence: ATTGGCTTTGATCTTGTGTAG

Forward U2-5'primer: CCGGCGATTGGCTTTGATCTT

RT-loop primer U2-5' sRNA: GTCGTATCCAGTGCAGGG TCCGAGGTATTCGCACTGGATACGACCT

Acknowledgments

This work was supported by funding from the European ASSEMBLE Plus (Association of European Marine Biological Research Laboratories Expanded, H2020-INFRAIA-1-2016-2017), the Fondation Bettencourt-Schueller and the "Initiative d'Excellence" program (Grant "DYNAMO," ANR-11-LABX-0011-01) to A.F.

References

1. Hüttenhofer A, Vogel J (2006) Experimental approaches to identify non-coding RNAs. Nucleic Acids Res 34(2):635–646
2. Carthew RW, Sontheimer EJ (2009) Origins and mechanisms of miRNAs and siRNAs. Cell 136(4):642–655
3. Guleria P, Mahajan M, Bhardwaj J et al (2011) Plant small RNAs: biogenesis, mode of action and their roles in abiotic stresses. Genom Proteom Bioinform 9(6):183–199
4. Falciatore A, Jaubert M, Bouly JP et al (2020) Diatom molecular research comes of age: model species for studying phytoplankton biology and diversity. Plant Cell 32(3):547–572
5. Bowler C, Allen AE, Badger JH et al (2008) The Phaeodactylum genome reveals the evolutionary history of diatom genomes. Nature 456:239–244
6. Rogato A, Richard H, Sarazin A et al (2014) The diversity of small non-coding RNAs in the diatom *Phaeodactylum tricornutum*. BMC Genom 15:698
7. Lopez-Gomollon S, Beckers M, Rathjen T et al (2014) Global discovery and characterization of small non-coding RNAs in marine microalgae. BMC Genom 15:697
8. Armbrust EV, Berges JA, Bowler C et al (2004) The genome of the diatom *Thalassiosira pseudonana:* ecology, evolution, and metabolism. Science 306:79–86
9. Mock T, Otillar R, Strauss J et al (2017) Evolutionary genomics of the cold- adapted diatom *Fragilariopsis cylindrus.* Nature 541: 536–540
10. Maumus F, Allen AE, Mhiri C et al (2009) Potential impact of stress activated retrotransposons on genome evolution in a marine diatom. BMC Genomics 22(10):624
11. Varkonyi-Gasic E, Wu R, Wood M et al (2007) Protocol: a highly sensitive RT-PCR method for detection and quantification of microRNAs. Plant Methods 3:12
12. Guillard RRL (1975) Culture of phytoplankton for feeding marine invertebrates. In: Smith WL, Chanley MH (eds) Culture of marine invertebrate animals. Springer, US, Boston, MA, pp 29–60
13. Vartanian M, De Sclés J, Quinet M et al (2009) Plasticity and robustness of pattern formation in the model diatom *Phaeodactylum tricornutum.* New Phytol 182:429–442
14. Livak KJ, Schmittgen TD (2001) Analysis of relative gene expression data using real-time quantitative PCR and the 2(-Delta C (T)) method. Methods 25(4):402–408

Chapter 18

Optimized Proteolistic Protocol for the Delivery of the Cas9 Protein in *Phaeodactylum tricornutum*

Monia Teresa Russo, Anna Santin, Alessandra Rogato, and Maria Immacolata Ferrante

Abstract

The CRISPR/Cas9 system coupled with proteolistics is a DNA-free nuclear transformation method based on the introduction of ribonucleoprotein (RNP) complexes into cells. The method has been set up for diatoms as an alternative to genetic transformation via biolistics and has the advantages of reducing off-target mutations, limiting the working time of the Cas9 endonuclease, and overcoming the occurrence of random insertions of the transgene in the genome. We present a point-by-point description of the protocol with modifications that make it more cost-effective, by reducing the amount of the enzyme while maintaining a comparable efficiency to the original protocol, and with an increased concentration of the selective drug which allows to reduce false positives.

Key words Genetic engineering, Diatom gene knockout, DNA-free nuclear transformation, CRISPR/Cas9

1 Introduction

Genetic engineering is a promising approach to obtain valuable insights into the diatom metabolism and physiology and to improve metabolic features for biotechnological applications [1, 2].

Recently, the search for new techniques for genome editing allowed the development of modern systems, such as TALEN (Transcription Activator-Like Effector Nucleases) and CRISPR (Clustered Regulatory Interspaced Short Palindromic Repeats)/Cas9 [3]. In particular, the CRISPR/Cas9 system, co-opted from a natural defence mechanism of bacteria, is a two components knockout approach based on the occurrence of a double-strand break (DSB) induced by the endonuclease Cas9, in correspondence of a specific genomic DNA target recognized by a specific single-

Monia Teresa Russo and Anna Santin contributed equally to this work.

Cinzia Verde and Daniela Giordano (eds.), *Marine Genomics: Methods and Protocols*,
Methods in Molecular Biology, vol. 2498, https://doi.org/10.1007/978-1-0716-2313-8_18,

strand guide RNA (gRNA). DSB damage is repaired by the cell through a nonhomologous end-joining (NHEJ) mechanism, which can induce insertions and/or deletions (INDELs) or single nucleotide replacements [3, 4].

The CRISPR/Cas9 system can induce the occurrence of unwanted off-target mutations [4], since the Cas9 can at times cut at sites with a certain degree of similarity to the target region, and unwanted perturbations due to random integrations in the genome of interest. This happens when using the election technique for diatom transformation, the biolistic method: exogenous DNA, encoding Cas9 and gRNA, attached to microspheres, is introduced by high pressure in target cells where it stably integrates in the endogenous genome [5, 6]. In 2015, an alternative method, bacterial conjugation, had been set up consisting in the introduction of a cargo DNA from *Escherichia coli* to diatoms throughout a conjugation process [7], showing a reduced occurrence of off-target mutations [8, 9]. Bacterial conjugation still requires the introduction of exogenous DNA in the cells in the form of an episome, which is not integrated in the diatom genome and can be easily removed by eliminating antibiotic selection from the medium [9]. The recently introduced proteolistic method [2] is a DNA-, bacterial-, and antibiotic-free technology. It is based on the use of ribonucleoprotein complexes (RNP), formed by the nuclease Cas9 and genomic target-specific gRNAs. The two most important advantages of this protocol are the reduction of off-targets, throughout the avoidance of long-term expression of the nuclease in the cell, and unwanted perturbations due to random integration of the transgene in the genome. An additional advantage of the technology is the selection system based on the simultaneous targeting of the gene(s) of interest along with the endogenous adenine phosphoribosyl transferase (APT) gene, that, when mutated, is able to confer resistance to 2-fluoroadenine (2-FA), a compound which is toxic for wild type cells [10].

We applied the method to the diatom *Phaeodactylum tricornutum*, using two gRNAs for APT as selection gene and two other gRNAs each for two target genes. We increased 2-FA concentration up to double compared to the original protocol [2], thus drastically decreasing the appearance of false positive colonies. Moreover, we modified protocol steps of RNPs complexes assembly being able to decrease of more than 80% the amount of Cas9 protein required for each shot, while maintaining a high transformation efficiency.

2 Materials

2.1 Culture and Medium

1. Cell culture flasks with vented filter cap.
2. Wild-type *P. tricornutum* culture.

3. Petri dishes.
4. Horizontal flow hood.
5. Plates with F/2 medium containing: deionized sterile water, 20 g/L Sea salts, 1× F/2 solution [11] without silicate and 10 g/L agar bacteriological grade.
6. Selective plates with F/2 medium containing: 20 μM 2-FA (2-fluoroadenine) and 5 mg/L adenine (*see* **Note 1**).

2.2 gRNA and Cas9 Components

1. CRISPR-Cas9 crRNAs (CRISPR RNA; *see* **Note 2**).
2. CRISPR-Cas9 tracrRNA (trans-activating CRISPR RNA; *see* **Note 2**).
3. HiFi Cas9 Nuclease V3 (10 g/L stock solution; kept at −20 °C; *see* **Note 2**).
4. Nuclease-Free Duplex Buffer: 30 mM HEPES, pH 7.5, 100 mM potassium acetate.
5. Nuclease-free water.
6. PBS (1×): 137 mM NaCl, 2.7 mM KCl, 8 mM Na_2HPO_4, and 2 mM KH_2PO_4.

2.3 Cas9 In Vitro Assay

1. Thermocycler.
2. Proteinase K.
3. Agarose gel: molecular biology Agarose and TAE buffer (0.5×), obtained from a TAE (50×) stock containing 50 mM EDTA sodium salt, 2 M Tris–HCl (adjust pH to 8.0) and 1 M acetic acid.
4. UV transilluminator.

2.4 Gold Nanoparticles

1. Gold particles (0.6 μm diameter).
2. Ethyl alcohol absolute.
3. Microfuge tubes.
4. Vortex.
5. High-speed microcentrifuge.
6. Sterile distilled water.

2.5 Gene Gun Equipment

1. Gene-gun PDS-1000/He System, with Helium tank, grade 4.5 (99.9% pure) pressurized to 2600 PSI and a vacuum source.
2. Ethyl alcohol absolute.
3. 1550 PSI rupture disks (*see* **Note 3**).
4. PDS-1000/He stopping screens (*see* **Note 3**).
5. PDS-1000/He macrocarriers (*see* **Note 3**).
6. PDS-1000/He macrocarrier holders (*see* **Note 3**).

3 Methods

3.1 Design of crRNAs

Identify Cas9 target sequences within your candidate gene using on line resources (i.e., CRISPOR (http://crispor.tefor.net/), CHOP-CHOP (https://chopchop.cbu.uib.no/), phytoCRISP-Ex (https://www.phytocrispex.biologie.ens.fr/CRISP-Ex/) paying attention to the following criteria.

1. They should preferably be located at the 5′ end of the gene, in correspondence of exons.
2. Sequences containing single nucleotide polymorphisms (SNPs) between alleles should be avoided to prevent targeting a preferential allele over the other.
3. Two target sequences for the same gene can be chosen to be used simultaneously in order to increase the probability to generate a frameshift as a result of an induced INDEL.
4. The input sequence in the crRNA order has to be the protospacer (20 bp), the protospacer adjacent motif (PAM) has to be excluded (*see* **Note 4**).

3.2 crRNA::tracrRNA Duplexes (gRNA)

1. Resuspend the lyophilized CRISPR-Cas9 RNA (crRNA) and CRISPR-Cas9 trans-activating RNA (tracrRNA) in nuclease-free duplex buffer mixing up and down to a final concentration of 100 μM (*see* **Note 5**).
2. Prepare a crRNA::tracrRNA solution containing 30 μM duplex mixing together 3 μL of 100 μM CRISPR crRNA, 3 μL of 100 μM CRISPR tracrRNA, 4 μL Duplex buffer (*see* **Note 6**). Incubate 5 min at 95 °C in a thermocycler and immediately remove from the block, cooling down at RT on the bench (Fig. 1).

3.3 Diatom Culture and Plating

1. Start a *P. tricornutum* culture, growing it at 18 °C under 12 h:12 h day–night cycles, 100 μmol photons m^{-2} s^{-1} white light during the day, in order to have 1.5×10^8 cells/shot collected during the exponential phase (around $1.0–2.0 \times 10^6$ cells/mL) on the day before the shots.
2. When the cells have reached the desired concentration, collect the required culture volume, to get 1.5×10^8 cells, by centrifugation at 1500 × *g* at 18 °C for 10 min, discard supernatant, resuspend the pellet with the bit of supernatant left, spread the cells on a plate, the spread has to be circular and around 4 cm in diameter. Let dry under the hood and then store at 18 °C overnight.

3.4 Assembly of RNP Complexes

1. Calculate the needed total volume of Cas9. Dilute the corresponding amount of HiFi Cas9 Nuclease V3 protein

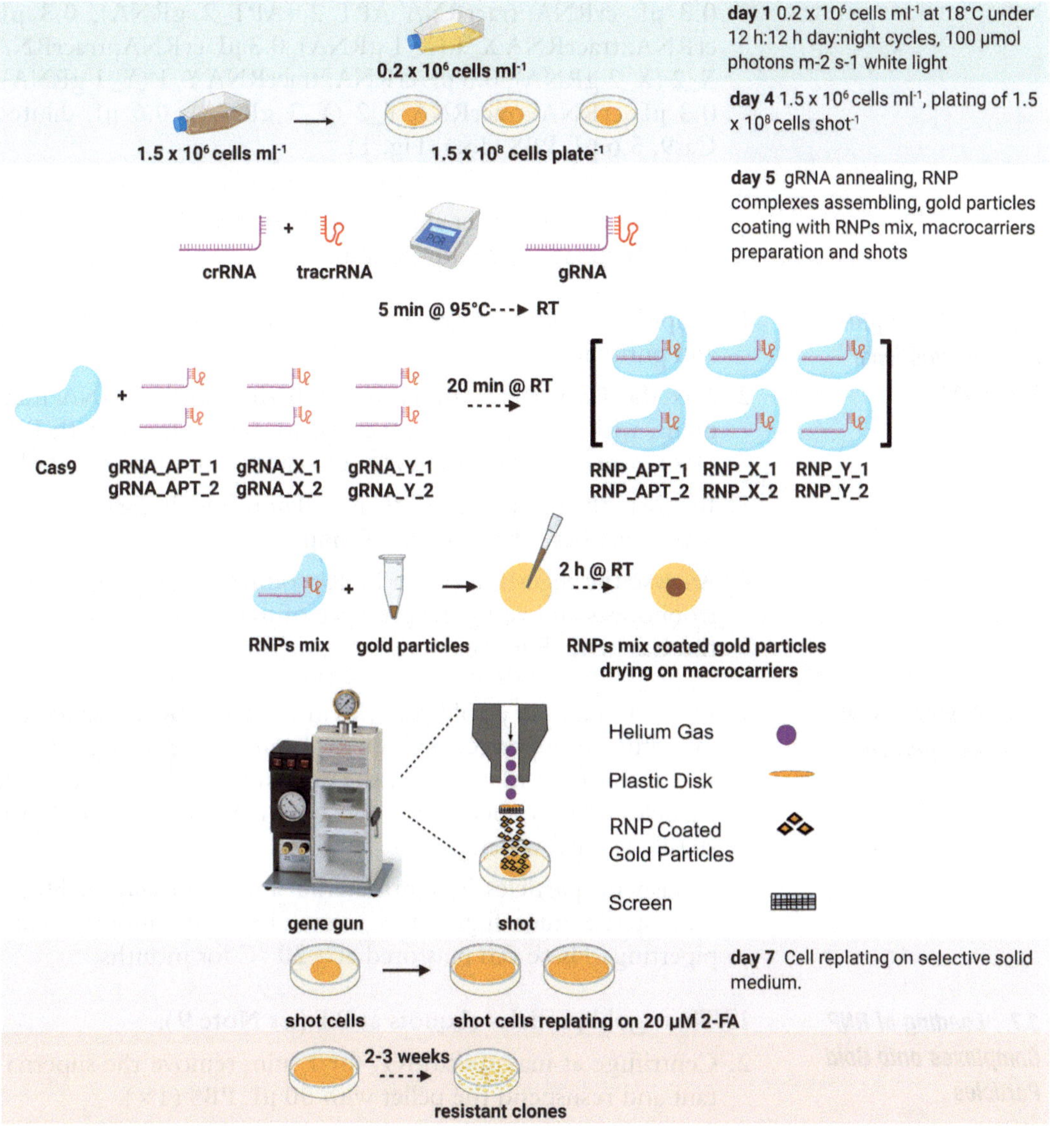

Fig. 1 Improved proteolistic protocol workflow. Main steps are indicated, with the approximate time required. (Created with BioRender.com)

from a first stock solution at 10 g/L (i.e., 65.5 µM), to a second stock solution at 2.5 g/L (i.e., 15.6 µM) using nuclease-free H_2O (*see* **Note** 7).

2. For each RNP complex to assemble, mix all the components in order to have 1.5 µg of Cas9 in a final volume of 8 µL (*see* **Note 8**), scaling up volumes based on your needs. In the following example, 2 gRNAs for each of 2 target genes (X and Y) and 2 gRNAs for the endogenous selection gene (APT) are included: 0.3 µL crRNA::tracrRNA APT_1 (APT_1 gRNA),

0.3 μL crRNA::tracrRNA APT_2 (APT_2 gRNA), 0.3 μL crRNA::tracrRNA X_1 (X_1 gRNA), 0.3 μL crRNA::tracrRNA X_2 (X_2 gRNA), 0.3 μL crRNA::tracrRNA Y_1 (Y_1 gRNA), 0.3 μL crRNA::tracrRNA Y_2 (Y_2 gRNA), 0.6 μL diluted Cas9, 5.6 μL PBS (1×) (Fig. 1).

3. For the negative control, with only Cas9, use the same amount of enzyme of the other shots: 0.6 μL diluted Cas9, 7.4 μL PBS (1×). Incubate at RT for 20 min.

3.5 Cas9 and gRNAs In Vitro Functional Validation

1. Amplify a genomic fragment encompassing the gRNA region and purify it.
2. Mix the Cas9-RNP complexes with the purified DNA fragment, with 10:1 molar ratio in 10 μL final volume with PBS (1×), considering to bring DNA final concentration to 50 nM.
3. Incubate the reaction at 37 °C for 60 min, add 20 μg proteinase K and incubate at 56 °C for 10 min.
4. Analyse the digestion of the PCR product by agarose gel electrophoresis and verify the presence of fragments deriving from the Cas9 cut (Fig. 2).

3.6 Preparation of Gold Nanoparticles

1. Place 60 mg of dry gold particles in 1 mL of 100% ethanol in a microfuge tube, vortex on high for 1 min and centrifuge at max speed, RT, for 1 min. Repeat the washing steps twice and pellet particles at max speed, RT, for 1 min. Wash twice in 1 mL sterile distilled water.
2. Resuspend particles in 1 mL sterile water and aliquot 50 μL into sterile microfuge tubes, vortexing continuously while pipetting. These can be stored at −20 °C for months.

3.7 Loading of RNP Complexes onto Gold Particles

1. Thaw gold particles aliquots at RT (*see* **Note 9**).
2. Centrifuge at max speed, RT, for 1 min, remove the supernatant and resuspend the pellet with 50 μL PBS (1×).
3. Repeat the washing steps twice, then resuspend the pellet with 50 μL PBS (1×) by flickering and vortexing.
4. Transfer 10 μL of the homogeneous gold particles solution into each RNP mixture, vortex and store on ice.
5. Under the hood, wash the macrocarrier holders and the macrocarriers with 100% ethanol, let them dry.
6. Place each macrocarrier into a macrocarrier holder, resuspend the RNP complex–coated particles by vortexing and spread the solution at the centre of the macrocarrier and let dry (≈2 h) under the hood (Fig. 1).

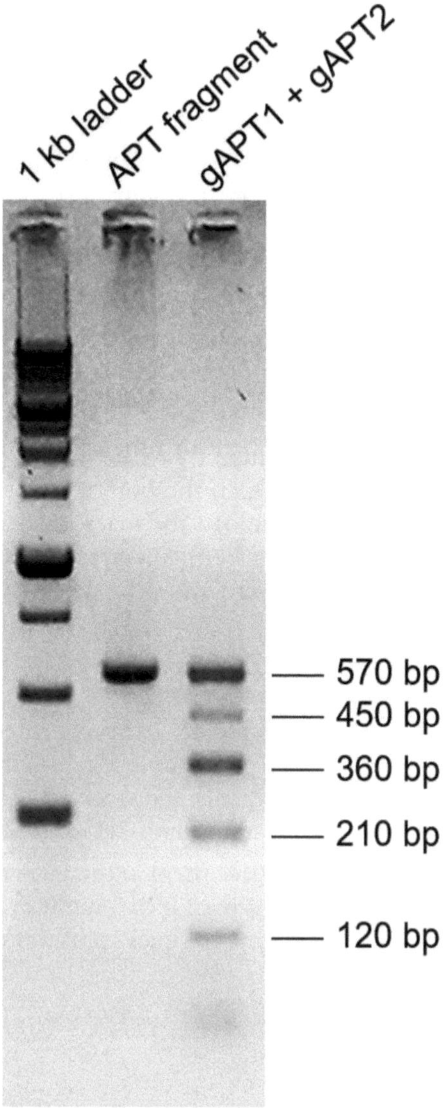

Fig. 2 Cas9 in vitro assay on the endogenous selection gene *APT*. *APT* is cut in shorter fragments by the Cas9 guided by both gRNAs, which divide the gene in 450 and 120 bp (gAPT1) and 360 and 210 bp (gAPT2) fragments. The presence of a residual uncut product is normal

3.8 Proteolistic Shots

1. Slowly open the main valve of the helium cylinder until gas pressure registers on the first gauge of the regulator.
2. Turn the regulator adjusting screw clockwise and adjust the pressure to 1700 psi (200 psi above the burst pressure of the selected rupture disk).
3. Put the 1550 psi rupture disk inside the assembly and screw it onto the acceleration tube. Gently tighten with torque wrench provided.

4. Assemble the macrocarrier holder with a sterile stopping screen and install it with the dried RNP complex–coated particles facing down.
5. Put the lid on and place the whole assembly in the second shelf slot in the chamber.
6. Place the Petri dish containing cells on the dish holder in the third shelf slot in the chamber and close the bombardment chamber door.
7. With the power on, start the vacuum pump and, when the vacuum gauge shows 23 in Hg, turn the switch to hold and press the fire switch continuously until the rupture disk bursts and the helium pressure gauge drops to zero.
8. Then turn the central switch to vent to release the vacuum.
9. Open the bombardment chamber door and remove the plate, unload the macrocarrier and stopping screen from the macrocarrier launch assembly, and unload the spent rupture disk.
10. Place the Petri dish in the growth chamber, cells facing down, at 18 °C, 12 h:12 h day–night cycles, 100 μmol photons $m^{-2} s^{-1}$.

3.9 Replating

1. After 48 h from proteolistic shots, resuspend the cells with 1 mL sterile F/2 medium without silicate and spread them equally between 2 selective plates containing 2-FA (*see* Subheading 2.1).
2. Place them in the growth chamber, cells facing down, at 18 °C, 12 h:12 h day–night cycles, 100 μmol photons $m^{-2} s^{-1}$. Resistant colonies appear after 2–3 weeks (Figs. 1 and 3).

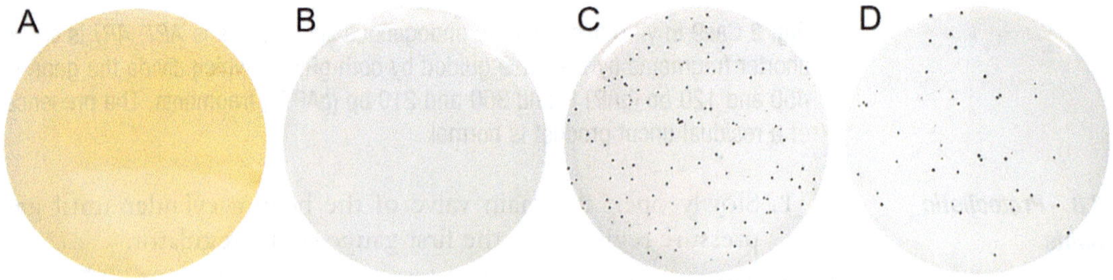

Fig. 3 Colonies obtained 2 weeks after shots. (**a**) positive control, transformed cells grown without selection; (**b**) negative control, untransformed cells grown in selection; (**c** and **d**) mutant colonies, transformed cells grown in selection

4 Notes

1. Working solution 2-FA was prepared starting from a 500× stock solution in DMSO, kept at 4 °C, and was stable over months. A 500× stock solution was prepared for adenine.
2. This material is currently provided exclusively by Integrated DNA Technologies (IDT).
3. This material is currently provided exclusively by Bio-Rad.
4. Before ordering the guides, amplify and sequence the candidate gene in your *P. tricornutum* strain since its sequence may significantly differ from that reported in the reference genome (https://protists.ensembl.org/Phaeodactylum_tricornutum/Info/Index?db=core).
5. Work in an RNase-free environment, use filter tips, wash bench-top with MilliQ H_2O and 60% ethanol, then spread RNaseZap.
6. Prepare the duplex in a PCR tube, mix the various components scaling-up volumes based on your needs. These can be stored at −20 °C for months.
7. This step will facilitate downstream work because the solution at 10 g/L is viscous and hard to manipulate [MCas9 = 1.6×10^5 g/mol].
8. Include all of the appropriate controls, comprising a negative control by transforming cells with Cas9 alone and a positive control with RNPs targeting the endogenous selection gene alone (*PtAPT*).
9. Each 50 μL aliquot contains 3 μg gold particles, sufficient for 5 proteolistic shots and can be stored at −20 °C for months.

Acknowledgments

A.S. is supported by a PhD fellowship funded by the Open University—SZN PhD program. The authors acknowledge the European Union's Horizon 2020 programme, grant agreement No. 730984, the Gordon and Betty Moore Foundation through grant #7978 to M.I.F., and the CNR project NUTR-AGE (FOE-2019, DSB. AD004.271).

References

1. Faktorová D, Nisbet RER, Fernández Robledo JA et al (2020) Genetic tool development in marine protists: emerging model organisms for experimental cell biology. Nat Methods 17: 481–494
2. Serif M, Dubois G, Finoux A-L et al (2018) One-step generation of multiple gene knockouts in the diatom *Phaeodactylum tricornutum* by DNA-free genome editing. Nat Commun 9: 3924

3. Nymark M, Sharma AK, Sparstad T et al (2016) A CRISPR/Cas9 system adapted for gene editing in marine algae. Sci Rep 6:24951
4. Kroth PG, Bones AM, Daboussi F et al (2018) Genome editing in diatoms: achievements and goals. Plant Cell Rep 37:1401–1408
5. Falciatore A, Casotti R, Leblanc C et al (1999) Transformation of nonselectable reporter genes in marine diatoms. Mar Biotechnol 1: 239–251
6. Poulsen N, Kröger N (2005) A new molecular tool for transgenic diatoms: Control of mRNA and protein biosynthesis by an inducible promoter-terminator cassette. FEBS J 272: 3413–3423
7. Karas BJ, Diner RE, Lefebvre SC et al (2015) Designer diatom episomes delivered by bacterial conjugation. Nat Commun 6:6925
8. Moosburner MA, Gholami P, McCarthy JK et al (2020) Multiplexed knockouts in the model diatom *Phaeodactylum* by Episomal delivery of a selectable Cas9. Front Microbiol 11:5
9. Russo MT, Aiese Cigliano R, Sanseverino W et al (2018) Assessment of genomic changes in a CRISPR/Cas9 *Phaeodactylum tricornutum* mutant through whole genome resequencing. PeerJ 6:e5507
10. Schaff DA (1994) The adenine phosphoribosyltransferase (APRT) selectable marker system. Plant Sci 101:3–9
11. Guillard RRL (1975) Culture of phytoplankton for feeding marine invertebrates. In: Smith WL, Chanley MH (eds) Culture of marine invertebrate animals. Springer, US, Boston, MA, pp 29–60

Chapter 19

Production of a Chimeric Mouse–Fish Monoclonal Antibody by the CRISPR/Cas9 Technology

Alessia Ametrano and Maria Rosaria Coscia

Abstract

The CRISPR/Cas9 system, a defense mechanism naturally occurring in prokaryotes, has been recently repurposed as an RNA-guided DNA targeting platform and widely used as a powerful tool for genome editing. Here we describe how to modify the carboxy-terminal region, called Fragment crystallizable (Fc) region, of a murine monoclonal antibody by replacing the heavy chain constant exons with those from a teleost fish antibody by the CRISPR/Cas9 system. We outline optimal conditions for knockout and knockin mechanisms to edit the Immunoglobulin heavy chain (IgH) constant region gene locus in a murine hybridoma cell line. A chimeric mouse–fish monoclonal antibody can be successfully produced by hybridoma cell lines engineered according to this protocol.

Key words CRISPR/Cas9, Monoclonal antibody, Fc region, IgH gene locus, Genome editing, Murine hybridoma, Teleost fish

1 Introduction

Antibody engineering is a powerful technology for high-affinity peptides/proteins discovery, receptor binding and epitope identification, or for investigating protein-protein interactions. In this field, hybridoma technology represents a significant milestone for monoclonal antibodies (mAbs) generation that has revolutionized biomedical research [1]. At present, mAbs can be produced not only by hybridoma cells, but also by means of a variety of expression systems, such as yeasts, bacteria, mammalian cells, or phages [2]. Although ensuring production of large amounts of antibodies, all these methods have some disadvantages such as the random integration of transgenes, unbalanced production of heavy and light chains, time-consuming procedures, and high costs.

In 2013, the CRISPR/Cas9 system has been used to perform, for the first time, genome engineering in mammalian cells. A synthetic guide RNA (gRNA), which is complementary to the DNA

Cinzia Verde and Daniela Giordano (eds.), *Marine Genomics: Methods and Protocols*,
Methods in Molecular Biology, vol. 2498, https://doi.org/10.1007/978-1-0716-2313-8_19,

target sequence, is required to direct the CRISPR-associated protein 9 (Cas9) nuclease to cleave the specific target region; a protospacer adjacent motif (PAM) sequence is essential for the recognition of target sites [3, 4]. The CRISPR/Cas9 system shows several potentials, for example, the capacity to use multiple guide RNAs to target multiple genes simultaneously in the same cell with high efficiency and specificity [5]. Apart from genome editing, the CRISPR/cas9 system has been used to regulate endogenous gene expression or label specific chromosomal loci in living cells [6]. Over the past few years, due to its simplicity and flexibility, the CRISPR/Cas9 technology has been also applied in the field of immunology to edit mouse and human Ig genes [7, 8]. Very recently, the CRISPR/Cas9 system has been used to produce site specifically modified antibody conjugates as therapeutic and diagnostic tools [9]. Overall, these studies demonstrate that the CRISPR-based editing offers a precise and inexpensive method to produce engineered mAbs.

These findings prompted us the idea to generate a chimeric mouse–fish mAb carrying part of the Fc region from a teleost fish antibody molecule (Fig. 1). As a first step, a DNA double-stranded break (DSB) is introduced into the target hybridoma IgH gene locus. In order to achieve highly specific knockout, avoiding off-targets in the genome, the gRNAs are designed and validated by using the E-CRISPR design tool. The gRNA showing the highest score of the Cas9 cleavage efficiency is cloned into the CRISPR/Cas9 plasmid, which contains the green fluorescence protein (GFP) that enables the detection of Cas9 expression. To generate gene knockout, the CRISPR/Cas9 plasmid is transfected into the hybridoma cells by applying the electroporation protocol. After isolating the GFP-positive cells by FACS, the effectiveness of Cas9 targeting in the mouse IgH genomic region is evaluated by PCR amplification, using primers that flank the DNA region of interest. Next step is knockin that consists in inserting the DNA donor template, which contains the fish IgH chain constant exons, into the target locus by repairing the DSB through the homology-directed repair (HDR) pathway. To construct the donor plasmid, the fish Ig insert is combined with the 5′ and 3′ homology arms by using the Gibson assembly protocol. In addition, the donor plasmid contains a fluorescent protein as selection marker for the correct sequence insertion. To perform gene knockin, the CRISPR/Cas9 plasmid and the donor plasmid are delivered into hybridoma cells by electroporation. The cells that carry the desired genome modification are fluorescent and can thus be easily isolated by FACS and maintained in culture for the production of mouse chimeric mAbs.

This protocol can represent a promising starting point for engineering the Fc region of a mAb with an innovative and versatile technology.

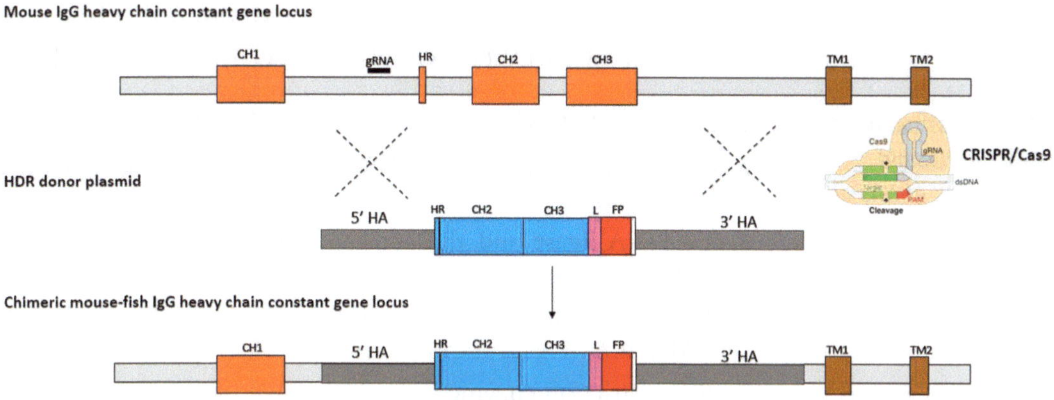

Fig. 1 Schematic representation of the generation of the chimeric mouse–fish monoclonal antibody by the CRISPR/Cas9 technology. The top scheme represents the target mouse IgG heavy chain constant gene locus comprising the CH1, CH2, and CH3 heavy chain constant exons, and the hinge region exon (HR) depicted in orange; TM1, TM2 (transmembrane) exons in brown. Within the intron between CH1 and HR, the guide RNA (gRNA) is depicted in black. It is one of the key elements required to edit the specific DNA target site by the CRISPR/Cas9 system (shown at the upper right-hand side of the figure). The HDR donor plasmid contains the fish Ig constant exons (HR, CH2, CH3) to be inserted, linked to a fluorescent protein (FP) through a linker peptide (L, purple). The stop codon is reported in white. The HDR donor plasmid is flanked on each side by a homology arm (HA). The bottom scheme represents the chimeric mouse–fish heavy chain constant region gene locus resulted from the knockin mechanism

2 Materials

Prepare and filter all solutions and buffers. Store all reagents according to the manufactory instructions. We recommend to use filtered and sterile tips, in particular for molecular and cell biology techniques. Before working with cell lines, check and clean cell hood and cell incubator.

2.1 Mammalian Cell Culture

1. Murine hybridoma cell line, secreting monoclonal IgG antibody (*see* **Note 1**).
2. Culture medium: RPMI 1640 medium supplemented with 10% heat inactivated fetal bovine serum (FBS), containing 100 U/mL penicillin and 100 μg/mL streptomycin (*see* **Note 2**).
3. 6-well, 24-well, flat bottom cell culture plates.
4. T25, T75 flasks.
5. Burker chamber cell counting.
6. Phosphate buffered saline (PBS; 1×): 10 mM Na_2HPO_4, 1.8 mM KH_2PO_4, pH 7.4, 137 mM NaCl, 2.7 mM KCl (*see* **Note 3**).
7. Electroporation apparatus and compatible electroporation cuvettes.

8. Sorting buffer: PBS (1×) supplemented with 0.1% FBS.
9. Fluorescence-activated cell sorting (FACS).

2.2 Construction of CRISPR Plasmid Containing gRNAs and HDR Donor Plasmid

1. Plasmid: pSpCas9(BB)-2A-GFP (pX458) [3] (*see* **Note 4**).
2. 0.1 nmol/mL oligos for gRNA construction and validation (*see* Subheading 3.1).
3. T4 DNA ligase and the accompanying 10× reaction buffer (10 μM ATP).
4. Annealing buffer (10×): 100 mM Tris–HCl, pH 7.4, 1 M NaCl.
5. *Bbs*I restriction enzyme.
6. TBE buffer (10×): 1 M Tris base, 1 M boric acid, 0.02 M EDTA, pH 8.0.
7. 0.8% agarose gel: 0.8 g agarose in 100 mL TBE (1×).
8. DNA gel extraction solutions. Gel dissolving buffer, 30% guanidine thiocyanate and sodium iodide-based buffer; DNA wash buffer, ethanol-based buffer.
9. DH5alpha chemically competent cells (*see* **Note 5**).
10. Ampicillin stock solution: 100 mg/mL of ampicillin in water, sterile filtered.
11. LB medium: 10 g tryptone, 10 g NaCl, 5 g yeast extract, 950 mL of distilled H_2O. Autoclave for 20 min.
12. LB agar: LB medium supplemented with 2% agar. Autoclave for 20 min. Cool the molten agar in a water bath at 56 °C prior to adding the antibiotic. Add ampicillin stock solution to LB agar to a final concentration of 100 μg/mL.
13. Plasmid DNA miniprep solutions.

    Buffer P1: 50 mM Tris–HCl, pH 8.0, 10 mM EDTA, supplemented with RNase A to a final concentration of 100 μg/mL.

    Buffer P2: 200 mM NaOH, 1% SDS.

    Buffer P3: 3 M potassium acetate, pH 5.2.
14. Plasmid DNA maxiprep solutions, Buffer P1, P2, and P3 prepared with endotoxin free deionized water. Wash buffer: 50 mM MOPS, pH 7.0, 1 M NaCl, supplemented with 15% isopropanol.
15. Isopropanol.
16. Ethanol.
17. UV spectrophotometer (NanoDrop).
18. U6 forward sequencing primer: 5′- ACTATCATATGCT TACCGTAAC-3′.

19. High-fidelity DNA polymerase and the accompanying buffer.
20. 10 mM dNTP set.
21. Thermocycler with programmable temperature stepping functionality.
22. Automated genetic analyzer.
23. pUC19 as donor plasmid (*see* **Note 6**).
24. Gibson Assembly Mix [10].

2.3 Evaluation of Genome Editing in Hybridoma Cells

1. Digestion buffer for genomic DNA extraction: 20 mM Tris–HCl, pH 8.0, 10 mM NaCl, 1 mM EDTA, 10% SDS, supplemented with Proteinase K, at a final concentration of 20 mg/mL, and RNase A, at a final concentration of 100 μg/mL.
2. Phenol–chloroform–isoamyl alcohol (25:24:1).
3. Taq DNA polymerase with standard Taq buffer.
4. TOPO cloning vector (pGEM).
5. M13/pUC forward sequencing primer: 5′- CCCAGTCAC GACGTTGTAAAACG-3′.

3 Methods

3.1 Design of gRNAs

1. Use the bioinformatics tool E-CRISP [11] for designing gRNAs. Select the organism and provide the name or sequence of the gene of interest (*see* **Note 7**).
2. Choose at least two gRNAs to test the knockout efficiency (*see* **Note 8**) (Table 1).
3. Synthesize two oligos for each selected gRNA, carrying a four-base overhang at the 5′ end (*see* below, in bold) compatible for cloning into *Bbs*I site in pX458. "N" and "n" represent forward and reverse primer complementary base pairs, respectively (*see* **Note 9**) (Fig. 2).

   Forward primer: 5′ -**CACC**GNNNNNNNNNNNNNNNNNNN- 3′.

   Reverse primer: 5′ -**AAAC**nnnnnnnnnnnnnnnnnnnC- 3′.

3.2 Cloning of gRNAs

1. Oligo annealing: reconstitute oligos in ddH_2O to a final concentration of 0.1 nmol/μL and vortex for a few seconds. Mix 11.2 μL of each oligo (forward and reverse) and 2.5 μL of 10× annealing buffer in a separate tube. Bring water to a boil in a beaker and make sure water temperature is at or above 95 °C. Place the oligo mixture in the water bath and allow to cool down naturally to 30 °C. Dilute the oligo mixture 1:400 with 0.5× annealing buffer (*see* **Note 10**).
2. Digest pX458 with *Bbs*I as follows:

Table 1
Outputs of the E-CRISP design tool for evaluation of the gRNAs quality

Name	Length	Start	End	Strand	Nucleotide sequence	S-Score	A-Score	E-Score	Number of Hits
mIgG_1	23	752	775	minus	GGTTGGCTGTCCAGCCTGGT NGG	108,6956	0	53,3108	1
mIgG_2	23	621	644	minus	GTTGCAGGTGACGGTCTCGC NGG	104,3478	0	58,8497	1
mIgG_3	23	562	585	minus	GAGTGTAGAGGTCA GATCC NGG	104,3478	0	57,0708	1
mIgG_4	23	220	243	plus	GCCTCAGGTCTCATCTGGTC NGG	88,6956	0	49,0717	2
mIgG_5	23	220	243	plus	GCCTCAGGTCTCATCTGGTC NGG	88,6956	0	49,0717	2
mIgG_6	23	946	969	minus	GTCTGGTGTGTGGTCATGCC NGG	81,7391	0	56,0988	3

Name: target site (mIgG, mouse IgG heavy chain); **Length**: target length; **Start**: target start relative to the input sequence; **End**: target end relative to the input sequence; **Strand**: orientation of sequence; **Nucleotide sequence**: target site nucleotide composition, including PAM sequence; **S-score**: Specificity score; **A-score**: Annotation score; **E-score**: efficacy score; **Number of Hits**: Cas9 off-target effects [11]

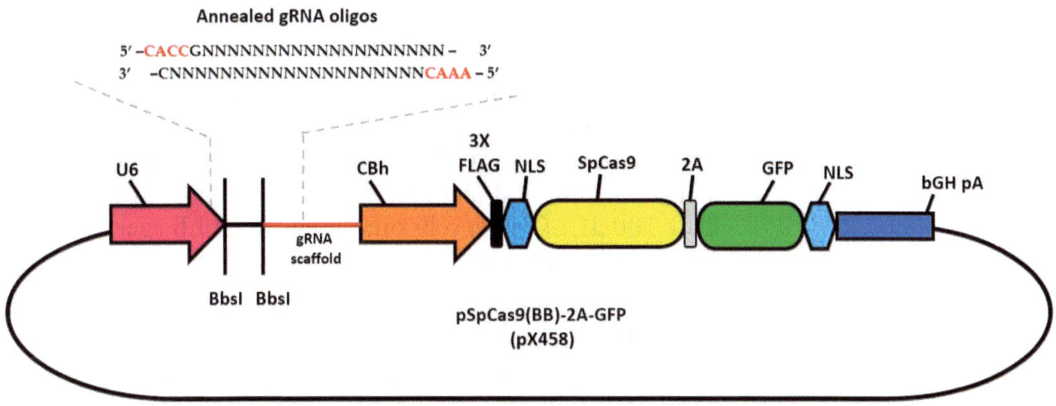

Fig. 2 Schematic representation of the CRISPR plasmid pX458 used for cloning the gRNA of choice. The gRNA oligos are shown on top. The top and bottom strand orientations contain overhangs (in red) for ligation into the pair of *Bbs*I sites at the U6 promoter insertion site (magenta arrow). The green fluorescent protein marker (GFP, green oval) is used to detect the expression of Cas9 (SpCas9, yellow oval) [3]

X μL pX458 (1–2 μg).

2 μL 10× *Bbs*I buffer.

1 μL *Bbs*I.

H_2O to final reaction volume of 20 μL.

Incubate the digestion reaction at 37 °C overnight. Run about 200 ng on 0.8% agarose gel for 1 h at 100 V to verify the complete digestion. Use 1 kb DNA ladder. Heat inactivate at 65 °C for 20 min (*see* **Note 11**). Gel-purify the digested plasmid. Run the digestion product on 0.8% agarose gel for 1 h at 100 V, followed by ethidium bromide staining.

3. Excise the plasmid band from the gel and incubate with Gel dissolving buffer (200 μL for 100 mg agarose gel) at 50 °C for 20 min. Load the dissolved gel up to 700 μL onto a spin column and centrifuge for 30 s at 11,000 × *g*. Wash the plasmid by adding DNA wash buffer to the column and centrifuge at the same conditions as described above. Elute the plasmid from the column with deionized water and determine the plasmid concentration by measuring the absorbance at 260 nm with an UV spectrophotometer (NanoDrop).

4. Set up ligation reaction as follows (*see* **Note 12**):

   1 μL gRNA oligo mixture.

   1 μL gel-purified digested pX458 (10–20 ng).

   1 μL 10× ligase buffer (10 μM ATP).

   1 μL T4 DNA ligase.

   H_2O to final volume of 10 μL.

Incubate at room temperature for 3 h. If not use immediately, store the reaction at −20 °C.

5. Transform DH5alpha competent cells with 7 μL of the ligation reaction. Incubate for 20 min at room temperature and heat shock for 45 s at 42 °C. Incubate on ice for 3 min, add 700 μL of LB medium and incubate at 37 °C while shaking for 1 h. Plate 100 μL of *E. coli* cells onto prewarmed LB agar/100 μg/mL ampicillin plates.
6. Pick single colonies and inoculate miniprep cultures in 5–6 mL of LB/100 μg/mL ampicillin.
7. Isolate miniprep DNA plasmid as follows. Pellet 5 mL of bacterial cultures by centrifugation for 5 min at 10,000 × *g*. Resuspend the pellet in 250 μL of Buffer P1 for cell lysis. Add 250 μL of Buffer P2 and incubate at room temperature for 5 min. Add 350 μL of Buffer P3 and centrifuge for 10 min at full speed. Discard the surnatant and wash the pellet with 100% ethanol and centrifuge in the same conditions as described above. Wash with cold 70% ethanol and centrifuge for 1 min at full speed. Resuspend the pellet with deionized water and determine the plasmid concentration by UV absorbance measurement.
8. Verify gRNA sequence by using the U6 sequencing primer.
9. After sequencing, retransform the plasmid into 100 μL of DH5alpha competent cells and perform the maxiprep protocol (*see* **Note 13**). Resuspend the bacterial pellet with 10 mL of Buffer P1. Add 10 mL of Buffer P2 and 10 mL of Buffer P3. Mix by inversion. Load the lysed cells onto spin column and then add 30 mL of wash buffer. Precipitate the plasmid with 10 mL of isopropanol and centrifuge at 15,000 × *g* for 30 min at 4 °C. Wash the pellet with 70% ethanol and centrifuge at 15,000 × *g* for 10 min at 4 °C. Resuspend the pellet with endotoxin-free water.

3.3 Gene Knockout

1. 24 h before transfection, seed the cells at a density of 5×10^5 in T25 flasks containing 5 mL of RPMI-1640 medium (*see* **Note 14**).
2. On the day of transfection, collect about 7×10^6 cells by centrifugation at 800 × *g* for 4 min. Wash the pellet twice with 5 mL of PBS and centrifuge again at the same conditions.
3. Resuspend the cells in 500 μL of PBS and add 5 μg of pX458 plasmid containing gRNAs.
4. Transfer the cell/DNA mixture into cuvette without introducing bubbles.

5. Immediately transform cells by electroporation according to the following general conditions for mammalian cells: 250 kV, 500 μF at the typical time constant of about 3 ms.
6. Keep cells in culture in 1 mL of RPMI-1640 supplemented with 20% FBS in 24-well plates at a density of 2×10^5 cells/well.
7. 24 h after electroporation, collect cells by centrifugation at $800 \times g$ for 4 min, resuspend in Sorting Buffer and perform cell sorting for GFP fluorescence.
8. Culture isolated GFP-positive cells in 24-well plates and progressively expand them into 6-well plates, and subsequently into T25 flasks.
9. Extract DNA from GFP-expressing hybridoma cells. Start from 5×10^5 cells and centrifuge for 5 min at $800 \times g$. Resuspend the pellet with 200 μL of digestion buffer and incubate for 10 min at 55 °C. Add 400 μL phenol–chloroform–isoamyl alcohol and centrifuge for 1 min at full speed. Transfer the aqueous phase to a fresh tube and add 1/10 volume of 5 M potassium acetate, pH 5.2, and 2 volumes of cold 100% ethanol. Incubate for 3 h at −20 °C. Centrifuge for 15 min at 4 °C at full speed. Discard the surnatant and add 500 μL of cold 70% ethanol and centrifuge for 5 min at 4 °C at full speed. Resuspend the pellet with deionized water.
10. Perform PCR amplification to validate genome editing (*see* **Note 15**) as follows.

    5 μL DNA polymerase reaction buffer.

    5 μL 2 mM dNTP.

    2.5 μL 10 μM forward primer (*see* **Note 16**).

    2.5 μL 10 μM reverse primer.

    X μL template DNA (250–500 ng).

    0.5 μL DNA Taq polymerase (*see* **Note 17**).

    H_2O to final volume of 50 μL.

    Thermocycler conditions for PCR amplification.

    Initial denaturation: 95 °C (3 min).

    Cycling (30 cycles): 95 °C (30 s); T_m-5 (30 s); 72 °C (1 min).

    Final Extension: 72 °C (5–15 min).

    Hold: 4–10 °C
11. Gel purify PCR products and clone into pGEM-vector. Isolate 5–10 positive clones by the blue–white screening method and sequence by sequencing forward primer M13/pUC (*see* **Note 18**).

3.4 Gene Knockin

1. Digest pUC19 with two polylinker restriction enzymes as follows.

   X μL pX458 (1–2 μg).

   5 μL 10X buffer.

   1 μL Enzyme 1.

   1 μL Enzyme 2.

   H_2O to final volume of 50 μL.

   Incubate the digestion reaction at 37 °C for 1 h. Run about 200 ng on agarose gel to verify the complete digestion. Heat-inactivate at 65 °C for 20 min and gel-purify linearized plasmid, as described above (Subheading 3.3).

2. Construct the donor plasmid by the Gibson Assembly protocol (*see* **Note 19**) (Fig. 3). Synthesize the fish Ig insert, containing overlapping ends for Gibson Assembly protocol (*see* **Note 20**) (Fig. 1). Design 5′ and 3′ homology arms (*see* **Note 21**) and amplify them by PCR (*see* **Note 22**) using a high-fidelity DNA polymerase as follows.

   5 μL 5× high-fidelity reaction buffer.

   0.5 μL 10 mM dNTP.

   1.25 μL 10 μM forward primer.

   1.25 μL 10 μM reverse primer.

   X μL template DNA (250–500 ng).

   0.25 μL high-fidelity DNA polymerase (*see* **Note 23**).

   H_2O to final volume of 25 μL.

   Thermocycler conditions for PCR amplification.

   Initial denaturation: 98 °C (30 s).

   Cycling (35 cycles): 98 °C (10 s); 50–72 °C (30 s); 72 °C (30 s).

   Final Extension: 72 °C (2 min).

   Hold: 4–10 °C.

3. Set up the Gibson Assembly reaction on ice as follows.

   10 μL Gibson Assembly Master Mix.

   X μL total amount of fragments (0.02–0.5 pmols) (*see* **Note 24**).

   H_2O to final volume of 20 μL.

   Incubate at 50 °C for 1 h. Store at −20 °C.

4. Transform the Gibson reaction as described in Subheading 3.2 (**steps 5, 6, 7, 9**) (*see* **Note 25**).

5. Perform the transfection protocol as described in Subheading 3.3, by adding 5 μg pX458 plasmid, containing the gRNA, and

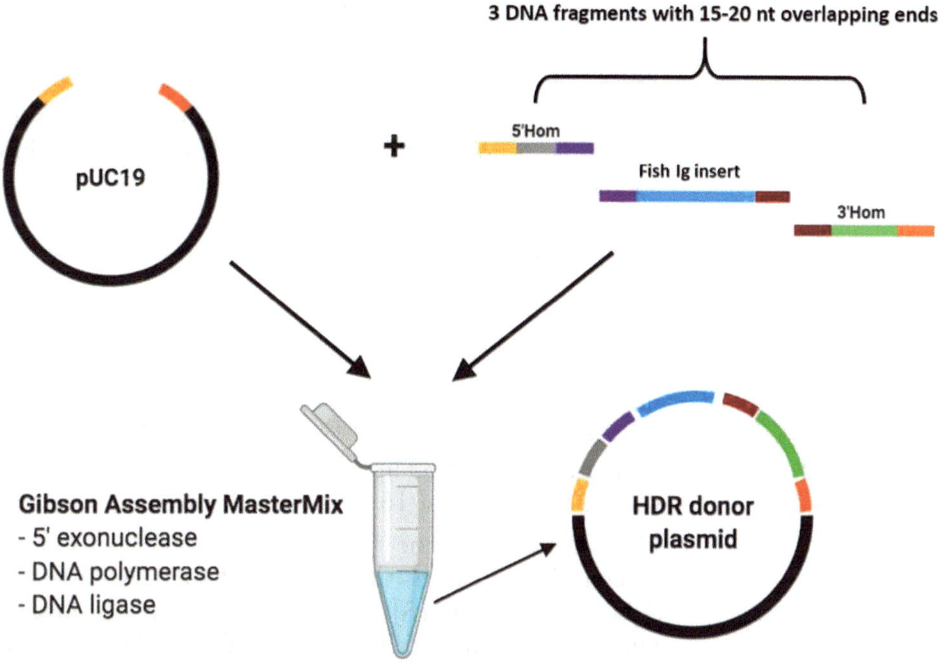

Fig. 3 Schematic representation of the construction of the donor plasmid using the Gibson Assembly protocol. First, overlapping ends, crucial for the correct assembly by using Gibson Assembly [10], are generated by PCR for each of the three DNA fragments to be cloned into pUC19. The overlapping ends are depicted: in magenta, for the 3′ end of the 5′ homology arm (5′Hom fragment) and for the 5′ end of the Fish Ig insert; in brown, for the 3′ end of the Fish insert and for the 5′ end of the 3′ homology arm (3′Hom fragment). This method allows multiple fragments to be joined in a single-step isothermal reaction based on the activity of three enzymes: the exonuclease chews back the 5′ ends of the DNA fragments, permitting the exposed complementary sequences to anneal; the DNA polymerase introduces nucleotides to fill in the gaps in the annealed regions; breaks are sealed and the DNA fragments linked together by the enzyme DNA ligase

5 μg of donor plasmid in a 1:1 ratio, to 10×10^6 hybridoma cells (*see* **Note 26**).

6. Culture transfected cells in 1 mL of RPMI-1640 supplemented with 20% FBS in 24-well plates at a density of 2×10^5 cells/well.
7. Twenty-four hours after electroporation, collect cells by centrifugation at $800 \times g$ for 4 min, resuspend in Sorting Buffer and perform cell sorting for GFP fluorescence.
8. Culture isolated GFP-positive cells in 24-well plates and progressively expand them into 6-well plates, and subsequently into T25 flasks.
9. After 2 weeks, starting from the GFP-positive cells, isolate the cells expressing the fluorescent protein inserted into the donor plasmid. This indicates that knockin occurred.

10. Evaluate the correct knockin by PCR amplifications following the conditions described in Subheading 3.3 (**steps 8** and **9**) (*see* **Note 27**).
11. Produce the chimeric mouse–fish mAb by culturing the engineered hybridoma cell line in T75 flasks with a hybridoma serum-free medium.

4 Notes

1. Maintain hybridoma cells at 37 °C with 5% CO_2 in 10 mL of culture in T25 flasks and passage every 48/72 h.
2. Filter all reagents in order to avoid mycoplasma contamination.
3. Dispense the solution into aliquots and sterilize them by autoclaving for 20 min at 15 psi. Store PBS at 4 °C.
4. Use pX458 (Addgene #48138). As an alternative approach, several investigators use pSpCas9(BB)-2A-Puro (pX459 Addgene #62988) [3], which contains the puromycin resistance selection marker for transfected cells. However, we recommend not to use this plasmid for the present protocol since murine hybridoma cells are particularly sensitive to antibiotic concentration.
5. In alternative, use Stbl3 chemically competent cells for cloning.
6. pUC19 is chosen as donor plasmid since it is suitable to incorporate large DNA inserts.
7. Carefully design the gRNAs by focusing on the intronic regions of the mouse IgG heavy chain constant region gene locus, to avoid insertions or deletions in CDS during knockin mechanism or unintended DNA modifications at nontargeted gene loci (Fig. 1).
8. E-CRISP provides design specifications and parameters on the basis of efficient prediction models' algorithms. For this purpose, choose gRNAs showing the number of hits 1, a value that indicates the highest level of target specificity. Since gRNAs having 100% homology to the target sequence may not have high cleavage efficiency, in addition to off-target activity, another important parameter to be considered is E-score, which evaluates on-target activity. E-score measures the percentage of CG in the target region, which is essential to maximize cleavage of the target sequence. It depends on the presence of at least two G or C nucleotides proximal to the PAM sequence, which is close to the selected target sequence [11]. Therefore, it is important to choose at least two gRNAs with the highest E-score value (Table 1) and test them for the best Cas9 cutting efficacy.

9. Make sure the PAM site is not included in the gRNA sequence. Add CACCG to forward oligo; if the 5′ end of the oligo has a G, just add CACC. We recommend to add two G bases to the 5′ end of the gRNA to improve the expression in mammalian cells by the U6 promoter in pX458 (Fig. 2). Design the reverse oligo by reverse-complement the sequence of the forward oligo, then add AAAC to 5′, C to 3′ end. Pay careful attention to the 5′ → 3′ orientations.
10. Store annealed and diluted oligos at −20 °C.
11. If not complete, add more enzyme and buffer and go longer. Rerun a gel to confirm digestion and when digest is complete, heat-inactivate.
12. Include only vector in ligation reaction as negative control to monitor vector self-ligation.
13. Prepare all solutions with endotoxin-free water in order to use the plasmid for cell transfection.
14. Do not plate more cells than the recommended density, since may reduce transfection efficiency.
15. Considering the heterogeneity of indels, several approaches have been described to validate genome editing. One of them is the Surveyor Nuclease assay, based on an enzyme mismatch cleavage [12]. PCR amplification in association with Next Generation Sequencing (NGS) is a good method not only to detect indels but also to determine the mutation frequencies [3].
16. Design PCR primers that flank the targeting region of gRNAs to confirm the presence of indels.
17. Use a DNA polymerase which generates 3′-dA overhangs, essential for cloning PCR products.
18. Sequence at least 5–10 clones to verify genome editing.
19. To generate the donor plasmid, the fish Ig insert is combined with the 5′ and 3′ homology arms using the Gibson Assembly protocol, which allows the joining of multiple DNA fragments in a single isothermal reaction. The DNA fragments are cloned into pUC19 vector through the overlapping ends (Fig. 3).
20. Include a fluorescent protein (e.g., mCherry, YFP) in the insert as selection marker for the correct sequence integration (Fig. 1).
21. Design homology arms in the intronic regions in order to avoid CDS knockin that can cause frameshift mutations. Evaluate the length of homology arms in relation with the insert size since both represent crucial features for the design of donor plasmids and thus for cloning efficiency [13]. For example, in the case of an insert of about 2 kb, the 5′ and 3′

homology arms should be 1.5-kb and 2.5-kb long, respectively, to ensure a successful integration. In addition, the homology arms must have polylinker restriction enzyme sites for cloning in pUC19 vector. Mutate the PAM sequence (NGG → NGA) in the 5′ homology arm that flanks the fish Ig insert, in order to prevent Cas9-cutting of the donor plasmid.

22. PCR primers for use in Gibson Assembly must possess the following features: (1) a 15–25 bases overlap sequence, required for the assembly of adjacent fragments; (2) a gene-specific sequence, essential for template priming during PCR amplification.
23. Use high-fidelity DNA polymerase to amplify homology arms in order to avoid DNA mismatch repair (MMR).
24. We recommend to use 0.005 pmols of both homology arms, 0.018 pmols of fish Ig insert and 50 ng of pUC19 vector.
25. Use M13/pUC sequencing primer to verify the correct assembly of donor plasmid and/or perform a DNA digestion by the restriction enzymes used for cloning the insert flanked by the homology arms.
26. To improve the knockin efficiency, use larger amounts of donor plasmid than the CRISPR plasmid (3:1 ratio).
27. Use PCR primers that flank homology arms.

References

1. Köhler G, Milstein C (1975) Continuous cultures of fused cells secreting antibody of predefined specificity. Nature 256:495–497
2. Frenzel A, Hust M, Schirrmann T (2013) Expression of recombinant antibodies. Front Immunol 4:217
3. Ran FA, Hsu PD, Wright J et al (2013) Genome engineering using the CRISPR-Cas9 system. Nat Protoc 8:2281–2308
4. Ran FA, Hsu PD, Lin CY et al (2013) Double nicking by RNA-guided CRISPR Cas9 for enhanced genome editing specificity. Cell 154:1380–1389
5. Cong L, Ran FA, Cox D et al (2013) Multiplex genome engineering using CRISPR/Cas systems. Science 339:819–823
6. Bikard D, Jiag W, Samai P et al (2013) Programmable repression and activation of bacterial gene expression using an engineered CRISPR-Cas system. Nucleic Acids Res 41: 7429–7437
7. Cheong TC, Compagno M, Chiarle R (2016) Editing of mouse and human immunoglobulin genes by CRISPR/Cas9 system. Nat Commun 7:10934
8. Pogson M, Parola C, Kelton WJ et al (2016) Immunogenomic engineering of a plug-and-(dis)play hybridoma platform. Nat Commun 7:12535
9. Khoshnejad M, Brenner JS, Motley W et al (2018) Molecular engineering of antibodies for site-specific covalent conjugation using CRISPR/Cas9. Sci Rep 8:1760
10. Gibson D, Young L, Chuang RY et al (2009) Enzymatic assembly of DNA molecules up to several hundred kilobases. Nat Methods 6: 343–345
11. Heigwer F, Kerr G, Boutros M (2014) E-CRISP: fast CRISPR target site identification. Nat Methods 11:122–123
12. Qiu P, Shandilya H, D'Alessio JM et al (2004) Mutation detection using surveyor nuclease. BioTechniques 36:702–707
13. Urnov FD, Miller JC, Lee YL et al (2005) Highly efficient endogenous human gene correction using designed zinc-finger nucleases. Nature 435:646–651

Chapter 20

Identification, Characterization, and Expression Analysis of Immunoglobulin Genes from Antarctic Fish by PCR Methods

Alessia Ametrano and Maria Rosaria Coscia

Abstract

We describe different applications of the polymerase chain reaction (PCR) method to amplify up to 2.5-kb DNA regions of the immunoglobulin genes from Antarctic teleost fishes and to evaluate their expression in different tissues. To enhance amplification of poorly expressed transcripts or difficult to amplify, for example, products generated by 5′ or 3′ RACE, we have set up optimal primer annealing conditions based on a three-step cycle strategy.

The protocols reported here along with materials required and general tips are modifications of conventional PCR.

Key words PCR, 5′ and 3′ RACE, qPCR, Antarctic teleost, IgH constant region gene, Cold-adapted gene

1 Introduction

Immunoglobulins (Igs) emerged with the earliest jawed vertebrates, cartilaginous fishes (Chondrichthyes), whose gene locus encoding the Ig heavy chain (IgH) is organized in multiple clusters [1]. Each cluster contains one V segment, a number of D segments, one J segment and one set of constant region exons (C) required for the production of a given Ig chain [2]. In contrast to cartilaginous fish, the IgH locus of bony fish (Osteichthyes) has a translocon-type organization [3], in which multiple V are located upstream of multiple J and D segments, followed by C genes. This type of configuration is highly conserved in all tetrapods. Change of the Ig-gene locus organization from the cluster to the translocon arrangement represents an evolutionary advantage, since enhancing the antibody diversity. Unlikely mammals, teleost fish possess a limited number of immunoglobulin classes. IgM and IgD are present in most bony fishes, suggesting that they are ancestral isotypes

Cinzia Verde and Daniela Giordano (eds.), *Marine Genomics: Methods and Protocols*,
Methods in Molecular Biology, vol. 2498, https://doi.org/10.1007/978-1-0716-2313-8_20,

[4]. Teleost IgM can be expressed on the surface of B cells or as soluble form. Interestingly, the membrane-bound form of teleost IgM lacks the fourth heavy chain constant exon (CH4), since the first transmembrane exon (TM) is spliced to the CH3 exon or, in the case of Antarctic fish, directly to CH2 [5]. IgD has been identified for the first time in channel catfish [6] and shows a peculiar feature of being a chimeric isotype, since having the first Cm domain followed by a various number of δ domains, according to the species considered [7]. IgT has been considered the oldest known Ig isotype specialized in mucosal immunity [8]. In most species, IgT H chain gene locus encodes four C domains [9], although some bony fishes possess a variable number of CH domains, for example, Antarctic fish have lost most second domain [10, 11].

Given the high sensitivity, affordability, and ease of use, the Polymerase Chain Reaction (PCR) technique has been applied in a wide variety of areas since its invention. Modifications of the original protocol have even allowed more applications. The ability to amplify a specific DNA sequence of a complex gene locus has made PCR particularly suited to perform genetic studies. Taking account of the heterogeneity of the Ig genes mentioned above, here we describe optimal conditions for PCR protocols, using thermostable DNA polymerases, for identifying Ig genes and for quantitative PCR (qPCR) for evaluating their expression in tissues from cold-adapted teleost fishes.

2 Materials

Prepare and filter all solutions and buffers. We recommend to use sterile tips, in particular for molecular biology techniques. Wear the gloves while handling RNA, since human skin is a source of RNase molecules.

2.1 DNA Extraction

1. Digestion buffer for genomic DNA extraction: 20 mM Tris–HCl, pH 8.0, 10 mM NaCl, 1 mM EDTA, 10% SDS, supplemented with Proteinase K, at a final concentration of 20 mg/mL, and RNase A, at a final concentration of 100 μg/mL.
2. Phenol–chloroform–isoamyl alcohol (25:24:1).
3. Taq DNA polymerase with standard buffer.
4. TBE buffer (10×): 1 M Tris base, 1 M boric acid, 0.02 M EDTA, pH 8.0.
5. 0.8% agarose gel, 0.8 g agarose in 100 mL TBE (1×).
6. DNA gel extraction solutions. Gel dissolving buffer, 30% guanidine thiocyanate and sodium iodide-based buffer; DNA wash buffer, ethanol-based buffer.

7. TOPO cloning vector (pGEM).
8. DH5alpha chemically competent cells (*see* **Note 1**).
9. LB medium, 10 g tryptone, 10 g NaCl, 5 g yeast extract, 950 mL of distilled H_2O. Autoclave for 20 min.
10. LB agar, LB medium supplemented with 2% agar.
11. Ampicillin 100 mg/mL, sterile filtered.
12. Plasmid DNA miniprep solutions:

    Buffer P1, 50 mM Tris–HCl, pH 8.0, 10 mM EDTA, supplemented with 100 mg/mL RNase A.

    Buffer P2, 200 mM NaOH, 1% SDS.

    Buffer P3, 3 M Potassium acetate, pH 5.2.
13. M13/pUC forward sequencing primer: 5′ - CCCAGTCAC GACGTTGTAAAACG- 3′.
14. UV spectrophotometer (NanoDrop).
15. Automated genetic analyzer.

2.2 RNA Extraction

1. RNase free 1.5-mL microcentrifuge tubes.
2. 1× Phosphate buffered saline (PBS), 10 mM Na_2HPO_4, 1.8 mM KH_2PO_4, pH 7.4, 137 mM NaCl, 2.7 mM KCl (*see* **Note 2**).
3. Sterile RNase-free water.
4. Isopropanol.
5. Ethanol.
6. Chloroform.
7. RNA lysis buffer, 4 M guanidinium thiocyanate, 25 mM sodium citrate, 0.5% *N*-laurosylsarcosine, 0.1 M 2-mercaptoethanol, pH 7.0.
8. DNase I.
9. 2% agarose gel, 2 g agarose in 100 mL TBE (1×).
10. UV spectrophotometer (NanoDrop).
11. Microcentrifuge.

2.3 cDNA Synthesis

1. DEPC water.
2. Reverse transcriptase and the accompanying 5× reaction buffer.
3. Oligo$(dT)_{15}$ primer.
4. Gene specific primers (forward and reverse).
5. dNTP set (10 mM).
6. Dithiothreitol (DTT).
7. RNase OUT recombinant ribonuclease inhibitor (40 U/μL).

Table 1
Primers for PCR and qPCR analyses of Ig genes from Antarctic fish

Oligo name	Domain	Sequence 5′ → 3′	Application
rtIgT2f fwr[a]	CH1	TCACGGTTGGTTGTCTTGC	DNA genomic amplification
G2IgT rev[a]	CH3	CCTTGCCTGGTTGTTGTTGAA	
GnIgT rev[a]	CH2	GTGATGGTAGCGTTAGTCTCTGG	5′ RACE as GSP1
Tufw fwr[a]	CH3	CCTTGCCTGGTTGTTGTTGAA	3′ RACE as GSP1
rtIgT1f fwr	CH1	ACCGGCACCTATAGCGATG	qPCR analysis
rtIgT1r rev	CH1	ATTGACTTGTTCCCACCAGC	

[a]Reference 11

2.4 5′ and 3′ Rapid Amplification cDNA Ends (RACE)

1. Adapter primer, 5′-GGCCACGCGTCGACTAGTACTTTTT TTTTTTTTTTTT-3′ (AP, 10 μM).
2. Abridged Universal Amplification Primer 5′-GGCCACGCGT CGACTAGTAC-3′ (AUAP, 10 μM).
3. Terminal deoxynucleotidyl transferase and the accompanying 5× reaction buffer.
4. dCTP (2 mM).
5. cDNA purification column.
6. Binding solution, 6 M sodium iodide.
7. 1× Washing buffer, ethanol-based buffer.
8. RNase H.
9. Thermocycler with programmable temperature stepping functionality and thin-wall PCR tubes.

2.5 qPCR Analysis*

1. Forward and reverse primers.
2. DNA intercalating dye-based qPCR mix.
3. RNase-free tubes.
4. qPCR thermal cycler with compatible optical 96-well PCR plates.

**See* Table 1 for primers details.

3 Methods

3.1 Identification of the IgH Constant Region Gene in Antarctic Fish

1. To extract genomic DNA, place the tissue sample (*see* **Note 3**) into a sterile microcentrifuge tube and add 1 mL of digestion buffer and incubate overnight at 55 °C until the lysis is complete. Add two volume of phenol–chloroform–isoamyl alcohol and centrifuge for 1 min at full speed. Transfer the aqueous phase to a fresh tube and add 1/10 volume of 5 M potassium

acetate, pH 5.2, and 2 volumes of cold 100% ethanol. Incubate for 3 h at −20 °C. Centrifuge for 15 min at 4 °C at full speed. Discard the surnatant and add 500 μL of cold 70% ethanol and centrifuge for 5 min at 4 °C at full speed. Resuspend the pellet with deionized water and determine the genomic DNA concentration by UV absorbance measurement. Run about 200 ng on 0.8% agarose gel for 1 h at 100 V, followed by ethidium bromide staining, to verify genomic DNA integrity. Use 1 kb DNA ladder for size determination.

2. Perform PCR amplification as follows.

   5 μL 10× DNA polymerase reaction buffer.

   1 μL 0.2 mM dNTP.

   2.5 μL 10 μM forward primer (*see* **Note 4**).

   2.5 μL 10 μM reverse primer.

   X μL template DNA (50–100 ng).

   0.5 μL DNA Taq polymerase (*see* **Note 5**).

   H_2O to final volume of 50 μL.

   Thermocycler conditions for PCR amplification (*see* **Note 6**).

   Initial denaturation: 95 °C (5 min).

   Cycling (40 cycles): 95 °C (30 s); Ta (30 s); 72 °C (1 min).

   Final Extension: 72 °C (10 min).

   Hold: 4–10 °C.

   Run about 200 ng on 1.5% agarose gel for 1 h at 100 V to verify the correct PCR amplification.

3. Gel-purify PCR product: excise the amplicon band from the gel and incubate with Gel dissolving buffer (200 μL for 100 mg agarose gel) at 50 °C for 20 min. Load the dissolved gel up to 700 μL onto a spin column and centrifuge for 30 s at 11,000 × *g*. Wash the amplicon by adding DNA wash buffer to the column and centrifuge at the same conditions as described above. Elute the amplicon from the column with deionized water and determine the amplicon concentration by measuring the absorbance at 260 nm with an UV spectrophotometer (NanoDrop).

4. Cloning the amplicon into pGEM plasmid. Set up ligation reaction as follows (*see* **Note** 7).

   X μL PCR product.

   1 μL pGEM vector (50 ng).

   5 μL 2× ligase buffer.

   1 μL T4 DNA ligase.

   H_2O to final volume of 10 μL.

Incubate at room temperature for 3 h. If not use immediately, store the reaction at −20 °C.

5. Transform DH5alpha competent cells with 7 μL of the ligation reaction. Incubate for 20 min at room temperature and heat shock for 45 s at 42 °C. Incubate on ice for 3 min, add 700 μL of LB medium and incubate at 37 °C while shaking for 1 h. Plate 100 μL of *Escherichia coli* cells onto pre-warmed LB agar/ 100 μg/mL ampicillin plates.
6. Pick single colonies and inoculate miniprep cultures in 5–6 mL of LB/100 μg/mL ampicillin.
7. Isolate miniprep DNA plasmid as follows. Pellet 5 mL of bacterial cultures by centrifugation for 5 min at 10,000 × *g*. Resuspend the pellet in 250 μL of Buffer P1 for cell lysis. Add 250 μL of Buffer P2 and incubate at room temperature for 5 min. Add 350 μL of Buffer P3 and centrifuge for 10 min at full speed. Discard the surnatant and wash the pellet with 100% ethanol and centrifuge in the same conditions as described above. Wash with cold 70% ethanol and centrifuge for 1 min at full speed. Resuspend the pellet with deionized water and determine the plasmid concentration by UV absorbance measurement.
8. Verify the sequence by using the M13/pUC forward sequencing primer.

3.2 5′ RACE of the IgH Constant Region Gene

1. To extract total RNA, homogenize tissue sample in 1 mL of RNA lysis buffer per 50 to 100 mg of tissue using a power homogenizer (*see* **Note 8**). Incubate the homogenized sample for 5 min at room temperature. Add 0.2 mL of chloroform per 1 mL of RNA lysis buffer. Shake tubes vigorously by hand for 15 s and incubate for 2 to 3 min. Centrifuge the samples at 12,000 × *g* for 15 min at 4 °C. After centrifugation, transfer the aqueous phase containing RNA to a fresh tube. Add 0.5 mL of isopropanol and incubate the sample at room temperature for 10 min and centrifuge at 12,000 × *g* for 10 min. Remove the surnatant and wash the RNA pellet by adding 1 mL of 75% ethanol. Mix the sample by vortexing and centrifuge at 7500 × *g* for 5 min at 4 °C. Dry the RNA pellet and dissolve it in DEPC-treated water and incubate for 10 min at 55 °C and determine RNA concentration by UV absorbance measurement. Run about 200 ng on 2% agarose gel for 1 h at 100 V, followed by ethidium bromide staining, to verify RNA integrity.
2. Perform cDNA synthesis as follows:

   2.5 μL 2.5 pmol gene specific primer (GSP1) (*see* **Note 9**).

   X μL sample RNA (*see* **Note 10**).

H_2O to final volume of 15.5 μL.

Incubate the reaction for 10 min at 70 °C. Chill 1 min on ice. Collect the content of the tube by rapid centrifugation and add the following reagents in the order given:

2.5 μL 10× Reverse transcriptase buffer.

2.5 μL 25 mM $MgCl_2$.

1 μL 10 mM dNTP.

2.5 μL 0.1 M DTT.

H_2O to final volume of 8.5 μL.

Mix gently and collect the reaction by centrifugation. Incubate for 1 min at 42 °C. Add 1 μL of reverse transcriptase. Mix and incubate for 50 min at 42 °C (*see* **Note 11**). Incubate at 70 °C for 15 min to finish the reaction. Add 1 μL of RNase H and incubate for 30 min at 37 °C.

3. Column-purify cDNA: equilibrate the binding solution to room temperature and 100 μL of sterilized water at 65 °C. Add 120 μL of binding solution to the cDNA mixture. Transfer the mixture to a cDNA column and centrifuge at 13,000 × *g* for 20 s (*see* **Note 12**). Discard the surnatant and add 0.4 mL of cold 1× wash buffer to the column and centrifuge at 13,000 × *g* for 20 s. Wash the column two times with 400 μL of cold 70% ethanol and centrifuge at 13,000 × *g* for 1 min. Transfer the column into a fresh sample recovery tube. Add 50 μL of sterilized water (*see* **Note 13**) and centrifuge at 13,000 × *g* for 20 s to elute the cDNA.
4. TdT tailing of cDNA, as follows (*see* **Note 14**):

5 μL 5× tailing buffer.

2.5 μL 2 mM dCTP.

10 μL purified cDNA sample.

H_2O to final volume of 24 μL.

Incubate for 3 min at 94 °C. Chill 1 min on ice. Collect the mixture of the tube by rapid centrifugation and place on ice. Add 1 μL TdT and incubate for 10 min at 37 °C (*see* **Note 15**). Heat inactivated the TdT for 10 min at 65 °C.

5. Perform PCR amplification of the dC-tailed cDNA as follows (*see* **Note 16**).

5 μL 10× DNA polymerase reaction buffer.

1 μL 0.2 mM dNTP.

2.5 μL 10 μM GSP1.

2.5 μL 10 μM AAP.

5 μL dC-tailed cDNA.

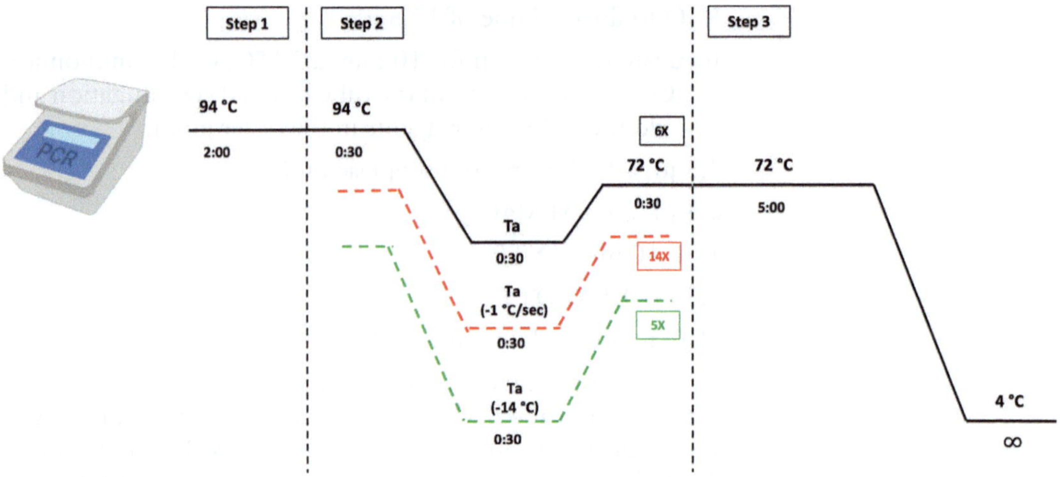

Fig. 1 Scheme of the thermocycler program to amplify 5′ and 3′ RACE products. Step 1: initial DNA denaturation. Step 2: first 6 cycles for primer annealing at the optimal temperature for the primer pairs used (solid black line); next 14 cycles for primer annealing at a decreasing temperature of −1 °C/s (dashed red line); final 5 cycles for primer annealing at a temperature decreased by 14 °C in respect to the initial annealing temperature (dashed green line). Step 3: final extension at 72 °C

0.5 μL DNA Taq polymerase.

H_2O to final volume of 50 μL.

Thermocycler conditions for PCR amplification (*see* **Note 17**) (Fig. 1).

Initial denaturation: 94 °C (2 min).

Cycling (6 cycles): 94 °C (30 s); Ta (30 s); 72 °C (30 s).

Cycling (14 cycles): 94 °C (30 s); Ta (*see* **Note 18**) (30 s); 72 °C (30 s)

Cycling (5 cycles): 94 °C (30 s); Ta (*see* **Note 19**) (30 s); 72 °C (30 s)

Final Extension: 72 °C (5 min).

Hold: 4–10 °C.

Cloning the amplified dC-tailed cDNA following the conditions described in Subheading 3.1 (**steps 3–8**).

3.3 3′ RACE of IgH Constant Region Gene

1. Extract RNA from tissue following the conditions described in Subheading 3.2 (**step 1**).
2. Perform cDNA synthesis as follows.

   1 μL 10 μM AP.

   X μL sample RNA.

   H_2O to final volume of 12 μL.

Incubate the reaction for 10 min at 70 °C. Chill 1 min on ice. Collect the content of the tube by rapid centrifugation and add the following reagentes in the order given.

2 μL 10× Reverse transcriptase buffer.

2 μL 25 mM $MgCl_2$.

1 μL 10 mM dNTP.

2 μL 0.1 M DTT.

H_2O to final volume of 7 μL.

Mix gently and collect the reaction by centrifugation. Incubate for 5 min at 42 °C. Add 1 μL of reverse transcriptase. Mix and incubate for 50 min at 42 °C. Incubate at 70 °C for 15 min to finish the reaction. Add 1 μL of RNase H and incubate for 30 min at 37 °C.

3. Perform PCR amplification of the target cDNA as follows (*see* **Note 16**).

   5 μL 10× DNA polymerase reaction buffer.

   1 μL 0.2 mM dNTP.

   2.5 μL 10 μM GSP1.

   2.5 μL 10 μM AUAP.

   5 μL target cDNA.

   0.5 μL DNA Taq polymerase.

   H_2O to final volume of 50 μL.

   Thermocycler conditions for PCR amplification as reported in Subheading 3.2 (**step 5**) (Fig. 1).

4. Cloning the amplified target cDNA following the conditions described in Subheading 3.1 (**steps 3–8**).

3.4 Ig Gene Expression Analysis by qPCR

1. Perform cDNA synthesis as follows.

   1 μL Oligo(dT)$_{15}$ primer.

   X μL sample RNA (1–500 ng).

   H_2O to final volume of 12 μL.

   Incubate the reaction for 5 min at 65 °C and quick chill on ice. Collect the content of the tube by rapid centrifugation and add the following reagents in the order given.

   2 μL 10× Reverse transcriptase buffer.

   2 μL 25 mM $MgCl_2$.

   1 μL 10 mM dNTP.

   2 μL 0.1 M DTT.

   H_2O to final volume of 7 μL.

Mix gently and collect the reaction by centrifugation. Incubate for 2 min at 42 °C. Add 1 μL of reverse transcriptase. Mix and incubate for 50 min at 42 °C. Incubate at 70 °C for 15 min to finish the reaction. Add 1 μL of RNase H and incubate for 30 min at 37 °C. Dilute and store cDNA samples at −20 °C.

2. Perform qPCR as follows.

   5 μL qPCR intercaling supermix.

   1 μL 0.5 mM forward primer (*see* **Note 20**).

   1 μL 0.5 mM reverse primer.

   H_2O to final volume of 8 μL.

   Pipet the reaction mixture on a PCR plate. Add 2 μL diluted cDNA (10 ng) per reaction.

3. Cover the PCR plate with optical seal and incubate the plate in a qPCR thermal cycler at the following temperatures.

   Initial denaturation: 95 °C (10 min).

   Cycling (40 cycles): 95 °C (15 s); Ta (15 s); 72 °C (1 min).

   Melting from 55 °C to 95 °C with steps of 0.5 °C increment each for 10 s (*see* **Note 21**).

4. As standardization of expression levels, reference genes need to be taken along for PCR (*see* **Note 22**).

4 Notes

1. In order to avoid plasmid recombination, we recommend to use Stbl3 chemically competent cells for cloning.
2. Dispense the solution into aliquots and sterilize them by autoclaving for 20 min at 15 psi. Store PBS at 4 °C.
3. For genomic DNA extraction, use 10–25 mg of tissue sample.
4. Design primers on conserved regions of IgH nucleotide sequences retrieved from genomic sequence databases available for teleost fishes (*see* Table 1).
5. Use a DNA polymerase that generates 3′-dA overhangs, essential for cloning PCR products.
6. We recommend to perform a second amplification, carried out in the same conditions as the primary PCR, in order to increase the amount of the specific amplicon.
7. Include only vector in ligation reaction as negative control to monitor vector self-ligation.
8. To make easier isolation of RNA from small quantities of samples, perform homogenization of samples in 0.8 mL of RNA lysis buffer and add 5 to 10 μg of glycogen as carrier.

9. Design GSP1 primer to anneal at least 300 bp from the mRNA 5′-end.
10. To perfom 5′ RACE, we recommend to use at least 1 μg of total RNA.
11. For short transcripts, 30 min incubation is usually sufficient; for longer transcripts, at least 50 min are required to synthesize enough cDNA.
12. Store the solution until the recovery of the cDNA is confirmed.
13. It is important that the distilled water is preheated to 65 °C in order to maximize recovery of DNA.
14. Given the high AT content of Antarctic fish Ig genes, we recommend to perform dC-tailing rather than dA-tailing.
15. Do not exceed the amount of TdT in the tailing reaction since may inhibit the following PCR amplification of tailed-cDNA.
16. If no specific PCR product will be generated, perform a nested PCR amplification using a gene specific primer (GSP2) that anneals to a site located within the primary PCR product.
17. Follow this thermocycler protocol in order to amplify low expressed transcripts (Fig. 1).
18. Set up the ramp down to Ta at −1 °C/s, for 14 cycles.
19. Perform the last 5 cycles at the temperature decreased by 14 °C, derived from the previous 14 cycles (*see* **Note 18**).
20. We recommend to test two to three different primer pairs, preferentially targeting different regions in the cDNA, except the untranslated regions. When RNA is not treated with DNase, make sure that primers span an intron of at least 700 base pairs.
21. Although the melting curve analysis indicates primer specificity, we recommend to analyze the PCR product by agarose gel electrophoresis. We recommend to run at least a duplicate for each sample.
22. For teleost fish immunoglobulin expression, use β-actin as reference gene.

References

1. Harding FA, Amemiya CT, Litman RT et al (1990) Two distinct immunoglobulin heavy chain isotypes in a primitive, cartilaginous fish, *Raja erinacea*. Nucleic Acids Res 18: 6369–6376
2. Hinds KR, Litman GW (1986) Major reorganization of immunoglobulin VH segmental elements during vertebrate evolution. Nature 320:546–549
3. Gally JA, Edelman GM (1972) The genetic control of immunoglobulin synthesis. Annu Rev Genet 6:1–46
4. Mashoof S, Criscitiello MF (2016) Fish immunoglobulins. Biology (Basel) 5(4):45. https://doi.org/10.3390/biology5040045
5. Coscia MR, Varriale S, De Santi C et al (2010) Evolution of the Antarctic teleost

immunoglobulin heavy chain gene. Mol Phylogenet Evol 55:226–233
6. Wilson M, Bengtén E, Miller NW et al (1997) A novel chimeric Ig heavy chain from a teleost fish shares similarities to IgD. Proc Natl Acad Sci U S A 94:4593–4597
7. Ramirez-Gomez F, Greene W, Rego K et al (2012) Discovery and characterization of secretory IgD in rainbow trout: secretory IgD is produced through a novel splicing mechanism. J Immunol 188:1341–1349
8. Sunyer JO (2013) Fishing for mammalian paradigms in the teleost immune system. Nat Immunol 14:320–326
9. Salinas I, Zhang YA, Sunyer JO (2011) Mucosal immunoglobulins and B cells of teleost fish. Dev Comp Immunol 35:1346–1365
10. Giacomelli S, Buonocore F, Albanese F et al (2015) New insights into evolution of IgT genes coming from Antarctic teleosts. Mar Genomics 24:55–68
11. Ametrano A, Gerdol M, Vitale M et al (2021) The evolutionary puzzle solution for the origins of the partial loss of the Cτ2 exon in notothenioid fishes. Fish Shellfish Immunol 116:124–139

Chapter 21

Physical Mapping of Repeated Sequences on Fish Chromosomes by Fluorescence In Situ Hybridization (FISH)

Laura Ghigliotti, Juliette Auvinet, and Eva Pisano

Abstract

The opportunity to map genes and noncoding DNA sequences on the chromosomes by fluorescence in situ hybridization (FISH) has greatly enhanced the potential for fish karyotyping and comparative cytogenetics. The use of FISH allowed for significant advances in our understanding of the fish genome architecture, especially when applied to the study of the repetitive component of the genome, that is generally underestimated in the bioinformatic assembly. Here we provide a step-by-step protocol for FISH of repeated sequences onto chromosomes of fish species.

Key words Repetitive sequence, Genome architecture, Fish, Ribosomal gene, Transposable element, Fluorescence microscopy

1 Introduction

Fluorescence in situ hybridization (FISH) is a technique developed in the late sixties for the physical detection of nucleic acids in their cell environment. It consists in hybridizing a labeled probe to its complementary DNA sequence on the chromosomes fixed on slides, and visualizing in situ the hybridization signals through fluorescence microscopy analysis. A major advantage of this technique resides in its ability to provide degree of resolution between molecular analysis and traditional cytogenetic investigation, while also retaining information at the single cell level.

Like most techniques in molecular biology and cytogenetics, FISH was initially developed in the field of mammalian research [1] and then gained widespread recognition in nonmammalian vertebrates, invertebrates and plants as in situ mapping technique to support increased karyotype resolution, comparative cytogenetics, as well as large-scale genome mapping and sequencing [2–7].

In fish, poor DNA compartmentalization resulting in weak chromosome banding [8] poses a major challenge to cytogenetic

Cinzia Verde and Daniela Giordano (eds.), *Marine Genomics: Methods and Protocols*,
Methods in Molecular Biology, vol. 2498, https://doi.org/10.1007/978-1-0716-2313-8_21,

research, and the opportunity to map genes and sequences on the chromosomes by FISH greatly enhanced the potential for fish karyotyping and comparative cytogenetics [9]. In particular, the in situ mapping of the repetitive component the genome provides relevant insights in the structure and organization of the genomes [10–12]. Although making up a major portion of the eukaryotic genomes [13], repetitive sequences are collapsed in bioinformatic assemblies, thus resulting masked and generally underestimated. The distribution of these sequence, as derived by FISH, integrates the sequence data, and allows to overcome the challenges of bioinformatic analyses, and this combined bioinformatic and physical mapping approach has contributed to the success of various fish whole genome deciphering projects (e.g., [14, 15]) or allowed the reconstruct of genomic regions of interest at high resolution [16, 17].

The method has been applied in fish chromosomes to investigate the position of repeated sequences of different types including multigene families (e.g., ribosomal genes, coding for the ribosomal RNA molecules), noncoding structural components of the chromosomes (e.g., repeated sequences found in the telomeres and centromeres), and dispersed repeats (e.g., transposable elements). The ribosomal genes are among the repetitive sequences most widely investigated by FISH (reviewed in [18–21]). The use of this technique allowed to step forward from the cytochemical approach, based on silver nitrate staining of the proteins associated to the nucleolar organizing regions (NORs), to a direct and more sensitive detection of this portion of the genome, opening the way to consistent comparative studies at interspecific and intraspecific level. The detection of large numbers of copies of the specific sequences at centromeres and telomeres by FISH supports a finer resolution of the chromosome structure, and holds potential to provide clues in support of chromosomal rearrangement hypotheses (reviewed in [22]). In situ hybridization has been fruitfully applied to characterize the distribution of transposable elements (TEs) in the landscape of the genome, leading to a better resolved fish genomic structure, and improved understanding of the evolutionary dynamics and plasticity of the genomes (e.g., [23, 24]).

In this chapter, we provide a step-by step protocol for FISH of repeated sequences onto the chromosomes of fish species. The text includes details on the preparation of spreads on the slide, probe labeling, hybridization, and washing procedures adapted for probes targeting repetitive DNAs. In fish the procedure is successfully used for the physical mapping of ribosomal genes, noncoding structural components of the chromosomes, and transposable elements (Fig. 1).

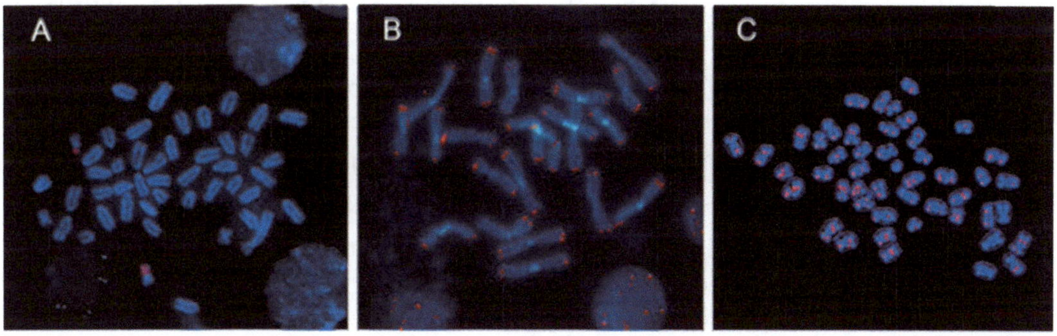

Fig. 1 Examples of FISH physical mapping of repetitive sequences on the chromosomes of fish species. (**a**) Physical mapping of major ribosomal genes on the chromosomes of the icefish *Chionodraco hamatus.* A 28S rDNA fragment prepared according to [25] was used as the probe. (**b**) Localization of telomeric sequences on the chromosomes of *Notothenia coriiceps.* The telomere detection probe was prepared by PCR in absence of template using primers $(TTAGGG)_5$ and $(CCCTAA)_5$ following [26], and DIG-labeled by PCR. (**c**) In situ mapping of Long Interspersed Nuclear Elements (LINEs) sequences on the chromosomes of *Pleuragramma antarctica.* The probe for FISH was obtained by PCR amplification from the genomic DNA of the Antarctic notothenioid fish *Dissostichus mawsoni*, cloned into a plasmid vector and nick translation labeled thereafter [27]

2 Materials

2.1 Preparation of Chromosome Spreads

1. Cell suspension in Carnoy fixative.
2. Cold Carnoy fixative (*see* **Note 1**): 3:1 methanol–glacial acetic acid (*see* **Note 2**).
3. Slide washing solution: 1% hydrochloric acid (HCl) 1 N in 96% ethanol (EtOH).
4. 10 mL round bottom centrifuge tubes.
5. Super frost slides.
6. Plastic Pasteur pipette.
7. Glass staining jar.
8. Paper towel.

2.2 Preparation of a Labeled Probe

1. DNA probe (1 μg) (*see* **Note 3**).
2. Biotin- or Digoxigenin-Nick Translation Mix.
3. Sterile double distilled water.
4. 0.5 M EDTA.
5. Ice-cold EtOH absolute.
6. 3 M sodium acetate (NaOAc), pH 5.2.
7. 70% EtOH.
8. Refrigerated microcentrifuge.
9. Standard agarose.

10. 1× TAE buffer: 40 mM Tris, 20 mM acetic acid, 1 mM EDTA, pH 8.
11. 6× Gel-loading buffer: 0.25 g bromophenol blue, 3 mL glycerol, 7 mL double distilled water (*see* **Note 4**).
12. DNA size standard (*see* **Note 5**).
13. Ethidium bromide (EtBr) (*see* **Note 6**).
14. Equipment for agarose gel electrophoresis with power supply device capable of 100 V.
15. Microwave oven.
16. UV transilluminator (or blue light LED instrument).

2.3 Fluorescence In Situ Hybridization (FISH)

1. 70% formamide in 2× Saline–Sodium Citrate Buffer (2×SSC): 0.3 M sodium chloride, 0.03 M sodium citrate, pH 7.0 (*see* **Note 7**).
2. Ice-cold 70% EtOH.
3. 80% EtOH.
4. 100% EtOH.
5. Hybridization solution: 0.2 μg of labeled DNA probe in Hybridization Buffer (50% formamide, 2× SSC, 50 mM potassium phosphate buffer pH 7, 0.1% SDS, 10% dextran sulphate) at a concentration of 10–20 ng/μL (*see* **Note 8**).
6. 50% formamide in 2×SSC, pH 7.0 (*see* **Note 7**).
7. 2×SSC, pH 7.0.
8. 4×SSC–Tween 20–0.6 sodium chloride, 0.006 sodium citrate, 0.5% Tween 20 v/v, pH 7.0.
9. 1/200 (v/v) Streptavidin-Cy3 (Amersham Bioscience) or anti-digoxigenin-rhodamine solution in Antibody buffer (5% bovine serum albumin, BSA, 4×SSC, 0.001× Tween 20).
10. 4′,6-Diamidine-2′-phenylindole dihydrochloride (DAPI) 1 μg/mL solution in 2×SSC.
11. Antifading solution: *p*-Phenylenediamine and glycerol in PBS: 137 mM NaCl, 2.7 mM KCl, 10 mM Na_2HPO_4, 1.8 mM KH_2PO_4.
12. Circulating water bath.
13. Thermoblock.
14. Moist chamber (*see* **Note 9**).
15. Coverslip (24 × 40).
16. Rubber cement.
17. Incubator at 37 °C.

3 Methods

3.1 Preparation of the Chromosome Spreads on Slides

1. Clean the slides in a solution made up of 100 mL of 96% EtOH and 1 mL of HCl and let them stay for about 10 min.
2. Transfer the cell suspension to 10 mL round bottom centrifuge tubes.
3. Centrifuge the cell suspension at 1000 × *g* for 10 min.
4. Discard the supernatant and resuspend the cells in a small volume of freshly prepared Carnoy fixative.
5. Carefully dry the slides with paper towel and drop the cell suspension onto the slides with a plastic Pasteur pipette.
6. Air-dry the slides and check the density of the nuclei on the slide under phase-contrast microscope. Good preparations should have well separated but not sparse nuclei (*see* **Note 10**).
7. Once verified the good nuclei concentration further slides could be prepared and dried at 37 °C overnight.
8. Select slides for FISH by analyzing the slides in phase-contrast under the microscope with regard to the number and quality of metaphases, and occurrence of metaphases in the central area of the slide (*see* **Note 11**).
9. Selected slides can be stored at −20 °C (*see* **Note 12**).

3.2 Preparation of Indirectly Labeled DNA Probe

The preparation of a well-labeled probe is key for the success of the FISH. We perform indirect labeling of probes with biotin or digoxigenin by nick translation. We recommend the use of commercial nick translation kits that, in our experience, results in the generation of properly labeled probes. The size of fragments generated by the nick translation procedure needs to be checked by gel electrophoresis after the labeling reaction. Optimal fragments for FISH should be in the range 300 to 700 bp. A DNA purification step after the indirect probe labeling is recommended to remove the un-incorporated dNTPs and reduce undesired background after hybridization.

1. Prepare 20 μL labeling reaction mix on ice.
2. Incubate at 15 °C for 90 min.
3. Prepare a 1% agarose gel (*see* **Note 13**) for electrophoresis.
4. Place the labeling reaction on ice.
5. Mix 3 μL of labeled DNA probe with sterile double distilled water, and 6× gel-loading buffer. Load the mixture to the gel and apply a voltage of 100 V. Let the DNA samples to migrate for 45–60 min. Examine the gel by UV or blue light to check the labeled probe fragmentation (*see* **Note 14**).

6. Stop the nick translation with 1 μL of 0.5 M EDTA.
7. Precipitate the probe by adding 2 v/v ice-cold ethanol absolute and 0.1 v/v 3 M NaOAc.
8. Incubate at −20 °C for 60 min or at −80 °C for 15 min.
9. Centrifuge at 14,000 × *g* for 30 min at 4 °C and discard the supernatant.
10. Fill the tube half way with 70% EtOH and centrifuge at 14,000 × *g* for 5 min at 4 °C.
11. Repeat **step 10**.
12. Air-dry the pellet.

3.3 FISH

1. Fill a Coplin jar with 70% formamide solution and prewarm at 68–70 °C (*see* **Note 15**).
2. Denature chromosomes on the slides by incubation in the warm 70% formamide solution for 15 s to 2 min (optimal time to be adjusted, *see* **Note 16**).
3. Stop the chromosome denaturation by immersion in ice-cold 70% EtOH for 5 min.
4. Dehydrate in ethanol series 70%–80%–100% (5 min each) and air-dry.
5. Denature the labeled probe by incubating the hybridization solution 10 min at 75 °C, and stop denaturation by placing the tube on ice for a few minutes.
6. Apply 20 μL of denatured probe solution onto the denatured-dehydrated chromosomes, cover with a 24 × 40 coverslip, and seal with rubber cement at the border of the coverslip.
7. Put the slides horizontal in a moist chamber (the slides should not be in contact with the tissue) and incubate at 37 °C overnight.
8. Carefully remove the rubber cement and coverslip with forceps and perform the following 5 min posthybridization washes at 45 °C: one wash in 2×SSC, two washes in 50% formamide, two washes in 2×SSC, and a final wash in 4×SSC–Tween 20.
9. Apply 150 μL of Streptavidin-Cy3 or anti-digoxigenin-rhodamine solution, cover with a 24 × 60 coverslip and incubate 45 min at 37 °C in a moist chamber.
10. Carefully remove the coverslip and wash the slides three times in 4×SSC–Tween 20 at 42 °C (5 min each), in the dark.
11. Transfer the slides in the DAPI solution and incubate 10 min, in the dark (*see* **Note 17**).

12. Put 35 μL of antifading solution on each slide, put a 24 × 50 coverslip and press firmly the slide between two sheets of filter paper to remove the excess of antifade solution and air bubbles that might be present under the coverslip (*see* **Note 17**).
13. Store the slides at 4 °C in horizontal position until the microscope analysis (*see* **Note 18**).

3.4 Image Analysis

For image acquisition, the typical workstation for analysis comprises a fluorescence microscope equipped with optical filters for DAPI, fluorescein isothiocyanate (FITC), and rhodamine; charge-coupled device (CCD) or complementary metal oxide semiconductor (CMOS) black and white digital camera; and an image acquisition software. Micro-Manager is a rather convenient open source microscopy software that allows to control automated microscopes, image acquisition and processing, together with the ImageJ open source software. If nonspecific binding of the probes is observed, the stringency of washing conditions can be increased by lowering SSC concentrations or by increasing the washing temperatures.

4 Notes

1. Carnoy's fixative should be prepared freshly.
2. Ethanol can also be used instead of methanol for preparing the fixative.
3. We usually clone repeated sequences in plasmid vectors and use pure plasmid DNA for nick translation. Good results are achieved with midi- and maxi-prep sized isolations. Backbones of plasmids do not need to be removed given the limited homology fish sequences. Usually, probes for telomeric and centromeric repeats are not can cloned into vectors, and can be directly labeled by PCR using a PCR labeling kit following the manufacturer's instructions.
4. Premade gel-loading buffers are commercially available. The DNA Gel-Loading Dye by Thermo Fisher Scientific allows two colors tracking of the DNA migration.
5. The DNA size standard should be chosen based on the expected size of the target DNA. As a routine, for checking the result of the nick-translation labeling procedure we use the 1 kb DNA Ladder or Lambda DNA Hind III Digest.
6. EtBr has historically been used as a fluorescent tag to detect DNA on agarose gel, however, in recent years, alternative dyes have been proposed as being less dangerous (noncarcinogenic and less mutagenic) than EtBr. We had good experience in using the Midori Green Advance stain, that can be visualized

by regular UV transilluminator as well as with LED based instruments.

7. To minimize pH changes, deionized formamide solutions should be stored in aliquots at −20 °C.
8. The hybridization solution can be stored at −20 °C.
9. We typically use plastic storage containers with a lid containing a tissue imbibed with distilled water at the bottom.
10. As an alternative method sometimes used to maximize the quality of the metaphase plates and reduce the number of overlapping of chromosome, the cell suspension is dropped on clean slides placed above a support on a water bath at 37 °C, and allowed to dry.
11. In fish species, cell cultures are uncommon and, typically, the chromosomes of fish species are obtained by direct preparation from hematopoietic tissues (mostly head kidney and spleen). As a consequence, the quality of the obtained chromosome suspensions varies among species, and even among individuals of the same species. The selection of high-quality slides for FISH is a critical step.
12. Once frozen the slides can be stored for years. Avoid repeated thawing and freezing.
13. Melt the agarose powder in TAE buffer in a microwave, let the molten gel cool to about 37 °C, and add EtBr to a final concentration of 0.5 μg/mL (or alternative DNA stain following the manufacturer's instructions). Pour the warm agarose solution in the mold and allow the gel to set. Cover the gel with TAE buffer 1× and remove the comb.
14. The nick translation labeled probe should appear as a smear of fragments between 300 and 500 bp, while unincorporated labeled dUTP appears as a band between 50 and 100 bp.
15. Be sure that the formamide solution has reached the required temperature, sharp.
16. Denaturation of chromosomal DNA is an essential step in FISH experiments, however the denaturation behavior of metaphase chromosome DNA differs from the pure DNA melting properties, and is associated to a number of variables. In our experience, aged chromosomes require longer denaturation time in respect to freshly prepared chromosomes. Additionally, a degree of variability in the duration of denaturation occur among species, related to the intrinsic structural characteristics of chromosomes in different taxa.
17. DAPI can be mixed with antifade as follows: 12 μL of a stock DAPI solution (100 μg/mL in 2×SSC) in 1 mL antifade. The

DAPI antifade solution can be stored in aliquots at −20 °C and directly spread on the slide before putting the coverslip.

18. Coverslips can be sealed with nail polish, and slides can be stored in the dark in a closed box at 4 °C for several weeks.

References

1. Pardue ML, Gall JG (1969) Molecular hybridization of radioactive DNA to the DNA of cytological preparations. Proc Natl Acad Sci U S A 64:600–604. https://doi.org/10.1073/pnas.64.2.600
2. Phillips RB (2007) Application of fluorescence *in situ* hybridization (FISH) to genome mapping in fishes. In: Pisano E, Ozouf-Costaz-C, Foresti F, Kapoor BG (eds) Fish cytogenetics. Science Publishers, Enfield, pp 455–471
3. Phillips RB, Keatley KA, Morasch MR et al (2009) Assignment of Atlantic salmon (*Salmo salar*) linkage groups to specific chromosomes: conservation of large syntenic blocks corresponding to whole chromosome arms in rainbow trout (*Oncorhynchus mykiss*). BMC Genet 10(1):1–11. https://doi.org/10.1186/1471-2156-10-46
4. Zhang P, Priebe B (2009) FISH on plant chromosomes. In: Fluorescence in situ hybridization (FISH)—application guide. Springer, Berlin, Heidelberg, pp 365–394. https://doi.org/10.1007/978-3-540-70581-9_32
5. Solinhac R, Leroux S, Galkina S et al (2010) Integrative mapping analysis of chicken microchromosome 16 organization. BMC Genomics 11(1):1–12. https://doi.org/10.1186/1471-2164-11-616
6. Abbasi FM, Khan MT, Perveen F et al (2010) Historical perspective of *in situ* hybridization for the analysis of genomic constitution of plants. Afr J Biotechnol 9(54):9142–9147
7. Jiang J (2019) Fluorescence *in situ* hybridization in plants: recent developments and future applications. Chrom Res 27(3):153–165. https://doi.org/10.1007/s10577-019-09607-z
8. Fortes GG, Bouza C, Viñas A et al (2007) Diversity in isochore structure and chromosome banding in fish. In: Pisano E, Ozouf-Costaz C, Foresti F, Kapoor BG (eds) Fish cytogenetics. Science Publishers, Enfield, pp 405–420
9. Pisano E, Ozouf-Costaz C, Foresti F, Kapoor BG (eds) (2007) Fish cytogenetics. Science Publishers, Enfield
10. Mazzuchelli J, Martins C (2009) Genomic organization of repetitive DNAs in the cichlid fish *Astronotus ocellatus*. Genetica 136(3):461–469. https://doi.org/10.1007/s10709-008-9346-7
11. Reichwald K, Lauber C, Nanda I et al (2009) High tandem repeat content in the genome of the short-lived annual fish *Nothobranchius furzeri*: a new vertebrate model for aging research. Genome Biol 10(2):1–17. https://doi.org/10.1186/gb-2009-10-2-r16
12. Majtánová Z, Moy KG, Unmack PJ et al (2019) Characterization of the karyotype and accumulation of repetitive sequences in Australian Darling hardyhead *Craterocephalus amniculus* (Atheriniformes, Teleostei). PeerJ 7:e7347. https://doi.org/10.7717/peerj.7347
13. Biscotti MA, Olmo E, Heslop-Harrison JS (2015) Repetitive DNA in eukaryotic genomes. Chromosom Res 23:415–420. https://doi.org/10.1007/s10577-015-9499-z
14. Crollius HR, Jaillon O, Dasilva C et al (2000) Characterization and repeat analysis of the compact genome of the freshwater pufferfish *Tetraodon nigroviridis*. Genome Res 10(7):939–949. https://doi.org/10.1101/gr.10.7.939
15. Freeman JL, Adeniyi A, Banerjee R et al (2007) Definition of the zebrafish genome using flow cytometry and cytogenetic mapping. BMC Genomics 8(1):1–10. https://doi.org/10.1186/1471-2164-8-195
16. Nicodemus-Johnson J, Silic S, Ghigliotti L et al (2011) Assembly of the antifreeze glycoprotein/trypsinogen-like protease genomic locus in the Antarctic toothfish *Dissostichus mawsoni* (Norman). Genomics 98(3):194–201. https://doi.org/10.1016/j.ygeno.2011.06.002
17. Zhuang X, Murphy KR, Ghigliotti L et al (2018) Reconstruction of the repetitive antifreeze glycoprotein genomic loci in the cold-water gadids *Boreogadus saida* and *Microgadus tomcod*. Mar Genomics 39:73–84. https://doi.org/10.1016/j.margen.2018.02.003
18. Pisano E, Ghigliotti L (2009) Ribosomal genes in notothenioid fishes: focus on the chromosomal organization. Mar Genomics 2:75–80. https://doi.org/10.1016/j.margen.2009.03.006

19. Gornung E (2013) Twenty years of physical mapping of major ribosomal RNA genes across the teleosts: a review of research. Cytogenet Genome Res 141(2–3):90–102. https://doi.org/10.1159/000354832
20. Rebordinos L, Cross I, Merlo A (2013) High evolutionary dynamism in 5S rDNA of fish: state of the art. Cytogenet Genome Res 141(2–3):103–113. https://doi.org/10.1159/000354871
21. Sochorová J, Garcia S, Gálvez F et al (2018) Evolutionary trends in animal ribosomal DNA loci: introduction to a new online database. Chromosoma 127(1):141–150. https://doi.org/10.1007/s00412-017-0651-8
22. Ocalewicz K (2012) Genomic distribution of telomeric DNA sequences—what do we learn from fish about telomere evolution. In: Li B (ed) Reviews on selected topics of telomere biology. InTech, Rijeka, pp 271–294
23. Auvinet J, Graça P, Ghigliotti L, Pisano E et al (2019) Insertion hot spots of DIRS1 retrotransposon and chromosomal diversifications among the Antarctic Teleosts Nototheniidae. Int J Mol Sci 20:701. https://doi.org/10.3390/ijms20030701
24. Schemberger MO, Nascimento VD, Coan R et al (2019) DNA transposon invasion and microsatellite accumulation guide W chromosome differentiation in a Neotropical fish genome. Chromosoma 128(4):547–560. https://doi.org/10.1007/s00412-019-00721-9
25. Ghigliotti L, Mazzei F, Ozouf-Costaz C et al (2007) The two giant sister species of the Southern Ocean, *Dissostichus eleginoides* and *Dissostichus mawsoni*, differ in karyotype and chromosomal pattern of ribosomal RNA genes. Polar Biol 30:625–634. https://doi.org/10.1007/s00300-006-0222-6
26. Ijdo JW, Wells RA, Baldini A et al (1991) Improved telomere detection using a telomere repeat probe (TTAGGG) n generated by PCR. Nucleic Acids Res 19(17):4780
27. Ghigliotti L, Cheng CHC, Ozouf-Costaz C et al (2020) Cytogenetic characterization of the Antarctic silverfish *Pleuragramma antarctica* (Boulenger 1902) through analysis of mitotic chromosomes from early larvae. Mar Genomics 52:100737

Chapter 22

Functional Genomics of Fish Erythrocytes

Mikko Nikinmaa and Amélie Crespel

Abstract

Fish red blood cells contain nucleus and therefore carry on active transcription throughout their life. As the steady state level of mRNA depends both on the production and breakdown of the mRNA, it is important to pay attention to sampling, cell separation, mRNA extraction, and transcript stability to conduct a comprehensive erythrocyte genomics study. In addition, particular attention should be made to tie the transcriptional changes to corresponding protein activities, as only those would impact cellular functions. Here, we thus describe the different steps to conduct comprehensive functional genomics studies with fish erythrocytes.

Key words Blood cell, Transcription, mRNA, Sequencing, Quantitative PCR, Stability

1 Introduction

Blood consists of different white blood cells (in this group we also include thrombocytes) and erythrocytes. Contrary to nonnucleated mammalian erythrocytes, fish red blood cells contain nucleus and carry on active transcription throughout their life in circulation. However, their activity is markedly smaller than that of white blood cells, whereby even though the proportion of leukocytes is only 1–5% of blood cells, they can contribute to more than 50% of the total oxygen consumption of blood [1]. Although the difference in transcriptional activity between fish leukocytes and erythrocytes has not been evaluated, it is probable that it is similar to that of oxygen consumption. In addition, the age of circulating erythrocytes varies, and both the transcriptional activity [2] and protein production [3] are dependent on erythrocyte age. Furthermore, the steady-state level of mRNA, given by RNA sequencing or microarray methodology, depends both on the production and breakdown of the mRNA [4]. Consequently, in a comprehensive erythrocyte genomics study one needs to pay attention to sampling and the treatment of samples. It is also important to evaluate the transcript stability. Finally, since transcriptional effects are meaningful with

Cinzia Verde and Daniela Giordano (eds.), *Marine Genomics: Methods and Protocols*,
Methods in Molecular Biology, vol. 2498, https://doi.org/10.1007/978-1-0716-2313-8_22,

regard to cellular functions only if they can be tied to protein functions, any functional genomics study should, in our opinion, include a component, which associates transcriptional change to protein levels and activities in phenomena, which the researcher is interested in. In the following, we present methodology for carrying out functional genomics studies with fish erythrocytes.

2 Materials

All the buffer and saline solutions can be stored for at least a month in refrigerator.

2.1 Blood Sampling

1. Blood Capillary.
2. Anaesthetic Solution: Ethyl 3-Aminobenzoate Methane Sulfonate (MS 222) 100 mg/L.
3. Heparinized syringe.

2.2 Cell Separation

1. Hematocrit centrifuge.
2. Physiological saline solution for rainbow trout blood [5]:128 mM NaCl, 3 mM KCl, 1.5 mM $MgCl_2$, 1.5 mM $CaCI_2$, and 20 mM Tris-HCl buffer.
3. Centrifuge with angled rotor at 45° going at least at 16,000 × *g*.
4. Polypropylene tubes: length 47 mm, diameter 4 mm, volume 0.5 mL.

2.3 RNA Extraction, Quantity and Quality Determination

1. TissueLyser for homogenization.
2. Stainless steel beads (diameter 2 mm, AISI 420).
3. NucleoZol (Macherey-Nagel).
4. Nanodrop spectrophotometer.
5. 1% agarose gel: The agarose gel is prepared by dissolving 1 g of agarose in 100 mL of 1× TAE buffer (Tris-acetate-EDTA buffer) with 5 μL of SYBR® Safe (DNA gel stain). The agarose solution is then heated for about 90 s in a microwave taking care of not boiling the solution. All the agarose needs to be dissolved after the 90 s of heat or additional 15 s of heat in the microwave should be added until the agarose is completely dissolved. The agarose solution is then cooled down for 5–10 min at room temperature after which the solution is transferred into the cassette with the well gel comb immediately inserted. The agarose solution will mould in about 30 min or until it forms a gel.
6. 1 kb ladder.

7. Gel electrophoresis system and power supply.
8. Bioanalyzer (Agilent systems).

2.4 Determining the Total Transcriptome: RNA Sequencing

1. Library preparation kit, for example, Illumina TruSeqStranded mRNA Sample Preparation Kit 2.4.2. HiSeq instrument for sequencing.
2. Bioinformatic analyses software: Bcl2fastq2 software; FastQC (www.bioinformatics.babraham.ac.uk/projects/fastqc/); Trimmomatic [6]; TopHat [7]; edgeR R/Bioconductor R package; subreads R package [8]; DEGseq R package [9]; BLAST2GO software [10]; BLASTx algorithm [11] in NCBI database (National Center for Biotechnology Information).

2.5 Picking Genes of Interest: Quantitative PCR

1. Primer3Plus software online.
2. Working solution of primers is prepared in the following way: first, a 10 mM stock solution is prepared by adding 10 volumes of water (in μL) to the quantity of primer (in nmol) directly in the primer tube. As an example, if the primer quantity is 20 nmol, 200 μL of water are added, and the solution is vortexed. Then the stock solution of the primer is diluted to get a working solution of 10 nM, that is, 10 μL of the stock solution is transferred into a new tube and 90 μL of water are added. The stock and working solutions are then stored at −20 °C until use.
3. cDNA Reverse Transcription Kit.
4. Evagreen® qPCR dye.
5. Bio-Rad CFX96.

2.6 Nuclear Run-On Assay Solutions

1. PBS solution: 140 mM NaCl, 2.7 mM KCl, 8.1 mM Na_2HPO_4, 1.5 mM KH_2PO_4.
2. Lysis buffer: 10 mM Tris-HCl pH 7.4, 3 mM $MgCl_2$, 10 mM NaCl, 1% Nonidet® P-40.
3. Glycerol storage buffer: 50 mM Tris-HCl pH 8.0, 30% glycerol, 2 mM $MgCl_2$, 0.1 mM EDTA.
4. Transcription buffer: 10 mM Tris-Cl pH 8.0, 5 mM $MgCl_2$, 300 mM KCl containing 10 μL of a 10× Biotin RNA labeling Mix (containing 3.5 mM biotin-16-UTP, 10 mM ATP, 10 mM CTP, 10 mM GTP).
5. Stopping reagents: 250 mM $CaCl_2$; 6 units of DNase I.
6. TriReagent LS.
7. Streptavidin-coated magnetic particles.
8. Binding buffer: 10 mM Tris-HCl, 1 mM EDTA, 100 mM NaCl, pH 7.5.

9. Washing buffer: 10 mM Tris-HCl, 1 mM EDTA, 1 M NaCl, pH 7.5.

2.7 mRNA Stability

1. Actinomycin D: 2.5 μg/mL.

3 Methods

3.1 Blood Sampling

The method for obtaining blood samples depends on the size of fish. The blood volume of an average teleost fish is about 3%, and since normally less than 1/3 of blood can be sampled, the amount of blood that can be obtained from approximately 5 g fish is 50 μL.

1. In the smallest fish a common method for blood sampling is cutting the tail and allowing blood to flow directly from the caudal vessel to a hematocrit capillary (Fig. 1). Although the method is crude, it suffices for a genomics study, where one is only interested in transcription in erythrocytes (*see* **Note 1**).
2. From fish over 10 g blood can be sampled with heparinized syringe from caudal vessel (Fig. 1), which is the best site for blood sampling in less than 100 g fish (and any flat fish). For genomic studies, fish are anaesthetized using, for example, MS 222, until fish stop to move in 2–3 min (*see* **Note 2**). The caudal vessels (both artery and vein) are situated adjacent to spine, and thus when one feels the spine, the needle can slip either above or below, and blood can be obtained.
3. In fish over 100 g with suitable shape also cardiac puncture is a good method for obtaining blood. Usually the fish is stunned with a blow on the head, and heart punctured between (and at the front end of) pectoral fins (Fig. 1).

3.2 Cell Separation

1. After blood sampling using hematocrit capillary, white and red blood cells are separated by centrifugation of the capillary tubes in a hematocrit centrifuge (usually over 2 min and 10,000 × *g*) and cutting the tubes well below the buffy coat. The weakness of obtaining blood samples from small fish is that it is impossible to separate the erythrocytes according to age, although there are marked differences in transcription between young and old erythrocytes [2].
2. With an adequate sample size (more than 0.1 mL blood) the erythrocytes can be separated to age fractions. This is done with density centrifugation, as the density of erythrocytes increases with age and leukocytes have lower density than erythrocytes. The increase in density is mainly due to the increase in the amount of haemoglobin in the cells with the age [12]. The erythrocytes are washed twice with physiological saline for fish, whereafter the cells are centrifuged to the bottom and an

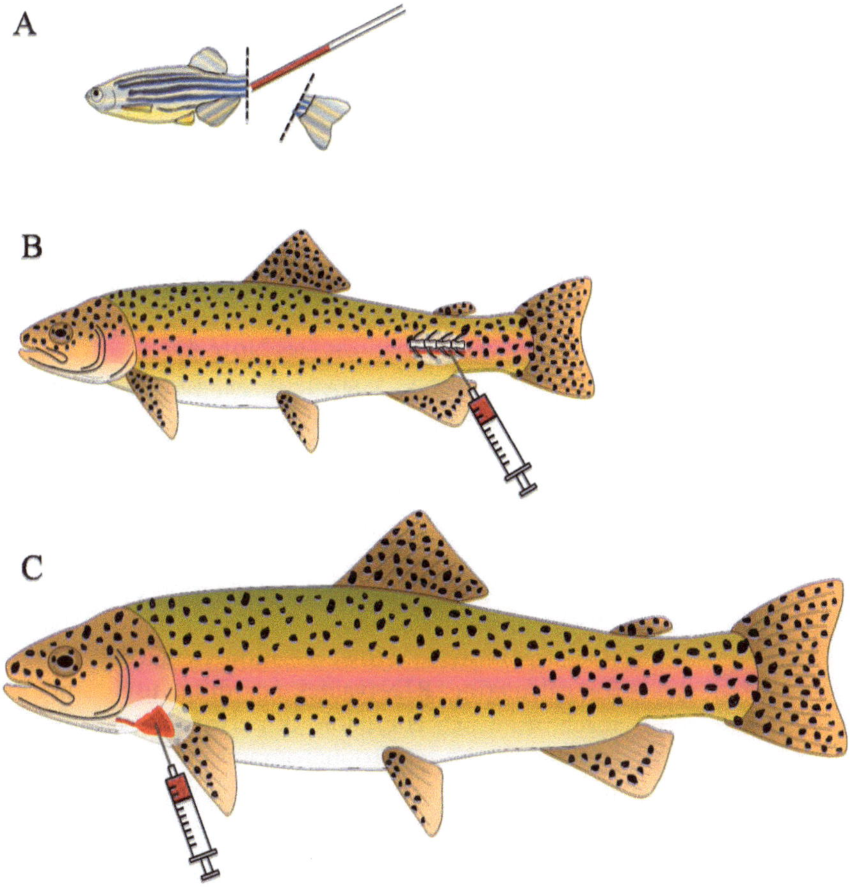

Fig. 1 Blood sampling depending on the size of the fish. (**a**) Small fish: blood sampling using a hematocrit capillary directly on the caudal vessel after cutting the tail of the fish. (**b**) Fish over 10 g: blood samples drawn from the caudal vessel using a heparinized syringe; once feeling the spine of the fish, blood can be obtained when needle slips either above or below the spine. (**c**) Fish over 100 g: blood sampling via cardiac puncture, the heart being situated between and at the front end of the pectoral fins

adequate volume of saline left on top of the erythrocyte fraction to give the hematocrit of ~80%. The erythrocyte suspension is then transferred into polypropylene tubes. Tubes are centrifuged in a fixed-angle rotor (45°) at 16,000 × *g* and 10 °C for 30 min. The tubes are then cut to give different age fractions. The topmost part contains the youngest and the bottommost part the oldest erythrocytes. To compare youngest and oldest erythrocytes only these two fractions (uppermost and bottommost 50 μL fraction) are retained, but depending on the purpose of the study one can include many intermediate fractions, as for example Speckner et al. [12]. The fractions are then transferred into new tubes, diluted with 100 μL saline solution and frozen at −80 °C.

3.3 RNA Extraction, Quantity and Quality Determination

1. Erythrocyte suspensions are homogenized (cells lysed) using TissueLyser with two stainless steel beads for 2 × 60 s at 30 Hz. From the homogenate RNA is isolated using Nucleo-Zol (Macherey-Nagel) according to the manufacturer's instructions (*see* **Note 3**).
2. The quality and quantity of the RNA obtained is checked using Nanodrop. One to two microliter of sample is added onto the Nanodrop membrane cell. Pure RNA has absorbance ratios at 260/280 nm and 260/230 nm around 2 (range between 1.9 and 2.1), and samples with ratios below 1.8 should not be used in further determinations (*see* **Note 4**).
3. The integrity of the RNA is checked by electrophoresis on 1% agarose gel. Once the 1% agarose gel is ready, 5 μL of 1 kb ladder and 5 μL of RNA samples with 1 μL of dye can be loaded into each different well. The gel is then run at 95 V for 1 h and UV light is used to see the migration at the end of the run. From the electropherogram bands for the 28S and the 18S are detected. As the simplest determination of RNA integrity, the 28S band should be twice as bright as the 18S band for intact RNA (*see* **Note 5**). However, more detailed electropherograms for calculating RIN values are done using the Bioanalyzer machine, which is capable of performing electrophoresis and generating the electropherograms; the algorithm calculating RIN is proprietary, whereby the RIN values necessarily refer to Agilent systems. If Bioanalyzer is available, the RIN values should be higher than 8.7 (max 10) for the RNA used in downstream analyses.

3.4 Determining the Total Transcriptome: RNA Sequencing

1. Sequencing libraries are usually prepared from the supplied RNA by commercial providers, using, for example, the Illumina TruSeqStranded mRNA Sample Preparation Kit. Sequencing is then done, for example, with a HiSeq 3000 instrument using single-end sequencing chemistry with 50 bp read length.
2. The sequence data then undergoes bioinformatic analyses. Base calling on the reads can be done using the Bcl2fastq2 software. The quality control of raw sequencing reads is performed with FastQC, and adapters and low quality bases are trimmed by Trimmomatic [6]. The read alignment is performed against the reference genome using TopHat [7]. The used reference genome depends on the species studied. There are already several fish genome sequences at different levels of resolution (*see* **Note 6**). The number of uniquely aligned reads can vary between 2 and 7 M per sample. The sample correlation values (Spearman's metrics) among replicates need to be above 0.9 for all samples. The gene-wise read counts are normalized using

the TMM normalization method of the edgeR R/Bioconductor R package before further statistical testing.

3. Differentially expressed genes (DEGs) are selected based on a fold change >2 and an FDR (Benjamini–Hochberg false-discovery-rate, adjusting *P* values for multiple hypothesis testing) <0.01. Sample pairing (for example young and old erythrocyte age classes of one individual) can be taken into account when building the linear model for statistical testing. The association between reads and known genes and the number of reads associated with each gene can be assessed using the subreads package [8]. Differential expression between groups is analyzed using MA-plots as implemented in DEGseq R package [9] in which the $\log_2$ fold change of each gene is plotted against the average $\log_2$ counts per million. The BLAST2GO software [10] is then used to predict Gene Ontology (GO) terms for all statistically significant DEGs (FDR < 0.01) and the most abundant genes. All pseudogenes (labeled with NA) and all RNAs can be removed from the list of DEGs, and the remaining list can subsequently be searched against the nonredundant NCBI protein database (NR database) using the BLASTx algorithm [11] with an *E*-value threshold of 10^{-5}, and a maximum of 20 "hit" sequences per query retained. GO annotations can be simplified to a smaller set of high-level GO terms using GO slim function implemented in BLAST2GO.

3.5 Picking Genes of Interest: Quantitative PCR

1. When one chooses the genes of interest, primers must be generated for them, so that they can be picked up by PCR. Primers can conveniently be designed using Primer3Plus software online: after copying the sequence of the gene of interest from NCBI into the software, primer sizes between 18–30 nucleotides with a product size range between 75 and 150 nucleotides are selected and the maximum of poly X should be 3. The temperatures of the forward and reverse primers should be similar (within 3 °C) and the CG content should be between 40 and 60% to favor binding.
2. The reverse transcription of the RNA into cDNA is done by using high capacity cDNA Reverse Transcription Kit. The reagents from the kit should be thawed on ice. For each RNA sample, a reverse transcription (RT) mix for a reaction to produce cDNA involving the enzyme reverse transcriptase (RT+) and a reaction without the enzyme to control later on for any DNA contamination in the samples (RT−) (*see* **Note 7**) are prepared (Table 1). The RT+ and RT− master mix for the total number of RNA samples are prepared by adding about 8–10% extra reactions to the mix. For example, for 48 samples, a mix for 52 reactions would be prepared. The different reagents from the kit are added by mixing them gently on ice.

Table 1
Volume of the different reagents in the reverse transcription (RT) mix for 52 reactions (48 samples)

	Volume per reaction (μL)	RT+ mix (μL)	RT− mix (μL)
10× RT buffer	2	104	104
25× dNTPMix (100 mM)	0.8	41.6	41.6
10× RT random primers	2	104	104
Reverse transcriptase	1	52	0
Nuclease free H_2O	0.2	10.4	62.4

Then, 6 μL of RT+ or RT− mix are added to 500 ng of RNA sample already diluted to a volume of 14 μL, the final volume of each reaction is now 20 μL. The tubes are then briefly spun down to favor the contact between the mix and the RNA samples and to eliminate any bubbles. The reverse transcription is performed into a thermocycler according to the instructions in the kit (10 min at 25 °C, 130 min at 37 °C, 5 min at 85 °C, ∞ at 4 °C). After the RT, 2 μL of each RT+ are combined to create a cDNA pool and all the samples are stored at −20 °C until use.

3. Four serial dilutions of the cDNA pool are prepared using five-fold dilution from 1/5 to 1/625 to be able to check the efficiency of the PCR reaction. For 1 volume of cDNA pool, 4 volumes of water are added (i.e., dilution 1/5). Then, 1 volume of the cDNA dilution 1/5 is diluted in 4 volumes of water (i.e., dilution 1/25). Then, 1 volume of cDNA dilution 1/25 is diluted in 4 volumes of water (i.e., dilution 1/125). And finally, 1 volume of cDNA dilution 1/125 is diluted in 4 volume of water (i.e., dilution 1/625). The serial dilutions are stored at −20 °C until use.

4. The gene of interest could then be amplified through quantitative real time PCR using Evagreen® dye (*see* **Note 8**). The dye needs to thaw on ice. For each gene of interest, an Evagreen® qPCR mix (Table 2) is prepared according to the sample number including samples RT+ (in triplicate), controls RT−, negative control (water instead of RT samples, also in triplicate), and the cDNA serial dilutions (in triplicate). Also, it is better to add 10% extra reactions to the mix to be sure to have enough mix for all the samples. For a total of 96 samples, a mix for 105 reactions is thus prepared.

   Then, 16 μL of the qPCR mix is added into the wells of the qPCR plate followed by 4 μL of each of your samples (RT+, RT−, negative controls, cDNA serial dilutions), the final volume of each reaction is 20 μL. The plate is sealed and spun down slowly. The amplification is performed into the Bio-Rad

Table 2
Volume of Evagreen® reagent and gene primers in the qPCR mix for 105 reactions (96 samples)

	Volume per reaction (μL)	qPCR mix (μL)
Evagreen (keep far from light)	10	1050
Primer solution Forward	0.5	52.5
Primer solution Reverse	0.5	52.5
H_2O	5	525

CFX96 following the protocol optimized for the Evagreen® dye, that is, 30 s at 95 °C, 30–45 cycles of 5 s at 95 °C and 5 s at 55–60 °C (depending on the optimal temperature of the primers), and it is important to add a melting curve (65 °C for 5 s followed by an increase of 0.5 °C every 5 s until 95 °C) to check for the quality of the amplification. The DNA denaturation induced by the melting curve should create a single distinct peak of fluorescence at a certain temperature, indicating only one product of amplification. The efficiency of the reaction and primers is then calculated by determining the slope of the cDNA serial dilutions. The efficiency of the reaction should range from 90 to 110% (*see* **Note 9**).

3.6 Rate of Transcription with Nuclear Run-On Assay

The nuclear run-on assay (NRO) is a useful method to quantify transcriptional activity in red blood cells. The method is described in Sambrook and Russell [13]. However, the method is optimized for mammalian cells, and several modifications are needed for studies with temperate fish. First, the RNA labeling temperature during the in vitro transcription is decreased to 20 °C. As the labeling temperature is decreased, the labeling time is prolonged to 1.5 h to achieve sufficient incorporation of the label into the transcript. Second, the radioactive label is replaced by biotin-16-dUTP which enables the determination of transcript using PCR [14].

1. Nuclei for the assay are prepared from approximately 5×10^7 red blood cells (30 μL blood sample). Red blood cells are washed 3 times by resuspension in ice-cold, sterile 1× PBS followed by centrifugation at 500 × *g* for 5 min at 4 °C. The washed pellet is resuspended in 4 mL ice-cold sterile lysis buffer, incubated for 5 min on ice, and nuclei collected by centrifugation (500 × *g*, 4 °C). The nucleus pellet is gently washed with lysis buffer devoid of Nonidet®P-40 and nuclei are collected again by centrifugation. After centrifugation, nuclei are resuspended in 100 μL ice-cold glycerol storage buffer and stored at −80 °C.
2. Frozen nuclei are thawed and one volume of 2 × transcription buffer is added to thawed nuclei. The mixture is incubated for

1.5 h at 20 °C with gentle agitation. The reaction is stopped by adding 6 μL of 250 mM $CaCl_2$ and 6 units of DNase I, and incubated for another 15 min at 20 °C. The nuclear run-on RNA is isolated using TriReagent LS according to the manufacturer's instructions. The resulting biotin-labeled RNA is resuspended in 40 μL of RNase-free water.

3. The biotin-16-UTP-labeled RNA represents the transcript formed in the 1.5 h incubation period. Its amount is evaluated with the help of its binding to magnetic particles (only the biotin-labeled nucleotide-containing RNA binds to the magnetic particles). Streptavidin-coated magnetic particles (40 μL) are washed 3 times in binding buffer and then mixed with an equal volume of biotin-labeled RNA. After incubation for 30 min at room temperature with gentle agitation, particles are washed twice in washing buffer and then eluted in 30 μL of RNase-free water by incubating them for 10 min at 80 °C. The concentration of the RNA samples is measured using a NanoDrop 2000 and nuclear run-on RNA is stored at −80 °C until the determination with PCR.

3.7 mRNA Stability

Stability of mRNAs (transcript half-life) is evaluated by preventing the synthesis of new mRNA, whereby the measurement of mRNA level as a function of time gives the decay rate of the mRNA. Synthesis of new transcripts is blocked by adding actinomycin D. The mRNA level at different time points (2 h intervals for most genes) is determined with quantitative PCR.

3.8 Tying Transcriptional Findings to Protein Function

In our opinion this is the most crucial step of functional genomic studies, since without understanding how transcriptional findings are reflected in functional changes, the findings are virtually meaningless. It is conceivable that there is a transcriptional change without a change in translation or that the transcriptional/translational change occurs as a response to a change in protein activity [15]. Another point that is important to remember is that, for example, the gene ontology terms are very general and, in the absence of appropriate delimiters, biased toward mammals/humans, because most information in gene banks is medically oriented. Because of this, the information in gene banks may be completely irrelevant, if the function of gene product is different in fish and rat. As a good example is the β_3-adrenergic receptor, which is involved in thermoregulation in mammals, but in regulation of ion transport in fish erythrocytes [16, 17].

Consequently, observing gene-level changes is just the beginning of a comprehensive functional genomics study, and should be followed by translational and functional studies [18]. When global transcriptional effects are evaluated by RNA sequencing, a first step should be evaluating how the functions of the products of the transcriptionally most affected genes vary. Thus, time-consuming enzyme activity, ion and gas transport determinations, and so on are needed. What is measured depends both on the transcriptional changes observed and on the interests of the researcher. This being the case, the methodology varies, and cannot be within the scope of this chapter apart from emphasizing the need for carrying out functional studies to relate genomics to protein function.

One significant methodological advancement which is starting to find itself to fish studies and is the result of next generation sequencing concerns the possibility to evaluate translation in real time. Ribosome profiling provides the means to quantify translational activity [19, 20]. In the method the samples are treated with ribonuclease so that all mRNAs except the ones inside ribosomes, that is, currently translated, are broken down. The mRNAs within the ribosomes are then sequenced. Detailed protocols using ribosome profiling including how the ribosomal RNA and the translated mRNAs are separated, are given in the seminal papers by Ingolia and his group [21, 22].

4 Notes

1. At present, there is no information about the possibility that sampling stress could affect the steady-state level of mRNAs. Because of the short duration it is not likely that significant changes can occur even if there is a significant transcriptional change in many genes. However, investigating possible stress-induced changes at the level of transcription and translation is certainly a fruitful direction of study.
2. Alternative methods for anesthesia in fish are described for example by Martins et al. [23].
3. Other protocols and kits could be used for the RNA isolation.
4. Lower ratios indicate the presence of contamination in the RNA samples, either with protein, phenol, or other contaminants that would interfere with the downstream RNA analysis.
5. Not intact RNA would create a smear without any bright bands and should not be used.
6. In our rainbow trout studies, we have used the Atlantic Salmon genome [ICSASG v2 genome; provided by the International Cooperation to Sequence the Atlantic Salmon Genome (ICSASG [24])] as a reference genome because of the higher coverage of the assembly compared to that of the rainbow trout.
7. It is important to have RT− to verify possible DNA contamination in your samples. Such DNA contamination could potentially modify your gene expression. In case of

contamination, the RNA samples should be treated with DNase before being converted again into cDNA.

8. Other dyes may be used and the mix volumes and cycles of the protocol for amplification should be adjusted accordingly. We preferred to use the Evagreen® dye as it is highly accurate and allows faster amplification than most dyes.
9. If the efficiency of the reaction is <90% or >110% there must either be some inhibitors or reagent contaminations in your RNA samples, the primers may not be optimal or some operator mistakes might have happened during the preparation of the qPCR plate (e.g., pipetting issues).

References

1. Wang T, Nielsen OB, Lykkeboe G (1994) The relative contributions of red and white blood cells to whole-blood energy turnover in trout. J Exp Biol 190:43–54
2. Gotting M, Nikinmaa MJ (2017) Transcriptomic analysis of young and old erythrocytes of fish. Front Physiol 8:1024
3. Lund SG, Phillips MCL, Moyes CD et al (2000) The effects of cell ageing on protein synthesis in rainbow trout (Oncorhynchus mykiss) red blood cells. J Exp Biol 203: 2219–2228
4. Gotting M, Nikinmaa M (2017) In vitro study on the regulation of cellular mRNA levels by changes in transcription rate and transcript stability in fish red blood cells. Comp Biochem Physiol B Biochem Mol Biol 213:35–44
5. Nikinmaa M (1983) Adrenergic regulation of haemoglobin oxygen affinity in rainbow trout red cells. J Comp Physiol B 152:67–72
6. Bolger AM, Lohse M, Usadel B (2014) Trimmomatic: a flexible trimmer for Illumina sequence data. Bioinformatics 30:2114–2120
7. Kim D, Pertea G, Trapnell C et al (2013) TopHat2: accurate alignment of transcriptomes in the presence of insertions, deletions and gene fusions. Gen Biol 14:R36
8. Liao Y, Smyth GK, Shi W (2014) featureCounts: an efficient general purpose program for assigning sequence reads to genomic features. Bioinformatics 30:923–930
9. Wang L, Feng Z, Wang X et al (2010) DEGseq: an R package for identifying differentially expressed genes from RNA-seq data. Bioinformatics 26:136–138
10. Götz S, García-Gómez JM, Terol J et al (2008) High-throughput functional annotation and data mining with the Blast2GO suite. Nucleic Acids Res 36:3420–3435
11. Altschul SF, Gish W, Miller W et al (1990) Basic local alignment search tool. J Mol Biol 215:403–410
12. Speckner W, Schindler JF, Albers C (1989) Age-dependent changes in volume and haemoglobin content of erythrocytes in the carp (Cyprinus carpio L.). J Exp Biol 141:133–149
13. Sambrook J, Russell DW (2001) Molecular cloning. A laboratory manual, 3rd edn. Cold Spring Harbor Laboratory Press, Cold Spring Harbor
14. Patrone G, Puppo F, Cusano R et al (2000) Nuclear run-on assay using biotin labeling, magnetic bead capture and analysis by fluorescence-based RT-PCR. BioTechniques 29:1012–1014, 1016–7
15. Nikinmaa M (2014) An introduction to aquatic toxicology, 1st edn. Academic, Waltham MA, p 240
16. Nickerson JG, Dugan SG, Drouin G et al (2003) Activity of the unique beta-adrenergic Na+/H+ exchanger in trout erythrocytes is controlled by a novel beta(3)-AR subtype. Am J Phys Reg Int Comp Phys 285:R526–R535
17. Nikinmaa M (2003) beta(3)-Adrenergic receptors - studies on rainbow trout reveal ancient evolutionary origins and functions distinct from the thermogenic response. Am Phys Reg Int Comp Phys 285:R515–R516
18. Nikinmaa M, Rytkonen KT (2011) Functional genomics in aquatic toxicology—do not forget the function. Aquat Tox 105:16–24
19. Zhao J, Qin B, Nikolay R et al (2019) Translatomics: the global view of translation. Int J Mol Sci 20(1):212
20. McGlincy NJ, Ingolia NT (2017) Transcriptome-wide measurement of translation by ribosome profiling. Methods 126: 112–129

21. Ingolia NT, Brar GA, Rouskin S et al (2012) The ribosome profiling strategy for monitoring translation in vivo by deep sequencing of ribosome-protected mRNA fragments. Nat Protoc 7:1534–1550
22. Ingolia NT (2010) Genome-wide translational profiling by ribosome footprinting. Methods Enzymol 470:119–142
23. Martins T, Valentim A, Pereira N et al (2019) Anaesthetics and analgesics used in adult fish for research: a review. Lab Anim 53:325–341
24. Lien S, Koop BF, Sandve SR et al (2016) The Atlantic salmon genome provides insights into rediploidization. Nature 533:200–205

Chapter 23

Stain-Free Approach for Western Blot Analysis of Zebrafish Embryos

Jianmin Xue and Paola Corti

Abstract

Western blots are employed to detect and characterize amounts of proteins in biological samples. Quantifications are traditionally carried out through data normalization by housekeeping protein method. This approach does not account for variations not intrinsically dependent on the sample such as different experimental conditions and type of samples. Zebrafish researchers often face the challenge of comparing embryos at different developmental stages or from different strains. Housekeeping protein amount can change in these conditions therefore adding an unwanted quantification error. Here we describe the method to analyze mutant zebrafish embryos at different stages by western blot using the Stain-Free technology for normalization. We present Globin X quantification at 2 and 5 days postfertilization in wild type and in the bloodless Vlad Tepes (*vlt*) zebrafish mutant that lack red blood cells.

Key words Stain-Free, Immunoblots, Zebrafish, Embryos, Larvae, Globin X, GbX

1 Introduction

Western blotting is a well-established technique widely utilized to measure relative protein amounts in biological samples [1]. In order to obtain quantitative data, the validation of the antibodies (Ab) is critical and the appropriate selection of a normalization method is crucial to assure that the reported protein quantification is not an artifact of the reference signal [2, 3]. Traditionally protein quantification has been carried out based on housekeeping proteins as reference signal but several challenges have to be taken into consideration when analyzing zebrafish tissues.

The analysis of zebrafish embryos often involves comparisons at different stages of development or samples with different genetic backgrounds. Because of natural variations of housekeeping proteins throughout development and between different strains, the housekeeping protein normalization method is often unreliable. Moreover, by their very own nature, fish research does not take

Cinzia Verde and Daniela Giordano (eds.), *Marine Genomics: Methods and Protocols*,
Methods in Molecular Biology, vol. 2498, https://doi.org/10.1007/978-1-0716-2313-8_23,

advantage of the numerous antibodies validated in mammalian models with a consequent lower availability of reliable sources. Additionally, zebrafish embryos contain limited amount of protein content because of their small size and differences between target protein expression among samples are typically very small. While it is necessary to load great amount of total protein lysate per lane to measure the target protein quantities, the housekeeping protein usually result to be grossly overloaded thus not serving the purpose of data normalization [4].

An alternative solution to the use of housekeeping protein for data normalization is the use of the Stain-Free technology based on the total lane density from the blot transferred proteins as a mean of data normalization. The Stain-Free technology described previously [5–7] is here utilized to analyze zebrafish embryos in order to determine the amount of Globin X (GbX) in wild type and *vlt* mutants carrying a mutation in the *gata1* gene, one of the earliest hematopoietic genes, at different stages of development.

2 Materials

2.1 Sample Preparation

1. Nuclease-Free water.
2. Pronase working solution: add 500 μL pronase stock solution (20 mg/mL) into 9.5 mL Egg water (E3) (*see* **Note 1**).
3. Glass vials: 15 × 45 mm Dram screw thread vial w/rubber lined cap opti-clear glass.
4. E3 embryos solution: add 1.2 g instant ocean sea salts to 1 L deionized water (*see* **Note 2**).
5. Radioimmunoprecipitation (RIPA) buffer: 1% NP-40, 1% sodium deoxycholate, 0.1% SDS, 150 mM NaCl, 50 mM Tris-HCl, 0.004% sodium azide, pH 7.4 (*see* **Note 3**). To prepare complete RIPA for use add the protease inhibitor mixture: combine 10 μL of 200 mM phenylmethylsulfonyl fluoride (PMSF) solution, 10 μL of 100 mM sodium orthovanadate solution, and 10 μL of protease inhibitor cocktail solution per mL of RIPA lysis buffer.
6. Cordless motor disposable pellet mixer and 1.5 mL microtubes with micro pestles (*see* **Note 4**).
7. Pierce BCA protein assay kit.

2.2 SDS Stain Free Gel

1. PAGE Mini-Protein system for electrophoresis.
2. Stain Free AnyKD precast gel (Bio-Rad patented).
3. SDS-PAGE running buffer: 0.025 M Tris-HCl, pH 8.3, 0.192 M glycine, 0.1% SDS.

4. SDS sample loading buffer (4×): 277.8 mM Tris-HCl (pH 6.8), 10% SDS, 44.4% (v/v) glycerol, 4.4% LDS, 0.02% bromophenol blue, store at room temperature.
5. β-Mercaptoethanol (β-ME) (*see* **Note 5**).
6. ChemiDoc MP imaging system (Bio-Rad).

2.3 Immunoblotting

1. Membrane PVDF: cut size as 5.5 × 9 cm (*see* **Note 6**).
2. Thick filter paper for blotting applications as part of the sandwich with gel and membrane. Precut size is 7.5 × 10 cm.
3. Western blot 10× transfer buffer: 25 mM Tris-HCl, 192 mM glycine (*see* **Note 7**).
4. Phosphate buffer saline (0.01 M phosphate buffered saline, 1× PBS): 0.138 M NaCl, 0.0027 M KCl in 1 L of deionized water, pH 7.4.
5. PBST: add 1 mL of Tween 20 to 1 L of 1× PBS to have 0.1% Tween 20 in PBS (*see* **Note 8**).
6. Blocking solution: 5% (w/v) nonfat dry milk in PBST (*see* **Note 9**).
7. Mini Trans-Blot cell for Western Blot (*see* **Note 10**).
8. Plastic container to fit the PVDF membrane, filter papers and sponges before protein transfer.
9. Plastic container to fit the PVDF membrane throughout the blotting procedure.
10. Rabbit anti GbX antibody: custom antibody against zebrafish GbX produced in rabbit.
11. Western Peroxide Reagent substrate.

2.4 Software

1. Software to image Stain-Free gel: Bio-Rad Image Lab software to calculate band intensity.
2. Microsoft Office Excel.

3 Methods

3.1 Sample Preparation

1. In order to have enough protein lysate, 30 embryos at 2 days postfertilization (dpf) and 15 embryos at 5 dpf are pooled in glass vials. The number of pooled embryos can be variable (*see* **Note 11**).
2. Add 1 mL pronase working solution to the 2 dpf samples and incubate for 10 min at room temperature (*see* **Note 12**).
3. Gently flick the vial and replace pronase solution with 2 mL of E3, repeat washing for 7–8 times.
4. Transfer the embryos into one 1.5 mL microtube by glass pipette. Remove all the solution from the embryos (*see* **Note**

13) and proceed to use. If needed, use the option to freeze the samples in liquid nitrogen and store them at −80 °C until application (*see* **Note 14**).

5. To prepare the samples for the electrophoresis, add 100 μL complete RIPA buffer to the frozen embryos pellet (*see* **Note 15**).
6. Homogenize samples using disposable pellet mixers with cordless motor for 1–2 min on ice. Incubate samples on ice for 60 min. Then centrifuge the samples at 13,000 × *g* at 4 °C for 10 min. Collect supernatant for electrophoresis.
7. Separate 10 μL of sample lysate to perform the protein assay (BCA protein assay) according to the following procedure:
   (a) Make serial dilution of the standard albumin (BSA) using the albumin stock provided within the kit (2 mg/mL). Use Nuclease-Free water for dilutions. The final volume for each dilution should be 25 μL to be loaded in a 96 well plate. Load standards from well 2–8. In well 1 load 25 μL of BSA stock solution and in well 9 load 25 μL of Nuclease-Free water as blank.
   (b) Dilute samples in Nuclease-Free water to a final volume of 40 μL. Use 25 μL for the assay (*see* **Note 16**).
   (c) Prepare master mix of BCA working reagent by mixing 50 parts of BCA reagent A with 1 part of BCA reagent B (*see* **Note 17**). Add 200 μL of the master mix to the samples and standards. Incubate the plate at 37 °C for 30 min and measure the absorbance at 562 nm on a plate reader. Use the Excel spreadsheet to calculate the concentration of samples.

3.2 Electrophoresis with Stain-Free Gel

1. Assemble electrophoresis cell with Any KD Stain-Free precast gel. Remove the tape and the comb following the manufacturer instructions. Wash the sample wells with running buffer.
2. To detect and measure GbX expression it is recommended to load 10–50 μg of protein per sample. Dilute the samples with Nuclease-Free water and 4× loading buffer in order to obtain 50 μg of protein in 12.5 μL volume (*see* **Note 18**). Heat samples at 95 °C for 5 min. Load 12.5 μL sample per lane.
3. Load 5 μL zebrafish recombinant GbX protein to gel (*see* **Note 19**).
4. Load 5 μL of standard molecular weight markers and 5 μL of Stain-Free molecular markers in separate lanes (*see* **Note 20**).
5. Run electrophoresis first at 50 mV for 5 min or until the samples has entered the gel, then increase the voltage up to 100–150 mV. Stop the electrophoresis when the dye front has reached the bottom of the gel. In these conditions this takes about 1 h.

6. Remove the gel cassette from the cell and open the cassette to release the gel. Carefully place the gel on the sample stage of the ChemiDoc MP Imaging System.
7. Start Image Lab software and capture the Stain-Free gel image using the following settings (*see* **Note 21**):
   (a) Application: Stain-Free gel.
   (b) Gel activation time: 1 min.
   (c) Imaging area: Bio-Rad Criterion gel.
   (d) Imagine exposure time: automatically optimized for the most intense bands.
8. Remove the gel from the sample stage and keep it in running buffer to avoid the gel to dry out.

3.3 Protein Transfer and Imaging

1. Immerse PVDF membranes in 100% methanol for 30 s. Transfer the membrane to 1× transfer buffer.
2. Use the Bio-Rad Criterion blotter and assemble the sandwich for protein transfer as follow: place sponge, thick filter paper, gel, PVDF, thick filter paper, sponge. Roll out bubbles.
3. Place the blotting chamber in a large container filled with ice and start the protein transfer to the membrane at 50 mV for 2 h.
4. Disassemble the blotting sandwich and place both the blotted membrane and the gel in a plastic container with deionized water.
5. To verify the transfer efficiency, capture an image of the post-transfer gel using ChemiDoc MP Imaging System. Use the same settings as previously but set the gel activation time to none. This image should show no protein bands.
6. Remove the gel from the sample stage and replace it with the posttransfer blot (*see* **Note 22**). Image the posttransfer blot (Fig. 1a, b *left images*) with the following settings:
   (a) Application: Stain-Free Blot.
   (b) Imaging area: Bio-Rad Criterion gel.
   (c) Image exposure time: automatically optimized for the most intense bands.
7. Remove the posttransfer blot from the sample stage and place it in a plastic container with PBST (0.1% Tween 20).

3.4 Antibody Incubation

1. Block the posttransfer blot in 5% no fat dry milk in PBST (0.1% Tween 20) at room temperature for 1 h with rotating.
2. Dilute the anti-GbX antibody 1: 500 in 0.5% BSA PBST. Incubate overnight at 4 °C with rotating.

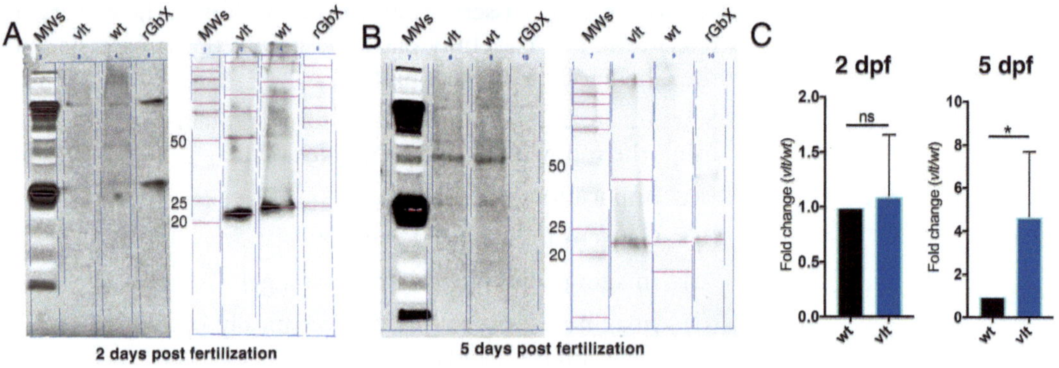

Fig. 1 Quantification of GbX by total protein normalization method. (**a**) Stain-Free representaive images of posttransfer blot and western blot at 2dpf. (**b**) Stain-Free representaive images of posttransfer blot and western blot at 5dpf. (**c**) Quantification of GbX in *vlt* compared to *wt*

3. Rinse the posttransfer blot three times for 5–10 min with PBST at room temperature with rotating.
4. Dilute the rabbit horseradish peroxidase (HRP)-conjugated secondary antibody 1:10,000 with 5% no fat dry milk in PBST. Incubate for 1 h at room temperature with rotating.
5. Rinse the western blot three times for 5–10 min with PBST with rotating.

3.5 Imaging and Analysis Using the ChemiDoc MP System

1. Keep the western blot in PBST while preparing the Clarity western ECL substrate mixture.
2. Mix the Clarity substrate kit components (A and B) in a 1:1 ratio. Incubate the membrane in the substrate solution for 1–5 min.
3. Place the western blot back on the sample stage of the Chemi-Doc MP Imager.
4. Start Image Lab software and capture the chemiluminescent signals on the blot (Fig. 1a, b *right images*) with the following settings:
   (a) Application: Chemi.
   (b) Imaging area: Bio-Rad Criterion gel.
   (c) Image exposure time: automatically optimized for the most intense bands (*see* **Note 23**).

3.6 Validate Total Protein Normalization

1. Overlay the two chemiluminescent and stain-free images of the blot (Fig. 1a, b *left and right images*) for normalization and proceed as follow:
   (a) Select the western blot image with the protein bands to be quantified. Select Normalization from Analysis tool box. Using default settings, lanes and bands are detected

automatically and bands density values are identified (*see* **Note 24**). Select Add Channel to have the image data added to the Sample Data field.

(b) Select the posttransfer blot image and add the data in the Normalization Data field.

(c) The protein band intensities in each sample are normalized to the total amount of protein per lane. To view the normalized target protein volumes, click the Analysis Table on the toolbar. All calculations will be performed automatically by the software, including the Normalization Factor and Normalized Volumes. The normalized values of protein amount can be exported into excel file.

2. Select molecular weights (MW) Analysis Tool and assign the MW values to each detected band in the MW lane. This step helps identifying the expected band to be quantified in the samples.
3. In order to calculate GbX differences between *wt* and *vlt* mutants for each experiment we compared the *vlt* band intensity previously normalized by total amount of protein loaded on the gel, to the *wt* normalized band. We represented the final result as the average of three experiment performed at each developmental stage (Fig. 1c).

4 Notes

1. Make pronase stock solution (20 mg/mL) in 1× E3 and store aliquots at −20 °C. Make fresh pronase working solution prior to use.
2. It is recommended to incubate the embryos in blue water for 1 day (1 mL methylene blue stock (1 g/L) into 1 L E3) to ensure they are not lost to fungal overgrowth. The methylene blue serves as an antifungal and antiparasitic reagent.
3. Store aliquots of 1 mL RIPA buffer at −20 °C. Thaw the aliquots needed for experiment immediately prior to use and quickly add the protease inhibitor mixture.
4. It is convenient to directly collect the embryos into 1.5 mL microtubes suitable for homogenization with micro pestles. To homogenize the embryos, immerse pestle operated by battery-powered motor into the lysis buffer. Completely homogenize embryos tissue with high-speed vortexing action for 1 min on ice.
5. Make complete SDS sample buffer adding β-ME prior to application. Add 100 μL β-ME into 900 μL of 4× sample buffer, mix well.

6. PVDF membranes must be wetted in 100% methanol and transferred to a container containing 1× transfer buffer prior to use. Carefully avoid the PVDF membrane to dry out. Filter paper and sponges should also be wetted in 1× transfer buffer prior to use.
7. To make 1× transfer buffer, add 100 mL of 10× transfer buffer, 200 mL of methanol and 700 deionized water to a 1 L graduated cylinder, mix well and store at room temperature.
8. To make PBST cut the end of a pipette tip to aspirate Tween 20 easily. Add Tween 20 to PBS, stir to dissolve completely and store at room temperature.
9. To prepare blocking solution add 5 g nonfat dry milk powder to 50 mL PBST in one conical sterile centrifuge tube and stir until dissolved. Stored on ice or 4 °C, use within 1–2 days.
10. The apparatus contains two gel holder cassettes for electrophoretic transfer of mini format gels. Color-coded cassettes and electrode ensure proper orientation of the gel during transfer.
11. Depending on the stage and the abundance of the protein to be quantified, the number of pooled embryos may vary. We found that pooling 30 embryos at 2 dpf and 15 embryos at 5 dpf works well for quantification of GbX.
12. This step can be substituted by manual dechorionation. At 5 dpf embryos are hatched therefore dechorionation is not necessary. When using pronase working solution it is highly recommended to use glass vials to avoid the embryos to stick to the walls after dechorionation.
13. To transfer and remove the whole solution from the embryos it is best to have the embryos in 1.5 mL microtubes and use a small tip to pipet out the solution without touching the embryos.
14. Storing the embryos at −80 °C for a few weeks is acceptable. However, for optimal results fresh preparation of samples is recommended.
15. When starting the lysis procedure, keep the frozen samples on ice and add complete RIPA buffer directly to the frozen samples.
16. For 15–30 embryos at 2–5 dpf, it is appropriate to dilute the samples 1:4 to be able to measure the absorbance within an acceptable range. To dilute 1:4 add 10 μL of supernatant to 30 μL water. This final volume will allow to have a little over the amount needed for the assay.
17. Prepare master mix of BCA working reagent. The volume needed is 200 μL master mix per 25 μL of sample. Calculate

the number of samples to be tested (including the number of standards) and prepare master mix accordingly. For example, to obtain the master mix for a total of 15 samples use 3 mL reagent A and 60 μL reagent B.

18. To prepare the sample at final concentration for loading, 80 μL sample supernatant is collected. If the resulting concentration of the sample by BCA assay is 5.4 mg/mL, add 27 μL of 4× loading buffer and water 1 μL to the 80 μL of sample in order to obtain a final protein concentration of 4 μg/μL. This concentration will allow to load on the gel 50 μg of protein in 12.5 μL volume for each sample. Only a portion of the diluted sample is used therefore the leftover sample can be stored at −20 °C for future applications. When needed, thaw the samples and reheat at 95 °C for 5 min prior to use.
19. Dilute the zebrafish recombinant GbX protein sample to a final concentration of 0.4 ng/μL. Use nuclease-free water and 4× loading buffer β-ME for dilutions and heat at 95 °C for 5 min.
20. It is best for imaging to use the MWs unstained protein standards. However, since these standard bands are not visualized on the blot after transfer, we found useful to load any regular MW standards in a separate lane as control to determine the success of protein transfer.
21. The gel needs to be activated by exposure to UV light for 1 min. This allows to make proteins fluorescent directly in the gel for the immediate visualization.
22. Keep the blot wet all the time; do not let it dry. This Stain-Free blot image is used for total protein loading control and normalization.
23. The image area for the stain-free and chemiluminescent images must be the same for follow-up total protein normalization by the Image Lab software.
24. Select and use Lanes and Bands tools to adjust the lanes and bands if necessary.

Acknowledgments

This work has been supported by AHA Career Development Award (18CDA34110344).

References

1. Towbin H, Staehelin T, Gordon J (1979) Electrophoretic transfer of proteins from polyacrylamide gels to nitrocellulose sheets: procedure and some applications. Proc Natl Acad Sci U S A 76(9):4350–4354
2. Taylor SC, Berkelman T, Yadav G et al (2013) A defined methodology for reliable quantification of Western blot data. Mol Biotechnol 55(3): 217–226

3. Posch A, Kohn J, Oh K et al (2013) V3 stain-free workflow for a practical, convenient, and reliable total protein loading control in western blotting. J Vis Exp 82:50948
4. Taylor SC, Posch A (2014) The design of a quantitative western blot experiment. Biomed Res Int 2014:361590
5. Gurtler A, Kunz N, Gomolka M et al (2013) Stain-Free technology as a normalization tool in Western blot analysis. Anal Biochem 433(2): 105–111
6. Li R, Shen Y (2013) An old method facing a new challenge: re-visiting housekeeping proteins as internal reference control for neuroscience research. Life Sci 92(13):747–751
7. Colella AD, Chegenii N, Tea MN et al (2012) Comparison of Stain-Free gels with traditional immunoblot loading control methodology. Anal Biochem 430(2):108–110

Chapter 24

Proteomics of Fish White Muscle and Western Blotting to Detect Putative Allergens

Liliana Anjos, Arsenios-Zafeirios Loukissas, and Deborah Mary Power

Abstract

A detailed workflow is provided for preparation from teleost fish white muscle of extracts for proteomics analysis. The protocol generates samples that can be analyzed by SWATH (Sequential Window data independent Acquisition of the Total High-resolution-Mass Spectra), a modern MS-based quantitative label free technology. The main steps for the extraction of three independent protein fractions, (1) soluble sarcoplasmic, (2) soluble myofibrillar, and (3) insoluble material, from fish white muscle are detailed. Coupled to the protein extraction protocol a Western blotting approach is outlined for detection of common fish allergens, in this case β-parvalbumin, in the white muscle sarcoplasmic protein fraction.

Key words Allergen, β-parvalbumin, European sea bass, Protein extraction, Proteomics, Western blot

1 Introduction

Fresh fish consumption has significant "health benefits" since they are a source of high-quality protein, omega-3 fatty acids, and trace nutrients. The white muscle of fish is the main commercial target of the fishing industry, since it is preferred for human consumption. High-throughput technologies for monitoring quality and safety of sea food products are needed and proteomics technology is a promising strategy to fill this gap. Proteomics is used in a panoply of applications under the catchy term "foodomics." In the case of sea food products, the main applications of proteomics techniques are for species authentication, monitoring of fish spoilage and quality during storage and processing and allergen detection [1–3]. The latter is of increasing importance since fish are one of the eight most common causes of allergy and there is an ambition to increase fish production and consumption to meet global protein needs [4]. Existing proteomics studies applied to sea food products are mainly based on conventional technologies and have identified a relatively small number of proteins. The application of high

Cinzia Verde and Daniela Giordano (eds.), *Marine Genomics: Methods and Protocols*,
Methods in Molecular Biology, vol. 2498, https://doi.org/10.1007/978-1-0716-2313-8_24,

throughput quantitative methodologies such as SWATH-MS (sequential window data independent acquisition of the total high-resolution-mass spectra), a quantitative label free technology can accelerate discovery and progress [5, 6].

The three main steps in the proteomics workflow are, (1) protein extraction from samples, (2) protein/peptide separation, and (3) protein identification/quantification. The first step is crucial and protein extraction should yield good quality proteins that are representative of the sample. This means methods should be sample-specific with the view of extracting proteins and peptides suitable for mass spectrometry. Three main protein fractions can be extracted from fish white muscle: 1—sarcoplasmic (around 30% of muscle protein, low salt soluble, enriched in enzymes); 2—myofibrillar (around 60% of muscle protein, high salt soluble, enriched in structural proteins); and 3—stromal (less abundant, enriched in connective tissue proteins). In the following section a detailed workflow for protein extraction from teleost fish white muscle that is compatible with SWATH proteomics analysis is outlined, using the European sea bass (*Dicentrarchus labrax*) as the example [7, 8]. The protocol provides the main steps for efficient extraction of the three main protein fractions. The method integrates mechanical disruption and successive protein fractionation steps to solubilize proteins by exploiting their biochemical properties. These kinds of approaches have been reported in a relatively restricted literature with a variable scope for several commercial fish species [9–13]. The method presented herein was successfully used for SWATH analysis of the sarcoplasmic and myofibrillar fractions of European sea bass white muscle and is compatible for use with white muscle from other fish species [7, 8].

Fish consumption is one of the eight most common causes of allergy. This is due to an IgE-mediated immune reaction, in a small proportion (<1% worldwide) [14] of people that consume fish, and it can be triggered by proteins such as parvalbumin [2, 14]. Parvalbumins are muscle sarcoplasmic calcium-binding proteins, that are resistant to high temperatures, denaturing agents, and proteolytic activity. The detection of fish parvalbumins is challenging compared to other food allergens. This may be due to isoform diversity and high biochemical and immunological variability, despite generally good amino acid sequence conservation of parvalbumin from different fish species [14, 15].

Using β-parvalbumin as an example of an allergen, we describe a second protocol based on Western blotting using European sea bass extracts of the white muscle sarcoplasmic protein fraction. Western blotting is a powerful and sensitive immunochemical technique developed in the seventies [16]. It is based on the use of specific antisera to detect proteins in a complex sample after separation by sodium dodecyl sulfate-polyacrylamide gel electrophoresis

(SDS-PAGE), [17] and transfer and immobilization on nitrocellulose membranes [18, 19].

2 Materials

Prepare and store all reagents at room temperature (unless indicated otherwise). Use ultrapure water (UP water, Milli-Q® Type 1 or equivalent), unless indicated otherwise. Diligently follow regulations when disposing of waste materials.

2.1 Global Protein Extraction from Fish White Muscle

To avoid contamination in MS proteomics analysis special care should be taken during protein sample preparation. Precautions include, the use of gloves and a lab coat, use of clean material, low protein binding tubes, and highly pure reagents. All protein preparation steps, when possible, should be done in a laminar flow cabinet or failing this in a clean environment.

2.1.1 Protein Extraction

1. Hammer, mortar, pestle, and aluminum foil.
2. Liquid nitrogen (*see* **Note 1**).
3. 2 mL micro tubes with cap, compatible with a tissue homogenizer in use (indicated below in **step 11**).
4. Glass beads (0.5 mm).
5. Micro spoon spatula (5 × 7 mm).
6. 6 M HCl (hydrochloric acid): Weigh 21.84 g of HCl and add dropwise to UP water to give a final volume of 100 mL and store at room temperature.
7. 1 M DTT: Weigh 1.54 g DTT (1,4-Dithiothreitol) and dissolve in sufficient UP water to give 10 mL. Prepare aliquots of 1 mL and store at −20 °C.
8. B1 - Sarcoplasmic protein extraction buffer: 10 mM Tris-HCl, pH 7.2.

   Place 800 mL of UP water in a 1 L glass beaker, add 1.21 g Tris-base and stir with a magnetic agitator until dissolved, and adjust to pH 7.2 with 6 M HCl and increase the volume to 900 mL with UP water (this will be made up to 1 L with the protease inhibitor solution). Store at 4 °C and immediately before use, add 100 μL/mL of a 10-times concentrated solution (10×) of protease inhibitor to the volume of buffer required for extraction.
9. B2 - Myofibrillar protein extraction buffer: 10 mM Tris-HCl, pH 7.2, 0.6 M NaCl (sodium chloride), 1.7% SDS (sodium dodecyl sulfate), 100 mM DTT. Add 600 mL of UP water to a 1 L glass beaker and add 1.21 g Tris-base and stir with a magnetic stirrer until dissolved, and adjust to pH 7.2 with

6 M HCl, then add 35.07 g NaCl and 17 g SDS and leave the solution standing at room temperature until the SDS is dissolved, only then gently mix with a magnetic stirrer taking care to avoid the formation of a foam. Make up to 800 mL in a volumetric flask by adding UP water. Store at 4 °C. Immediately before use, add 100 μL/mL of 10× protease inhibitor, and 100 μL/mL of 1 M DTT to the volume required for extraction.

10. B3 - Insoluble protein extraction buffer: 8 M Urea, 50 mM Tris base pH 8, 100 mM NaCl, 10 mM EDTA (Ethylenediaminetetraacetic acid). Add 300 mL of UP water to a 1 L glass beaker and add 3.04 g of Tris-base and stir with a magnetic stirrer until dissolved and adjust to pH 8 with 6 M HCl. To the Tris base solution gradually add 240.24 g of urea, 2.922 g of NaCl, 14.62 g of EDTA and mix until fully dissolved (this may take 1–2 h). Make up the final volume of the solution to 500 mL in a volumetric flask by adding UP water and store at room temperature.
11. Tissue homogenizer: "Tissue lyser II."

2.1.2 Protein Quantification

1. Bradford protein assay kit and bovine serum albumin standard.
2. 96-well flat bottomed multiwell plates.
3. Microplate reader.

2.1.3 1D SDS-PAGE

1. 40% polyacrylamide–bis solution (37.5:1) (*see* **Note 2**).
2. Resolving gel buffer: 1.5 M Tris-HCl, pH 8.8. Dissolve 36.34 g of Tris base in 200 mL of UP water and adjust to pH 8.8 with 6 M HCl and store at 4 °C.
3. Stacking gel buffer: 1 M Tris-HCl pH 6.8. Dissolve 24.23 g of Tris base in 200 mL of UP water and adjust to pH 6.8 with 6 M HCl and store at 4 °C.
4. Glass plates (8 × 10 cm), combs (15 wells), spacers (1.5 mm), and assembly chamber.
5. Mini-PROTEAN® Tetra Cell or equivalent system for gel casting and electrophoresis.
6. 10% (m/v) SDS in UP water.
7. Ammonium persulfate (APS): 10% (m/v) solution in UP water (*see* **Note 3**).
8. TEMED (N,N,N′,N′-Tetramethyl ethylenediamine).
9. Whatman paper.
10. 5× SDS loading buffer, (five times concentrated): 250 mM Tris-HCl, pH 6.8, 10% SDS (m/v), 50% glycerol (v/v), 0.25% bromophenol blue (w/v), 500 mM DTT.

In a 15 mL falcon tube mix 2.5 mL of stacking gel buffer (1 M Tris-HCl, pH 6.8), 1 g SDS, 5 mL glycerol, 0.025 g bromophenol blue, and 0.77 g of DTT. Leave to stand for 10 min and then vortex to mix all reagents and adjust the volume to 10 mL with UP water. Prepare 10 aliquots of 1 mL and store 9 at −20 °C and leave one aliquot at 4 °C for use.

11. SDS-PAGE running buffer: 25 mM Tris-HCl, pH 8.3, 192 mM glycine, 0.1% SDS. Add 800 mL of UP water to a 1 L glass beaker and add 3.03 g of Tris-HCl and 14.4 g of glycine and stir with a magnetic stirrer until dissolved. Add 1 g of SDS to the surface of the solution and leave for 5 minutes (min) to dissolve and then make up to 1 L with UP water and store at 4 °C.
12. Prestained Protein Ladder (range 10–200 kilodaltons, kDa).
13. Power supply: Power Pack Universal Power Supply.
14. Coomassie blue (CBB) staining solution: 40% methanol, 10% acetic acid, 0.1% Coomassie brilliant blue R-250, ddH_2O (double distilled water).
15. Destaining solution: 40% methanol, 10% acetic acid, ddH_2O.

2.2 Western Blotting to Detect β-Parvalbumin in the White Muscle Sarcoplasmic Fraction

1. Primary antibody: monoclonal anti-parvalbumin clone PARV-19, mouse ascites fluid. Store in 5 μL aliquots at −20 °C.
2. Secondary antibody: anti-mouse IgG (whole molecule)—peroxidase-labeled antibody, produced in rabbit. Store in 5 μL aliquots at −20 °C.
3. Nitrocellulose membrane 0.2 μm.
4. TBS (tris-buffered saline): 20 mM Tris base, 150 mM NaCl, deionized water and adjust to pH 7.6 with 6 M HCl.
5. TBS-T: TBS, 1% Tween 20 (*see* **Note 4**).
6. Blocking buffer: TBS-T, 5% skimmed milk powder (w/v) (*see* **Note 5**).
7. Transfer buffer: 25 mM Tris base, 192 mM glycine, 10% methanol, 0.01%. SDS, ddH_2O.
8. Ponceau S staining: 1.3 mM (0.1%) Ponceau S tetrasodium salt, 0.874 M acetic acid, deionized water.
9. Whatman paper.
10. Roller, pencil, scissors, smooth-ended tweezers, ruler.
11. Plastic or glass container for "transfer unit" preparation and small plastic container for incubations (e.g., square petri dishes).
12. Western blotting chemiluminescent detection reagent.

13. Western blotting transfer tank (wet transfer) system with plastic cassettes and sponges.
14. Power supply: power pack, universal power supply.
15. CCD (charge-coupled device) camera that detects chemiluminescence.

3 Methods

3.1 Global Protein Extraction of Fish White Muscle for Proteomic Analysis

Sea bass white muscle (1 cm^3) was sampled with a clean scalpel, just below the dorsal fin and approximately mid-point of the standard length. Skin was dissected off, and the muscle wrapped in tin foil and immediately immersed in liquid nitrogen and stored at −80 °C until protein extraction. For the protein extraction protocol all steps should be carried out at 4 °C unless otherwise indicated. A schematic workflow of the procedure is presented in Fig. 1.

3.1.1 Protein Extraction

1. Add approximately 340 mg of silica beads (0.5 mm) with a micro spoon spatula to a 2 mL micro tube with a cap and note the weight (*see* **Note 6**).
2. With a sterile scalpel quickly cut-off approximately 350 mg of white muscle, register the weight, wrap in aluminum foil, dip in liquid nitrogen and smash with a hammer.
3. Place the muscle fragments in a cold mortar, add liquid nitrogen and create a fine power using the pestle (keep frozen throughout by adding liquid nitrogen as necessary) (*see* **Note 7**).
4. Add approximately 125 mg of powdered muscle to the tube containing the silica beads, weigh and note the weight of added tissue. Hold the tubes at −80 °C until sample extraction.
5. Add the sarcoplasmic protein extraction buffer (B1); for 0.5 g of tissue add 2 mL buffer.
6. To extract the sarcoplasmic proteins (FSc) insert the sample tubes into a tissue homogenizer and program 2 cycles of 20 seconds (s) with an interval of 30 s between each cycle (*see* **Note 8**).
7. Centrifuge at 28,000 × *g* for 15 min at 4 °C.
8. Collect the FSc supernatant into a new tube. Take an 80 μL aliquot of FSc and store at −20 °C for immediate use and store the remaining FSc extract at −80 °C.
9. Solubilize the pellet containing the myofibrillar protein (FM) fraction with extraction buffer (B2) (using 2 mL for 0.5 g pellet). Resuspend the FM fraction by gently pipetting up and down and vortexing for 1 min.

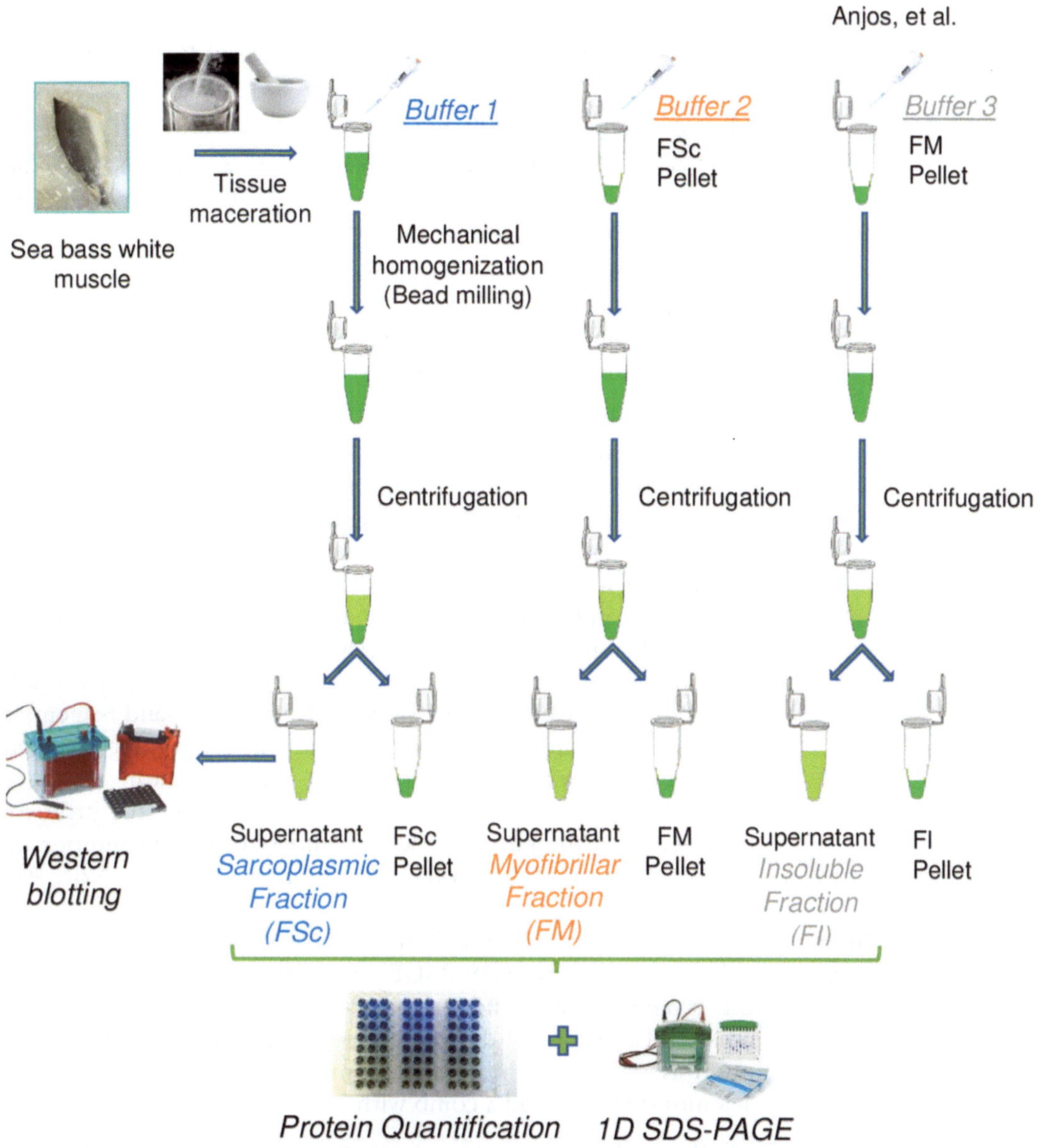

Fig. 1 Overview of the general experimental workflow used for global protein extraction of white muscle from fish for proteomic analysis and detection of allergens by Western blotting. The white muscle used for extraction in the study was from the European sea bass (*Dicentrarchus labrax*). Following tissue maceration (combining liquid nitrogen and mechanical homogenization by bead milling), the sarcoplasmic (FSc—Blue), myofibrillar (FM—orange), and insoluble (FI—gray) protein fractions were sequentially extracted using appropriate buffers (Buffer 1, 2, and 3). The extraction buffers were chosen taking into consideration the biochemical characteristics of each of the muscle protein fractions. The protein concentration of the different fractions was quantified using the Bradford method and the protein extracts were analyzed by 1D SDS-PAGE (12%). The sarcoplasmic fraction (FSc) was also analyzed by Western blotting to identify β-parvalbumins

10. Heat the FM protein extract at 95 °C for 5 min, vortex briefly and then heat at 95 °C for 5 min. Allow to cool to room temperature.
11. Centrifuge at 28,000 × *g* at 22 °C for 15 min.
12. Collect the FM protein supernatant into a new tube. Take an 80 μL aliquot of FM and store at −20 °C for immediate use and store the remaining FM extract at −80 °C.
13. Solubilize the insoluble muscle protein in the pellet with extraction buffer B3 (using 1 mL for 0.5 g pellet). Resuspend the insoluble muscle protein pellet by gently pipetting up and down and vortexing for 1 min.
14. Centrifuge at 28,000 × *g* for 15 min at 20 °C.
15. Collect the insoluble muscle protein supernatant (FI) into a new tube. Take a 50 μL aliquot of FI and store at −20 °C for immediate use and store the remaining FI extract at −80 °C.

3.1.2 Protein Quantification

To proceed with subsequent protein quantification, a colorimetric assay based on the Bradford method can be used [20]. Several kits are available and the Quick Start™ Bradford protein assay kit is appropriate for measuring the FSc, FM and FI protein extracts prepared from sea bass white muscle. To save time and reagents a 96-well microplate assay can be used, and a standard curve prepared using bovine serum albumin (BSA) with a linear range of 0.125–2 mg/mL is included in each assay. Follow the protocol provided with the kit used and remember to include appropriate controls for each of the protein extraction protocols (*see* **Note 9**).

3.1.3 Proteome Profile Analysis by 1D SDS-PAGE

One-dimensional (1D) sodium dodecyl sulphate-polyacrylamide gel electrophoresis (SDS-PAGE) is used to assess the quality of the protein extracts and the proteome profile of sea bass white muscle. The protocol described below uses a vertical gel electrophoresis system, with a standard mini gel setup (8 × 10 cm), 1.5 mm thickness and a comb with 15 wells.

1. Warm the reagents to room temperature and ensure the glass plates, combs, and spacers are clean and dry. Prepare the glass gel cassettes according to the instructions of the manufacturer.
2. Prepare the resolving gel (12%) by mixing 3 mL of polyacrylamide, and 4.296 mL water in a 50 mL beaker with a magnetic stirrer. Degas with a water pump for 10 min and then add 100 μL of SDS, 100 μL of APS, and 4 μL of TEMED and rapidly cast the gels and overlay with isobutanol or water (*see* **Note 10**).
3. Prepare the stacking gel (5%) by mixing 0.630 mL of resolving buffer, 0.625 mL of polyacrylamide, 3.640 mL water, 50 μL of SDS, 50 μL of APS, and 5 μL of TEMED and cast on top of the

polymerized resolving gel from which the isobutanol or water has been removed. Immediately after casting the stacking gel insert the comb without introducing air bubbles and only remove it when the gel has polymerized.

4. For sample preparation mix 15 μg of total protein from FSc, FM, and FI samples (*see* Subheading 3.1.2) with 5× SDS loading buffer. To facilitate loading make up all samples to the same final volume (e.g., 10 μL) using the appropriate extraction buffer for samples of each of the muscle fractions.
5. Denature the samples by boiling at 95 °C for 5 min in a dry bath, centrifuge at 14,000 × *g* at room temperature for 3 min and leave at room temperature until loaded onto the polyacrylamide gel (*see* **Note 11**). Do not add lysis buffer to the prestained protein ladder or subject it to heat.
6. Place the polyacrylamide gels into an electrophoresis chamber (*see* **Note 12**), fill the upper and lower buffer chamber with the running buffer and wash-out the sample wells of the stacking gel with running buffer.
7. Load the protein samples into the wells of the stacking gel (one sample/well); add to one of the wells the protein ladder for sizing the proteins after electrophoresis.
8. Run the polyacrylamide gel at a constant current (30 mA) until the dye front reaches the bottom (takes around 90–100 min). Run the electrophoresis at 4 °C to avoid overheating, since this reduces electrophoresis resolution.
9. After electrophoresis, remove the polyacrylamide gels and release them from the glass-cassette with a flat spatula. Rinse the gels in water for approximately 5 min and then stain them with Coomassie blue (CBB) (50 mL/gel) by shaking for 1 h at room temperature.
10. Substitute the CBB with the destain solution and incubate the polyacrylamide gels in several changes of destain until blue staining protein bands are clearly visible. Document with an imager. A typical protein profile of different fractions of fish white muscle extracts is presented in Fig. 2a.

3.2 Western Blotting to Detect β-Parvalbumin in a Fish White Muscle Sarcoplasmic Fraction

Prior to Western blotting, a titration assay can be performed to optimize the working dilutions of primary and secondary antibodies using the Dot Blot technique (not described here). Plan your experimental work in advance (define samples and antibody dilutions to be tested, membrane type and working buffers) (*see* **Note 13**). It may be difficult to obtain commercial antibodies that specifically detect allergenic proteins in fish as few are available in the market.

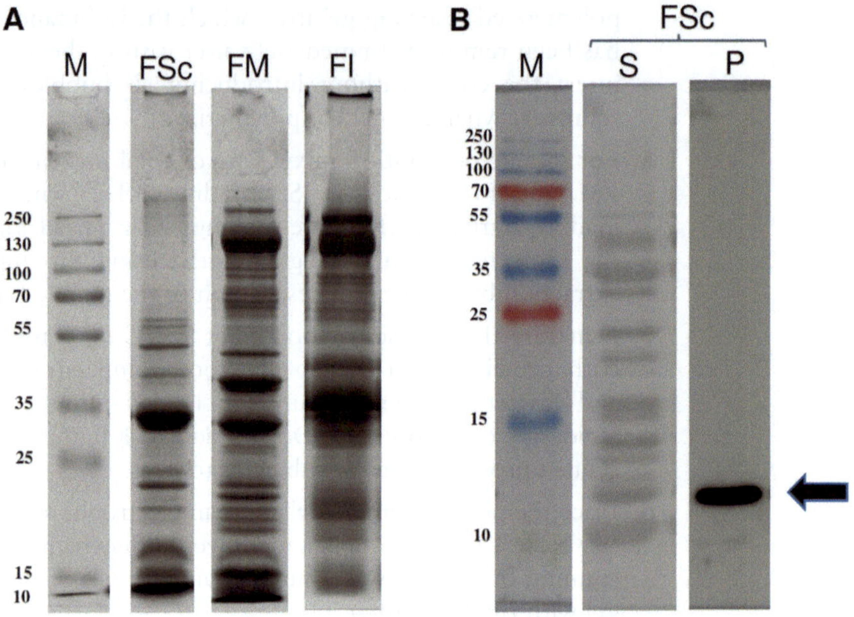

Fig. 2 (**a**) Coomassie blue stained 1D SDS-PAGE (12% polyacrylamide) of the proteins extracted from the sarcoplasmic fraction (FSc), myofibrillar fraction (FM), and insoluble fraction (FI) of the European sea bass white muscle. (**b**) Western blot of β-parvalbumin (P) in soluble sarcoplasmic protein extracts of European sea bass white muscle. Proteins in FSc were transferred to the membrane after electrophoresis by 1D SDS-PAGE (15% polyacrylamide). The FSc total protein profile was identified by *Ponc*eau S staining (S). *M* molecular weight marker (in kDa). Note the strong immunoreaction highlighted by a black arrow

1. Prepare and run the samples (FSc extracts and the prestained protein ladder) on a 1D SDS-PAGE gel (*see* Subheading 3.1.3, **steps 1–8**).
2. On completion of electrophoresis rinse the gel (still supported by one of the glass plate) in UP water to remove traces of SDS-PAGE running buffer.
3. Trim the gel to the appropriate size, removing the stacking gel and any excess gel without protein and incubate it in 50 mL of transfer buffer for at least 10 min.
4. Accurately measure the dimensions of the gel and prepare a piece of nitrocellulose membrane and 8 pieces of Whatman paper (N° 1) with the same dimensions (*see* **Note 14**).
5. Equilibrate the nitrocellulose membrane with the transfer buffer by soaking it for 5 min in 50 mL of distilled water followed by 10 min in transfer buffer (*see* **Note 15**).
6. Soak the transfer cassette, sponges and Whatman paper (N° 1) in transfer buffer. Build the "sandwich" by placing the gel on the 4-sheet stack of Whatman paper, then overlay the gel with the nitrocellulose membrane and then the remaining 4 sheets

of Whatman paper. Gently pass a roller over the "stack" to remove air bubbles that will impede protein transfer.

7. Place your "sandwich" in the cassette between the two wet sponges and lock the cassette (*see* **Note 16**).
8. Fill your transfer tank with transfer buffer at 4 °C, add a magnetic stirrer to agitate the buffer during transfer and then introduce the cassette (*see* **Note 17**).
9. Put the lid on the apparatus (pay attention to the orientation of the electrodes to ensure that proteins will transfer to the nitrocellulose membrane) and initiate transfer at 4 °C under a constant current (300 mA) for 1 h (*see* **Notes 18** and **19**).
10. After protein transfer from the polyacrylamide gel to the nitrocellulose membrane remove the gel and stain it in Coomassie blue (*see* Subheading 3.1.3) to confirm that protein transfer occurred.
11. Cut a very small portion off the right-hand corner of the membrane (to facilitate identification of membrane orientation in subsequent steps). Then wash the membrane for 5 min in UP water at room temperature (RT) with gentle agitation (90 rpm) (*see* **Note 20**). From this step onward, all the membrane incubation steps can be done at room temperature unless indicated otherwise.
12. To confirm that protein transfer occurred to the nitrocellulose membrane incubate it in 50 mL of Ponceau S for 5 min and wash it quickly in deionized water until the protein bands became visible. Take a picture of the nitrocellulose membrane and eliminate the Ponceau S staining by washing in deionized water for 5 min or until the red bands disappear.
13. Incubate the nitrocellulose membrane in TBS-T for 5 min and then place it in blocking buffer at 4 °C overnight or for 1 h at room temperature (*see* **Notes 21** and **22**).
14. Incubate the nitrocellulose membrane with the primary antisera at an appropriate dilution (β-parvalbumin, 1:5000 in TBS) and incubate for 1 h at room temperature and then remove excess antisera by washing the membrane twice in TBS-T for 5 min, followed by 5 min in TBS (*see* **Notes 22** and **23**).
15. Incubate the nitrocellulose membrane with the secondary antibody conjugated to horse radish peroxidase (1:80,000 in TBS) for 1 h at room temperature and then wash twice in TBS for 5 min.
16. Develop with Western blotting chemiluminescent detection reagent following the manufacturer's instructions (*see* **Note 24**) and detect the chemiluminescence with a CCD camera (exposure time is adjusted as required, in the case of β-parvalbumin, with the antisera dilutions used 3 min exposure

worked well). Figure 2b presents the results of Western blotting for β-parvalbumin in the sea bass white muscle FSc fraction.

4 Notes

1. 9 L of liquid nitrogen is sufficient for approximately 60 samples.
2. Respect safety rules and consult the safety data sheets (MSDS) of each reagent prior to use, for example, acrylamide as a monomer is considered toxic, directly affecting the nervous system, and it may be a carcinogen. Acrylamide is readily absorbed through intact skin from aqueous solutions. Wear gloves and a laboratory coat when preparing the acrylamide gel solution.
3. The APS should be made fresh.
4. Tween 20 is viscous; it may be easier to use a measuring cylinder to measure the necessary volume instead of a pipette.
5. Ensure the milk powder is well dissolved by mixing it with a magnetic stirrer for 10 min. You can store the blocking buffer at −20 °C and reuse it several times.
6. Approximately 340 mg of silica beads (0.5 mm) were weighed by adding 3 spoons of silica beads.
7. Use safety glasses and gloves.
8. Take a photograph of one tube.
9. To determine the protein concentration: calculate the average of the readings obtained for the blank and subtract this value from the readings obtained for the standards and samples. Plot the standard curve by plotting the 595 nm values (*y*-axis) versus their concentration in mg/mL (*x*-axis). Determine the unknown sample concentration using the absorbance and reading off the corresponding concentration from the standard curve. If the samples were diluted before protein determination, adjust the final concentration of the unknown samples by multiplying by the dilution factor (e.g., 1/10).
10. The APS should be made fresh immediately before use and the TEMED should be added immediately before pouring the gel (both reagents are initiators of the acrylamide polymerization). Agitate the solution slowly while adding the reagents to avoid bubble formation. Water, isobutanol or 80% isopropanol in ddH_2O can be used to cover acrylamide gels undergoing polymerization, to prevent contact with atmospheric oxygen (which inhibits acrylamide polymerization).

11. Centrifuging samples prior to SDS-PAGE helps remove insoluble debris, which can lead to streaking in the gels (evident when stained with Coomassie blue).
12. Always follow the instructions of the manufacturer.
13. Always consider the recommendations of the producer of the antibodies as well as the nitrocellulose membranes (or suitable alternatives) and if possible, take into account the characteristics of the antigens (e.g. irrespective of protein structure the antibody recognizes the antigen epitope).
14. Use a carbon pencil and a ruler and do not touch the membrane with bare hands.
15. In our experiment we used a nitrocellulose membrane, but (Polyvinylidene difluoride) PVDF membranes can also be used. If a PVDF membrane is used soak it for 10 s in 100% methanol, followed by distilled water for 5 min followed by TBS for 10 min.
16. You can leave it the whole time, always carry out the transfer at 4 °C.
17. Check that there are no bubbles inside the tank to ensure a steady and homogeneous flow of electricity through the polyacrylamide gel.
18. Always introduce the membrane on the positive facing side of the transfer cassette, as the proteins in the polyacrylamide gel are negatively charged and migrate toward the positive pole.
19. This current intensity is effective for the transfer of parvalbumin, the transfer current can be calculated as 2 $mA/h/cm^2$. The time of transfer should be optimized for the proteins under study.
20. This wash step is very important to remove residual SDS, which is an inhibitor of Ponceau S staining.
21. In our experiments the working buffer used was TBS, but you can also use phosphate buffered saline (PBS) or phosphate buffer (PB), unless alkaline phosphatase is used for signal detection.
22. The volume of the washing buffer and blocking buffer can be calculated as approximately 2 mL/cm^2 of nitrocellulose membrane.
23. Primary antibody dilutions should be prepared in sterile TBS and under sterile handling conditions so that these working solutions can be reused and to minimize microbial contamination.
24. Development with ECL plus start kit according to the instructions, remember (a) remove the kit from the fridge 20 min

before development and (b) remove excess TBS buffer by blotting with Whatman paper.

Acknowledgments

PO MAR2020, Portugal2020, EU - Fundo Europeu dos Assuntos Marítimos e das Pescas (FEAMP), Project SEAFOODQual (MAR-01.03.01-FEAMP-0050) (Program: MAR2020-P01M03). Additional funds came from H2020-MSCA-RISE-2019, ICHTHYS (n° 872217). FCT provided CCMARs core funding (UIDB/04326/2020) and additional funds came from the operational programs CRESC Algarve 2020, COMPETE 2020, to EMBRC.PT ALG-01-0145-FEDER-022121. L.A. has a contract with the University of Algarve, funded by FCT, with reference L57/2016/CP1361/CT 0011.

References

1. Carrera M, Cañas B, Gallardo JM (2013) Proteomics for the assessment of quality and safety of fishery products. Food Res Int 54(1): 972–979
2. Carrera M, Pazos M, Gasset M (2020) Proteomics-based methodologies for the detection and quantification of seafood allergens. Foods 9(8):1134
3. Mazzeo MF, Siciliano RA (2016) Proteomics for the authentication of fish species. J Proteome 147:119–124
4. FAO - Food and Agriculture Organization of the United Nations (2014) The state of world fisheries and aquaculture: opportunities and challenges. FAO
5. Anjo SI, Santa C, Manadas B (2015) Short GeLC-SWATH: a fast and reliable quantitative approach for proteomic screenings. Proteomics 15(4):757–762
6. Gillet LC, Navarro P, Tate S et al (2012) Targeted data extraction of the MS/MS spectra generated by data-independent acquisition: a new concept for consistent and accurate proteome analysis. Mol Cell Proteomics 11(6):1–17
7. Anjos L, Pinto PIS, Tsironi T et al (2019) Experimental data from flesh quality assessment and shelf life monitoring of high pressure processed European sea bass (*Dicentrarchus labrax*) fillets. Data Br 26:1–9
8. Tsironi T, Anjos L, Pinto PIS et al (2019) High pressure processing of European sea bass (*Dicentrarchus labrax*) fillets and tools for flesh quality and shelf life monitoring. J Food Eng 262:83–91
9. Carrera M, Fidalgo LG, Saraiva JA et al (2018) Effects of high-pressure treatment on the muscle proteome of hake by bottom-up proteomics. J Agric Food Chem 66(17):4559–4570
10. Etienne M, Jérôme M, Fleurence J et al (2000) Identification of fish species after cooking by SDS-PAGE and urea IEF: a collaborative study. J Agric Food Chem 48(7):2653–2658
11. Ortea I, Rodríguez A, Tabilo-Munizaga G et al (2010) Effect of hydrostatic high-pressure treatment on proteins, lipids and nucleotides in chilled farmed salmon (*Oncorhynchus kisutch*) muscle. Eur Food Res Technol 230(6):925–934
12. Pazos M, Méndez L, Gallardo JM et al (2014) Selective-targeted effect of high-pressure processing on proteins related to quality: a proteomics evidence in atlantic mackerel (*Scomber scombrus*). Food Bioprocess Technol 7(8): 2342–2353
13. Pazos M, Méndez L, Vázquez M et al (2015) Proteomics analysis in frozen horse mackerel previously high-pressure processed. Food Chem 185:495–502
14. Kuehn A, Swoboda I, Arumugam K et al (2014) Fish allergens at a glance: variable allergenicity of parvalbumins, the major fish allergens. Front Immunol 5:1–8
15. Carrera M, Cañas B, Vázquez J et al (2010) Extensive de Novo sequencing of new parvalbumin isoforms using a novel combination of bottom-up proteomics, accurate molecular mass measurement by FTICR-MS, and

selected MS/MS ion monitoring. J Proteome Res 9(9):4393–4406

16. Burnette WN (1981) "Western Blotting": electrophoretic transfer of proteins from sodium dodecyl sulfate-polyacrylamide gels to unmodified nitrocellulose and radiographic detection with antibody and radioiodinated protein A. Anal Biochem 112(2):195–203
17. Laemmli UK (1970) Cleavage of Structural Proteins during the Assembly of the Head of Bacteriophage T4. Nature 227(5259): 680–685
18. Renart J, Reiser J, Stark GR (1979) Transfer of proteins from gels to diazobenzyloxymethyl-paper and detection with antisera: a method for studying antibody specificity and antigen structure. Proc Natl Acad Sci U S A 76(7): 3116–3120
19. Towbin H, Staehelin T, Gordon J (1979) Electrophoretic transfer of proteins from polyacrylamide gels to nitrocellulose sheets: procedure and some applications. Proc Natl Acad Sci U S A 76(9):4350–4354
20. Bradford MM (1976) A rapid and sensitive method for the quantitation of microgram quantities of protein utilizing the principle of protein-dye binding. Anal Biochem 72(1): 248–254

Chapter 25

In Vitro Assays for the Bifunctional Acylpeptide Hydrolase (APEH) Enzyme from Antarctic Fish

Carmela Fusco, Bruna Agrillo, Marta Gogliettino, Gianna Palmieri, and Ennio Cocca

Abstract

The bifunctional enzyme acylpeptide hydrolase (APEH) is involved in important metabolic processes both as an exopeptidase and as an endopeptidase. Hence, the growing interest in the study of this protein and the need to set up in vitro assays for its characterization. This chapter describes two in vitro assays able to detect the activities of APEH, one for the exopeptidase activity and one for the endopeptidase activity. In particular, these assays have been set up on the two APEH isoforms from Antarctic fish, characterized by a distinct functionality and marked exo- and endopeptidase activities.

Key words Acylpeptide hydrolase (APEH), Oxidized protein hydrolase (OPH), Antarctic fish, In vitro assay

1 Introduction

APEH (acylpeptide hydrolase or acylaminoacyl peptidase) is a cytosolic serine protease that plays a key role in protein metabolism, as it catalyzes as exopeptidase the hydrolysis of N-acylated peptides and proteins, releasing an N-acyl-amino acid and a peptide/protein with a free N-terminus [1]. Furthermore, APEH is also known as OPeH (oxidized protein hydrolase) because performs an important function as an endopeptidase, degrading proteins that have undergone strong oxidative damage, thus contributing to the elimination of cytotoxic proteins [2].

An enzyme with these characteristics has become strategic in the evolutionary processes of notothenioid fish, which are endemic to Antarctic waters. In an increasingly colder environment, up to today's subzero temperatures, the oxidative pressure due to the greater solubility of oxygen has become ever increasing. These organisms have therefore adopted various stratagems to increase antioxidant defenses, and among these, the development of a well-

Cinzia Verde and Daniela Giordano (eds.), *Marine Genomics: Methods and Protocols*,
Methods in Molecular Biology, vol. 2498, https://doi.org/10.1007/978-1-0716-2313-8_25,

equipped APEH system able to participate effectively in the process of protein turnover and in the elimination of oxidized proteins. Moreover, while mammals possess only one APEH gene, fish have APEH1 and APEH2 isoforms, both exhibiting exopeptidase activity that in APEH2 is accompanied by a very pronounced endoprotease function [3, 4].

In vitro assays are used in biochemistry to detect the reaction catalyzed by an enzyme by following the variation in the quantity of a specific substrate or of a product of the reaction itself, and mainly are based on spectroscopic techniques [5, 6]. This chapter describes methods of in vitro analysis developed for the characterization of the two APEH isoforms of fish, but applicable for each component of this enzyme family.

The first assay regards the determination of the exopeptidase activity of APEH in a protein extract, based on the excitation and light emission of a specific fluorimetric substrate. The second assay is aimed at the detection of the endoproteolytic activity of purified APEH from hemolysates, based on the use of an oxidized protein model as substrate, whose degradation is detected by SDS-PAGE [7, 8].

2 Materials

All buffers and solutions are prepared with deionized water and analytical grade reagents, and are stored after use at the temperatures specifically indicated.

2.1 Tissue Collection

1. Fish sampling and stabilization in aquarium.
2. Tricaine methanesulfonate (MS 222, 300 mg/L).
3. Heparinized syringes.
4. Dissection tools.

2.2 Total Protein Extraction

1. Extraction buffer solution: 25 mM Tris-HCl, pH 7.5, prepared from 1 M Tris-HCl, pH 7.5 diluted in deionized water. Add about 100 mL water to a 1 L glass beaker. Weigh 121.14 g Tris-HCl and transfer to the beaker. Add water to a volume of 900 mL. Mix and adjust pH with HCl (*see* **Note 1**). Make up to 1 L with water. Appropriately dilute one in 40 the amount of buffer 1 M Tris-HCl, pH 7.5 in the desired volume. Store at 4 °C.
2. Centrifuge.
3. Ultracentrifuge.
4. Phenylmethylsulfonyl fluoride (PMSF) 1 mM (*see* **Note 2**), diluted from a 60 mM stock solution in 100% isopropanol.
5. Glycerol 5%, diluted from 100% in deionized water.

2.3 APEH Exopeptidase Fluorescent Assay

1. Buffer assay solution: 50 mM Tris-HCl, pH 7.5, prepared from 1 M Tris-HCl, pH 7.5 as described in Subheading 2.2.
2. Stock solution of 9.57 mM Ac-Met-AMC (acetyl-methionine-7-amino-4-methylcoumarin) substrate, prepared dissolving 50 mg Ac-Met-AMC in 15 mL of dimethylsulfoxide (DMSO) solution in the dark, to prevent autodegradation. Aliquot in 1 mL fractions and store at −20 °C (*see* **Note 3**).
3. Spectrofluorometer equipped with a thermostated cuvette compartment and quartz microcuvette to house the reaction mix for the assay.

2.4 Batch-Based Hydrophobic Chromatography of Hemolysates

1. Equilibration buffer: 25 mM Tris-HCl, pH 7.5 (diluted from 1 M stock solution described in Subheading 2.2) and 1 M $(NH_4)_2SO_4$ (prepared dissolving 132.14 g in 800 mL H_2O), make up to 1 L with water and autoclave.
2. Elution buffer: 25 mM Tris-HCl, pH 7.5.
3. Phenyl Sepharose High Performance chromatography resin.

2.5 Gel Filtration Chromatography

1. Equilibration buffer: 25 mM Tris-HCl, pH 7.5 and NaCl 50 mM, prepared from a 4 M stock solution. Weigh 233.76 g NaCl in 800 mL H_2O. Make up to 1 L with water and autoclave. Appropriately dilute one in 80 the amount of 4 M NaCl in the desired volume.
2. Superdex 200 column attached to a chromatography system (*see* **Note 4**).

2.6 10% Sodium Dodecyl Sulfate–Polyacrylamide Gel Electrophoresis (SDS-PAGE)

1. Stacking gel buffer: 0.5 M Tris-HCl, pH 6.8. Weigh 60.6 g Tris-HCl in 900 mL of deionized water, mix and adjust pH with HCl. Make up to 1 L with water. Store at 4 °C.
2. Resolving gel buffer: 1.5 M Tris-HCl, pH 8.8. Weigh 181.7 g Tris-HCl, add water to a volume of 900 mL. Mix and adjust pH with HCl. Make up to 1 L with water. Store at 4 °C.
3. 40% acrylamide/bis solution (19:1) (*see* **Note 5**).
4. 10% ammonium persulfate (APS) solution in water (*see* **Note 6**).
5. *N*,*N*,*N*,*N*′-tetramethyl-ethylenediamine.
6. Glasses (1.5 mm thick) for gel casting.
7. 2-propanol.
8. Absorbent paper.
9. Running buffer: 25 mM Tris-HCl, 192 mM glycine and 10% sodium dodecilsulfate (SDS) (*see* **Note** 7). Weigh respectively: 3 g, 14.4 g and 10 g in 1 L of deionized water.
10. Electrophoresis chamber suitable for 10 × 10 cm gels, connected to a power supply.

11. Sample buffer (SDS reducing buffer): 62.5 mM Tris-HCl, pH 6.8, 20% Glycerol, 2% SDS, 5% β-mercaptoethanol and 0.5% (w/v) bromophenol blue, in water.
12. Coloring solution: 50% methanol, 10% acetic acid and 0.25% Coomassie Brilliant Blue G-250 powder dissolved in deionized water.
13. Decoloring solution: 10% acetic acid and 25% methanol in deionized water.

2.7 Oxidation of Bovine Serum Albumin (BSA)

1. Bovine serum albumin (BSA).
2. Hydrogen peroxide (H_2O_2), 9 M stock solution.
3. Horse radish peroxidase (HRP), 0.2 mg/mL stock solution in deionized water.
4. Thermostated bath.

2.8 Densitometric Analysis

1. Gel Imager.
2. Program for densitometry.

3 Methods

3.1 Tissue Collection

In fish, blood and liver are among the most interesting tissues to study APEH, because the former was the first one where the endoproteolytic activity of this enzyme was identified [9], and liver is one of the tissues where APEH is well expressed.

1. After sampling and stabilization in aquarium, fish are anesthetized with tricaine methanesulfonate (MS222, 300 mg/L) for at least 30 min.
2. Blood and liver are sampled as described in the next sections.

3.2 Exopeptidase Activity

3.2.1 Preparation of Total Protein Extracts from Fish Blood

1. Use a fresh sample of 1–3 mL of blood drawn from the caudal vein of an anesthetized fish with heparinized syringes or slowly thaw on ice a corresponding amount of blood stored at −80 °C.
2. Resuspend blood in hypotonic buffer (25 mM Tris-HCl, pH 7.5) and leave the suspension for 30 min on ice (*see* **Note 8**).
3. Centrifuge the sample at 9200 × *g* for 40 min at 4 °C to separate the soluble fraction.
4. Recover the surnatant in a free tube and discard the pellet.
5. Add 5% glycerol as cryopreservant and 1 mM phenylmethylsulfonyl fluoride (PMSF) as serine protease inhibitor to preserve proteins.

6. Quantify the total protein amount by Bradford assay or equivalent methods to check if the procedure was successful.
7. Keep an aliquot of hemolysate for subsequent control on SDS-PAGE (*see* Subheading 3.3.3).
8. Proceed to in vitro assay to test APEH exopeptidase activity, as described in Subheading 3.2.3.

3.2.2 Preparation of Protein Extract from Fish Liver

1. Perform the truncation of the spinal cord of an anesthetized fish to dissect a fresh portion of liver from the abdominal pack or slowly thaw on ice a sample stored at −80 °C.
2. Homogenize it in four volumes of ice-cold homogenization buffer, composed of 10 mM Tris/HCl, pH 7.5, containing 150 mM NaCl (*see* **Note 9**).
3. Centrifuge the homogenate at 288,000 × *g* for 1.5 h at 4 °C.
4. Recover the surnatant in a free tube and discard the pellet.
5. Add 5% glycerol as cryopreservant and 1 mM PMSF as serine protease inhibitor to preserve proteins.
6. Quantify the total protein amount by Bradford assay or equivalent methods to check if the procedure was successful.
7. Proceed to in vitro assay to test APEH exopeptidase activity, as described in Subheading 3.2.3.

3.2.3 Exopeptidase Activity by Spectrofluorometric Assay

1. Mix in a microcuvette an aliquot of protein extract in exopeptidase assay buffer solution to get a total volume of 950 μL and preincubate at 37 °C for 3 min.
2. Add 50 μL of 10 mM of Ac-Met-AMC substrate (*see* **Note 10**) and insert the microcuvette containing the reaction mixture in a spectrofluorometer thermostated at 37 °C.
3. Follow for 10 min the fluorescence generated at 460 nm (*see* **Note 11**) by hydrolysis of the substrate due to APEH activity. The release of the AMC fluorescent group during the assay give rise to a graph that represents the variation of intensity of light in function of time (Fig. 1).
4. Expect to obtain a region of the graph with constant slope, which indicates that the enzymatic reaction is taking place under optimal conditions for measuring activity. The slope of the line gives a delta value of fluorescence intensity that is utilized to measure the activity of the enzyme in the sample in Unit/min (*see* **Note 12**).

3.3 Endoprotease Assay

The assay for the endoprotease activity of APEH (OPeH activity) consists in verifying the degradation of oxidized BSA, chosen as a model substrate, by a time-course evaluation. The assay is semi-quantitative as the degradation of BSA is measured by

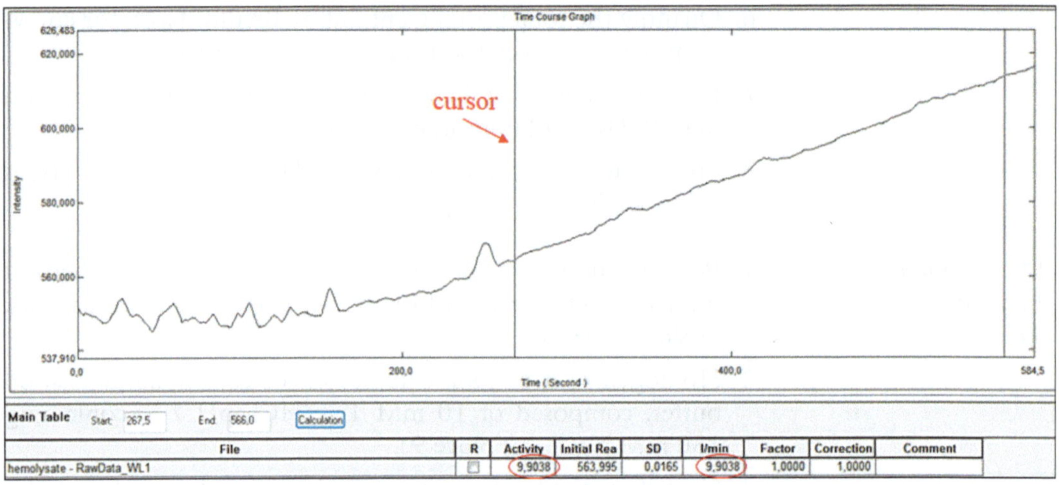

Fig. 1 Spectrofluorometer software interface calculates the slope of the graph that represents the variation of fluorescence released by the hydrolyzed substrate in function of time. In the example, the cursor is put at the beginning of the region with a constant slope, which corresponds to an enzymatic activity of 9.9038 Unit/min

densitometric analysis of the protein bands separated by SDS-PAGE. The activity is expressed as a relative percentage value, since it is not possible to measure the enzymatic units in absolute way. Since various enzymes with endoprotease activity are present in a protein extract, it is necessary to purify APEH to specifically test its endo activity. Blood is the tissue of choice for this purification as it is the one with the main APEH endoprotease activity. Two chromatographic steps are performed, the first one consisting in a hydrophobic batch-based separation step, and the second one in a gel filtration chromatography on Superdex 200.

3.3.1 Batch-Based Hydrophobic Interaction Chromatography (HIC)

1. Dialyze hemolysate in 1 M $(NH_4)_2SO_4$ at 4 °C for 2 h.
2. Mix 0.5 mL of hemolysate (prepared as Subheading 3.2.1) with 1 mL of phenyl Sepharose high-performance chromatography resin for 10 min at room temperature, under gentle agitation in a 2 mL microtube.
3. Centrifuge the mix at 15,700 × *g* for 10 min at 4 °C and discard the supernatant.
4. Equilibrate the resin treated with hemolysate with 0.5 mL of 500 mM $(NH_4)_2SO_4$ in 25 mM Tris-HCl, pH 7.5, for 10 min.
5. Centrifuge at 15,700 × *g* for 10 min at 4 °C and discard the supernatant.
6. Equilibrate the resin with 25 mM Tris-HCl, pH 7.5, for 10 min at room temperature under gentle agitation.
7. Centrifuge again at 15,700 × *g* for 10 min at 4 °C and recover the supernatant enriched with APEH.

8. Test supernatant by the spectrofluorometric assay described in Subheading 3.2.
9. Keep an aliquot of supernatant for subsequent control on SDS-PAGE (*see* Subheading 3.3.3).
10. Dialyze in 25 mM Tris-HCl, pH 7.5 and proceed with gel filtration chromatography, as described in the next section.

3.3.2 Gel Filtration Chromatography on Superdex 200

1. Use a chromatographic column calibrated with molecular weight standards (*see* **Note 13**).
2. Load the fraction from batch-based HIC on a Superdex 200 column connected to an FPLC system.
3. Elute the bound proteins with 25 mM Tris-HCl, pH 7.5, containing 50 mM NaCl at a flow rate of 0.1 mL/min (Fig. 2).
4. Test fractions by the spectrofluorometric assay, described in Subheading 3.2, and collect and pool the active ones.
5. Dialyze in 25 mM Tris-HCl, pH 7.5 and store in the same buffer supplemented with 5% glycerol.
6. Load the active fractions on 10% SDS-PAGE, together with the previously saved aliquots of hemolysate and batch-based HIC extract and molecular weight markers, in order to evaluate the purification fold (Fig. 3).
7. Proceed to in vitro assay to test APEH endoprotease activity, as described in Subheading 3.3.5.

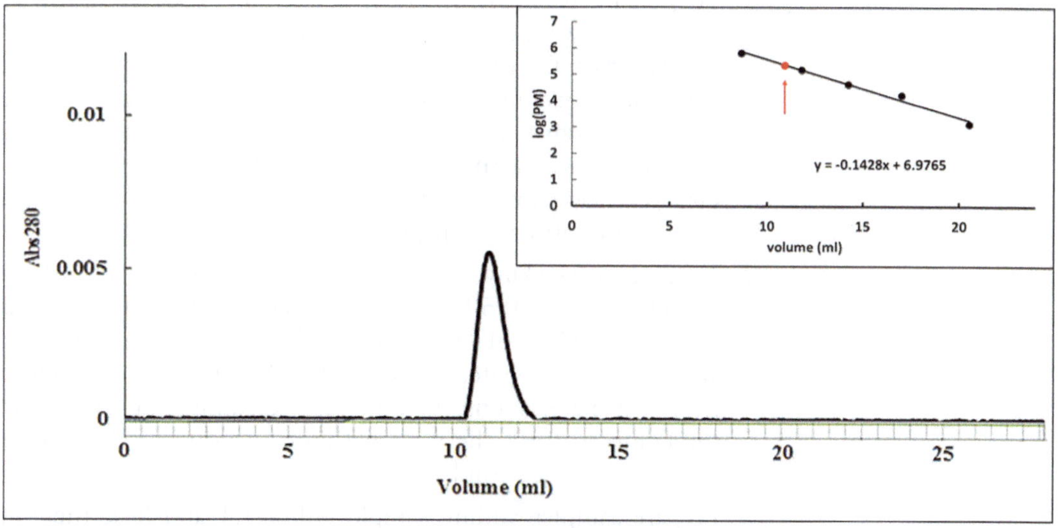

Fig. 2 Gel filtration chromatography profile on a Superdex 200 Increase 10/300 column of batch-based HIC fraction. Absorbance is measured at 280 nm. Insert: calibration curve obtained with the protein standards Thyroglobulin (670 kDa), γ-globulin (158 kDa), Ovalbumin (44 kDa), Myoglobin (17 kDa), Vitamin B12 (1.35 kDa). The elution peak corresponds to a molecular weight of approximately 250 kDa, which can be assigned to the homotetramer of Antarctic fish APEH-2 isoform

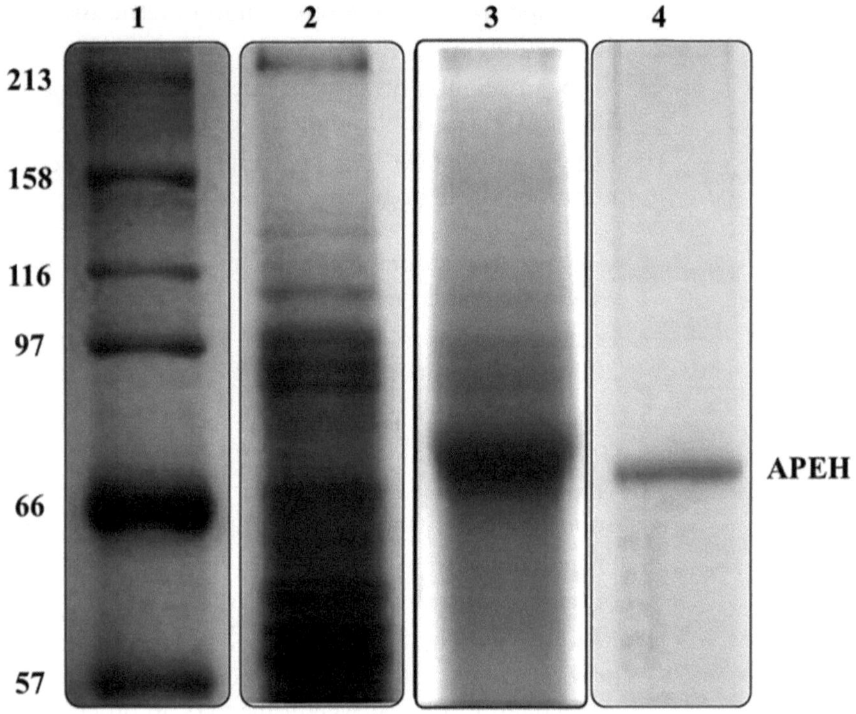

Fig. 3 SDS-PAGE analyses of APEH purification steps from blood. Lane 1: molecular weight markers. Lane 2: hemolysate. Lane 3: fraction from hydrophobic batch purification. Lane 4: purified APEH from gel filtration chromatography on Superdex 200 column

3.3.3 10% SDS-PAGE

1. To prepare resolving gel mix in a 15 mL tube 4.34 mL of deionized water, 2.50 mL of acrylamide mixture, 2.5 mL of 1.5 M Tris-HCl, pH 8.8, 100 μL of 10% SDS. Add 50 μL of ammonium persulfate, 10 μL of TEMED and cast gel within a 10 cm × 10 cm × 1.5 mm gel cassette. Allow space for stacking gel gently cover with 2-propanol or water to remove air bubbles. After polymerization, remove 2-propanol or water with a thin layer of absorbent paper.
2. To prepare stacking gel mix in a 15 mL tube 3.17 mL of deionized water, 0.50 mL of acrylamide mixture; 1.25 mL of 0.5 M Tris-HCl, pH 6.8; 50 μL of 10% SDS. Add 25 μL of ammonium persulfate, 5 μL of TEMED. Pour the solution into gel cassette containing the polymerized running gel and insert the comb to form the wells without introducing air bubbles (we use a 10-well comb).
3. Prepare samples adding sample buffer 1:1 and heat them at 95 °C for 5 min. Centrifuge at 3000 × *g* for 30 s to bring down the condensate and load samples and a Molecular Weight Marker on the gel.
4. Run the gel in denaturing condition with a constant voltage (*see* **Note 14**).

5. After the electrophoresis, wash the gel in deionized water for 15 min and stain with coloring solution for 2 min.
6. Use the decoloring solution to visualize protein bands on the gel.

3.3.4 BSA Oxidation

The method of Fujino et al. [8] is followed to oxidize BSA. Prepare a reaction mix composed of 0.5 mg/mL BSA in Tris-HCl pH 8.0, 20 mM H_2O_2, 1 μg/mL HRP and incubate at 37 °C for 20 h.

3.3.5 APEH Endoprotease Assay (OPeH Activity)

Time-Course Incubation

1. Prepare two sets of three mixtures in 1.5 mL tubes, one for unoxidized BSA and one for oxidized BSA, mixing 60 μg/mL of unoxidized or oxidized BSA, respectively, with 30,000 APEH exopeptidase Units (*see* **Note 15**) in 30 μL of 25 mM Tris-HCl pH 8.0.
2. Cover the cap of the tubes with parafilm, place them in a float in a thermostated bath. Incubate at 37 °C for 0, 24 and 48 h (*see* **Note 16**).
3. Centrifuge the mixes and analyze them on 10% SDS-PAGE (Fig. 4).

Densitometric Analysis

1. Take a photo of the gel and analyze it with a software for densitometric analysis. The electrophoretic pattern of BSA consists of a unique band at 66 kDa. The degradative activity of APEH produces a gradual reduction of the 66 kDa band, which increases with the incubation time, and the concomitant appearance of bands at lower molecular weights due to BSA fragmentation (Fig. 4).
2. To express APEH endoprotease activity as a relative percentage value at a chosen reference incubation time (for example 48 h), consider as "zero" the value obtained for the 66 kDa band for unoxidized BSA and consider as "100" the densitometric value obtained when this band is completely degraded by APEH (Fig. 4). Therefore, calculate a relative percentage value of APEH endoprotease activity for intermediate situations by dividing the respective densitometric values by that obtained for unoxidized BSA. In the example in Fig. 4, APEH with strong and weak endoprotease activity are compared, from Antarctic fish blood (APEHafb) and from pig liver (APEHpl), respectively.

4 Notes

1. At first use concentrated HCl (12 N) to shorten the difference from the starting to the required pH. Reached a value close to the desired one (one-two units of difference), it would be

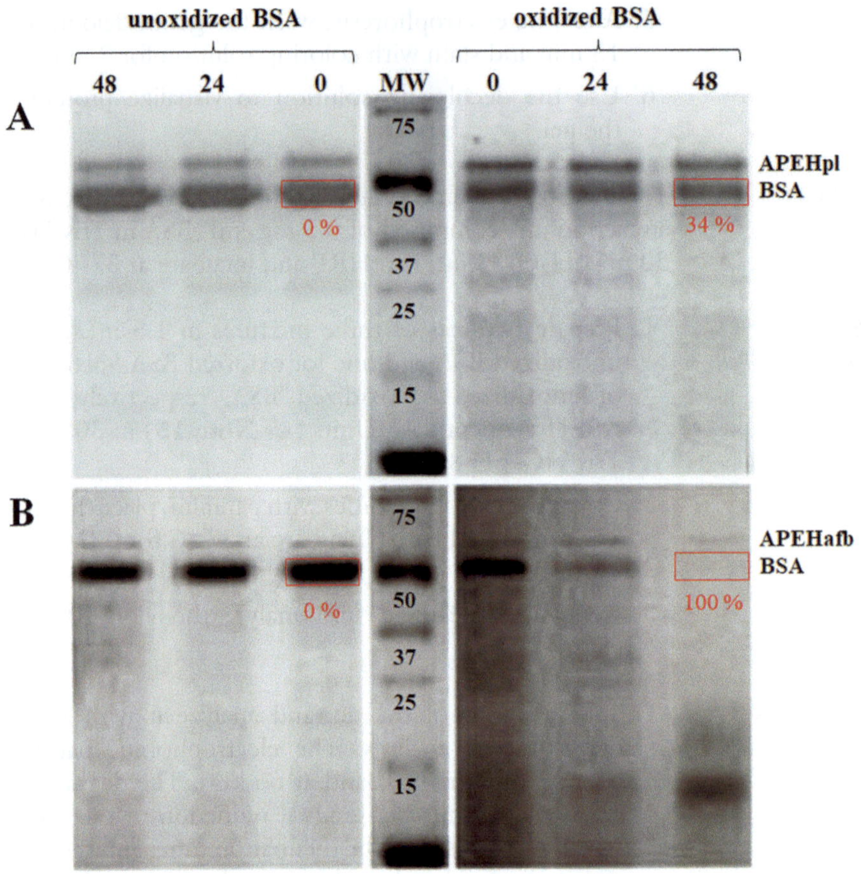

Fig. 4 SDS-PAGE profiles of unoxidized BSA (left) and oxidized BSA (right) treated with (**a**) commercial APEH from porcine liver (APEHpl) and (**b**) APEH-2 isoform from Antarctic fish blood (APEHafb), at pH 8.0 and 37 °C for the indicated period (h). The bands at 66 kDa corresponding to nondegraded BSA and those of APEH used in the assay are indicated, as well as the molecular weight (kDa) of the markers. The gel regions analyzed by densitometry and the relative percentage values of endoprotease activity are in red

better to use a diluted HCl (e.g., 1 N) to avoid a sudden drop in pH below the required one.

2. As phenylmethylsulfonyl fluoride could inactivate APEH enzyme, we tested that at 1 mM concentration it is ineffective for APEH but works on other serine protease.
3. Powdery substrate are electrostatic and difficult to handle, especially in small amounts; for this reason, to avoid losing of dust is recommended to dissolve the entire aliquot of substrate (50 mg of Ac-Met-AMC Bachem®) in 15 mL of DMSO. The solution and its aliquots must be shielded from light as much as possible, to avoid self-degradation of the substrate. All the tubes have to be wrapped in aluminum foil and then stored at −20 °C.

4. We use a Superdex 200 Increase 10/300 GL column attached to an AKTA chromatography system.
5. Caution: acrylamide is a neurotoxin, it is absorbed through the skin and it is toxic by inhalation. Always wear gloves and mask when preparing the solutions. To protect yourself, prepare all solutions and gel under a chemical hood.
6. Dissolve 1 g APS in 8 mL H_2O. Adjust volume to 10 mL with H_2O. Solution is stable at 4 °C for 2 weeks. In our lab we prefer to prepare 1 mL aliquots that can be stored at −20 °C for 4 months.
7. Dissolve 10 g SDS in 90 mL H_2O. Heat to 68 °C to solubilize crystals. Adjust pH to 7.2 with HCl (about 50 μL). Adjust volume to 100 mL with H_2O. Caution: crystals of SDS disperse easily; it is important to wear a mask when weighing SDS and clean the weighing area and balance after use.
8. Do vortex pulses every 10 min.
9. To improve homogenization it is advisable first to crumble the tissue with a cutter.
10. The substrate should be add by pipetting slowly several times in order to obtain a homogeneous solution.
11. Excitation and emission wavelengths of Ac-Met-AMC are 380 nm and 460 nm, respectively.
12. Enzymatic activity is expressed in arbitrary units (U), where one U (μmol/min) is defined as the amount of the enzyme that catalyzes the conversion of one micromole of substrate per minute under the specified conditions of the assay method [2].
13. A calibration curve is drawn measuring the elution volumes of several protein standards and plotting them versus the decimal logarithm of their molecular weight. The molecular weight of an unknown protein can be determined from the calibration curve once its elution volume is measured.
14. We prefer use a 30 mA current at 100 V.
15. The reaction mixture is composed of a standard concentration of unoxidized and oxidized BSA and a standard concentration of APEH enzymes. To normalize APEH concentration we utilize the assay for exopeptidase APEH activity described in Subheading 3.2.
16. Long incubation times are required to detect APEH endoprotease activity on SDS-PAGE.

References

1. Perrier J, Durand A, Giardina T et al (2005) Catabolism of intracellular N-terminal acetylated proteins: involvement of acylpeptide hydrolase and acylase. Biochimie 87:673–685
2. Shimizu K, Kiuchi Y, Ando K et al (2004) Coordination of oxidized protein hydrolase and the proteasome in the clearance of cytotoxic denatured proteins. Biochem Biophys Res Commun 324:140–146
3. Gogliettino M, Riccio A, Balestrieri M et al (2014) A novel class of bifunctional acylpeptide hydrolases—potential role in the antioxidant defense systems of the Antarctic fish *Trematomus bernacchii*. FEBS J 281(1):401–415. https://doi.org/10.1111/febs.12610
4. Riccio A, Gogliettino M, Palmieri G et al (2015) A new APEH cluster with antioxidant functions in the Antarctic hemoglobinless icefish *Chionodraco hamatus*. PLoS One 10(5):e0125594. https://doi.org/10.1371/journal.pone.0125594. PMID: 25946123; PMCID: PMC4422685
5. Brand L, Johnson ML (eds) (1997) "Fluorescence Spectroscopy" methods in enzymology, vol 278. Academic, San Diego
6. Bisswanger H (2011) Practical enzymology. Wiley-VCH Verlag. https://doi.org/10.1002/9783527659227
7. Vandooren J, Geurts N, Martens E et al (2013) Zymography methods for visualizing hydrolytic enzymes. Nat Methods 10:211–220. https://doi.org/10.1038/nmeth.2371
8. Fujino T, Kojima M, Beppu M et al (2000) Identification of the cleavage sites of oxidized protein that are susceptible to oxidized protein hydrolase (OPH) in the primary and tertiary structures of the protein. J Biochem 127(6):1087–1093. https://doi.org/10.1093/oxfordjournals.jbchem.a022702
9. Fujino T, Ishikawa T, Inoue M et al (1998) Characterization of membrane-bound serine protease related to degradation of oxidatively damaged erythrocyte membrane proteins. Biochim Biophys Acta 1374:47–55

INDEX

Cinzia Verde and Daniela Giordano (eds.), *Marine Genomics: Methods and Protocols*,
Methods in Molecular Biology, vol. 2498, https://doi.org/10.1007/978-1-0716-2313-8,

M

N

O

P

Q

R

S

T

U

V

Z

MIX
Papier aus verantwortungsvollen Quellen
Paper from responsible sources
FSC® C105338

If you have any concerns about our products,
you can contact us on
ProductSafety@springernature.com

In case Publisher is established outside the EU,
the EU authorized representative is:
Springer Nature Customer Service Center GmbH
Europaplatz 3, 69115 Heidelberg, Germany

Printed by Libri Plureos GmbH
in Hamburg, Germany